Sustainable Natural Fiber Composites

Edited by

Anish Khan[1,2], A. Manikandam[3], M. Ramesh[4], Imran Khan[5]. Abdullah M Asiri[1,2]

[1]Center of Excellence for Advanced Materials Research, King Abdulaziz University, Jeddah 21589, Saudi Arabia

[2]Chemistry Department, Faculty of Science, King Abdulaziz University, Jeddah 21589, Saudi Arabia

[3]Department of Chemistry, Bharath Institute of Higher Education and Research, Chennai - 600073, India

[4]Department of Mechanical Engineering, KIT-Kalaignarkarunanidhi Institute of Technology, Coimbatore-641402, Tamil Nadu, India

[5]Humanities section, ZH College of Engineering and Technology, Aligarh Muslim University, Aligarh India

Published by **Materials Research Forum LLC**
Millersville, PA 17551, USA

Published as part of the book series
Materials Research Foundations
Volume 122 (2022)
ISSN 2471-8890 (Print)
ISSN 2471-8904 (Online)

Print ISBN 978-1-64490-184-7
eBook ISBN 978-1-64490-185-4

Distributed worldwide by

Materials Research Forum LLC
105 Springdale Lane
Millersville, PA 17551
USA
https://www.mrforum.com

Manufactured in the United States of America
10 9 8 7 6 5 4 3 2 1

Table of Contents

Preface

All-Cellulose Composites Derived from Natural Plant Fibers
Juan Francisco Delgado, Jimena Bovi, María Laura Foresti, Celina Bernal 1

Innovations in Natural Fibre Reinforced Composites for Improved Performance
N. Gokarneshan, PG Anandhakrishnan, R. Sasirekha, Obulam Vidya Sagar, D. Haritha, and K.M. Pachiyappan 37

Next Generation Biodegradable Material for Dental Teeth and Denture base Material: Vegetable Peels as an Alternative Reinforcement in Polymethylmethacrylate (PMMA)
Anirudh Kohli, Arun Y Patil, Mahesh Hombalmath 77

Microstructure Characteristics of Cellulose Fibers in Cement Paste and Concrete - A Review
Syed Habibunnisa, Ruben Nerella and Madduru Sri Rama Chand 96

Application of Surface Modification Routes to Coconut Fiber for its Thermoplastic-Based Biocomposite Materials
Ümit Tayfun, Mehmet Doğan 11

Durability Against Fatigue and Moisture of Natural Fibre Composite
Bidita Salahuddin, Zhao Sha, Shazed Aziz, Shaikh N. Faisal, Mohammad S. Islam .. 128

Resol–Vegetable Fibers Composites
Wei Ni, Lingying Shi 154

Measurement of Thermal Conductivity of Natural Fiber Composites
K. Ramanaiah, Srinivas Prasad Sanaka, A.V. Ratna Prasad, K. Hemachandra Reddy 199

Natural Fiber and Biodegradable Plastic Composite
Manoj Mathad, Anirudh Kohli, Mrutyunjay Adagimath, Arun Patil, Anish Khan 209

Polymer Natural Fibre Composites and its Mechanical Characterization
Manju Sri Anbupalani, Chitra Devi Venkatachalam, P. Manjula 238

Glass and Natural Fiber Composite: Properties and Applications
Guravtar Singh Mann, Anish Khan, Abdullah M Asiri 256

Pineapple Natural Fibre Composite: Extraction, Mechanical Properties and Application
Anirudh Kohli, Manoj Mathad, Arun Patil, Anish Khan.. 282

Keyword Index

About the Editors

Preface

Natural fiber composite research is an emerging area in the field of polymer science with tremendous growth and using a green approach. The book is as a one stop reference book for researchers working in the area of natural fiber composites for engineering applications. The book reviews all the important natural fibers and its composites for engineering applications in a widespread way. It will help beginners as well as young researchers by exploring cutting edge research on natural fibre composites which can provide future direction and opportunities. The state of the art demonstrates the potential of various natural fibres and its composite to reach all potential improvements achievable for high-demanding engineering applications. The target of this book is to give a comprehensive demonstration and to show the ability to improve the hydrophobic nature of natural fiber composites. Extensive survey for various engineering applications on the development of natural fiber composite are given in the book. This book also focusses on modern technologies and hybrid natural fiber composites able to be used as alternatives in structural components subjected to severe conditions.

We are highly thankful to contributors of the different book chapters who provided us their valuable innovative ideas and knowledge in this edited book. We attempt to gather information related to natural fiber composites for environmental application from diverse fields around the world and finally complete this venture in a fruitful way. We greatly appreciate contributor's commitment to their support to compile our ideas in reality. We are highly thankful to the MRF team for their generous cooperation at every stage of the book production.

Editors

Anish Khan, A. Manikandan, M. Ramesh, Imran Khan, Abdullah M. Asiri

Dedication
Editors are honoured to dedicate

this book to

My

Grandfather

Late Mr Rasheed

Sustainable Natural Fiber Composites   Materials Research Forum LLC
Materials Research Foundations **122** (2022) 1-36   https://doi.org/10.21741/9781644901854-1

Chapter 1

All-Cellulose Composites Derived from Natural Plant Fibers

Juan Francisco Delgado[#], Jimena Bovi[#], María Laura Foresti[a*], Celina Bernal[b*]

Instituto de Tecnología en Polímeros y Nanotecnología (ITPN-UBA-CONICET), Facultad de Ingeniería, Universidad de Buenos Aires, Argentina

[a]mforesti@fi.uba.ar, [b]cbernal@fi.uba.ar

[#]authors contributed equally to this work

Abstract

A promising approach to prepare high performance cellulose-reinforced biodegradable materials is the production of innovative composites made completely from cellulose according to the concept of single-polymer composites. All-cellulose composites (ACCs) are distinguished for the excellent interfacial adhesion between the matrix and the reinforcement which results in outstanding mechanical properties, and for their enhanced recyclability. In this context, the current chapter will focus on the most important processing routes and the main properties of ACCs and ACNCs (all-cellulose nanocomposites) totally or partially derived from natural plant fibers, using either the entire fibers or the cellulose isolated from them.

Keywords

Natural Plant Fibers, Cellulose, Biocomposite Materials, All-Cellulose Composites and Nanocomposites, Processing Methods, Properties

Contents

All-Cellulose Composites Derived from Natural Plant Fibers 1

1.   Introduction..2

2.   Cellulose fundamentals 3

3.   Isolation of cellulose and nanocellulose from natural fibers................5

4.   All-cellulose composites (ACCs)..**8**

    4.1   Cellulose dissolution...8

    4.2   Processing routes and derived ACCs properties 12

    4.2.1 Partial dissolution method 13

    4.2.2 Impregnation method..16

5.   All-cellulose nanocomposites (ACNCs) **21**

Conclusions and future trends...**26**

References **27**

1.   Introduction

Over the last decades, the use of natural fibers as reinforcing phase of polymer composites has been the subject of much research, and a huge number of articles on the topic have been published. According to a general definition, natural fibers are fibers derived from natural sources, such as plants (e.g. jute, sisal, hemp, ramie, flax, cotton, kenaf, wheat straw), wood (softwood and hardwood) and certain animals (e.g. wool, silk). In spite of the different uses of the term found in the literature, in this chapter the term *natural fibers* will refer to *natural plant fibers*. Provided their high tensile strength, stiffness, low density, biodegradability, renewable origin, small carbon footprint and good thermal and acoustical insulation properties; natural fibers have been recognized as attractive alternatives to other traditional synthetic reinforcements such as glass or carbon fibers.

The main constituent of plant fibers is cellulose, which is found in the plant cell wall together with hemicellulose, lignin, pectins and some proteins. Given its availability, biodegradability, biocompatibility, non-toxicity, and high tensile strength and stiffness, cellulose finds plenty of uses in the paper and cardboard, textile, packaging, construction and pharmaceutical industries, among others. In 2018 the global market of cellulose was valued in almost 220 billion USD, out of which paper production accounted for 42-43% [1]. Besides, cellulose is also used as reinforcement of several polymeric matrices, especially biodegradable and/or biobased ones, for the production of innovative biocomposites. Particularly, cellulosic reinforcements with dimensions in the nanoscale are recognized for their good reinforcing capability due to their high aspect ratio and stiffness of the crystal regions. However, the incompatibility between hydrophilic cellulose and many existing hydrophobic resin systems used in the industry results in poor fiber-matrix interfacial adhesion and very low load transfer efficiency. This issue is a great challenge which has been often addressed by physical adsorption of surfactants or polyelectrolytes, polymer grafting and chemical modification of the cellulosic reinforcement/nano-reinforcement [2].

Another recent promising approach to prepare high performance cellulose-reinforced biodegradable materials is the production of innovative composites made completely from cellulosic materials according to the concept of "single-polymer composites" (SPCs). In these composites the matrix and the reinforcement belong to the same polymer family, thus being a very promising alternative to overcome the incompatibility limitations already described for traditional composite materials, while enhancing their recyclability. Using this concept, Nishino *et al.* (2004) coined the term "all-cellulose composites" (ACCs), although ACCs development has frequently involved not only highly pure cellulose sources (e.g. filter paper, microcrystalline cellulose, bacterial cellulose, etc.) but also natural plant fibers such as ramie, hemp and cotton, among others [3,4,5]. ACCs will be the focus of the current chapter, with particular emphasis on those totally/partially derived from natural plant fibers.

2.   Cellulose fundamentals

As already described, cellulose is found in natural plant fibers along with hemicellulose and lignin as main components. Cellulose fibrous structure enables the maintenance of plants structural integrity, providing them strength and stability. On the other hand, hemicelluloses fulfill various roles in mechanical support, reserve storage and development [6]. Lignin plays a role as a binder between hemicelluloses and cellulose fibrils, giving support and structure to the plant, and also some resistance to microbial degradation [7]. The cell wall of plant fibers has often been described as a natural composite material in which stiff and strong semicrystalline cellulose microfibrils are embedded in a matrix of hemicellulose and lignin [8].

In terms of structure, cellulose is a linear polymer formed by glucose monomers linked by β ($1 \rightarrow 4$) bonds. The result of the glycosidic bond is a unit known as cellobiose. Depending on the source analyzed, the polymerization degree of cellulose varies widely, from 300 to 10,000 units [4,9]. Cellulose is characterized by a hierarchical structure, involving cellulose chains organized in single subunits, elementary fibrils and microfibrils; the latter containing crystalline and amorphous regions [10].

Based on the preponderance of accessible hydroxyl functional groups, cellulose has strong affinity for water. Interactions with water result in cellulose swelling (i.e. volume increase due to liquid uptake), which is especially meaningful in its amorphous regions [11]. Moisture affinity is recognized as one of the main drawbacks of cellulose fibers which could limit the outdoor applications of the derived composite materials, since cellulosic fiber swelling may result in significant dimensional instability.

On the other hand, hydroxyl groups of cellulose are also involved in intra- and intermolecular hydrogen bonds, i.e. intrachain H-bonds between neighboring monomers of a single cellulose polymer chain and interchain H-bonds between adjacent chains. The hydrogen bonding patterns in cellulose are considered as one of the most influential factors on its physical properties [12]. Strong hydrogen bonding systems in cellulose play a key role on its limited solubility in most solvents, on the reactivity of the hydroxyl groups, on the crystallinity of cellulose samples, on their mechanical properties, and they are also the major interaction to stabilize the hierarchical architecture of higher plants. Hydrophobic areas of cellulose (around the C atoms) also influence its overall properties [13].

Hydrogen bonding in cellulose results in various ordered crystalline arrangements. Naturally produced crystalline cellulose corresponds to Cellulose I, and it can be found in two polymorphs: I_α, which has a triclinic cell unit and I_β which has a monoclinic cell unit. Both forms are found in native cellulose, although the relationship between them depends on the cellulose source [10]. On the other hand, Cellulose II can be produced by a process of mercerization or by regeneration of cellulose from a previous dissolution. Cellulose III can be obtained from Cellulose I or Cellulose II by treatment with liquid ammonia at high pressures [14]. Cellulose IV is obtained by heating Cellulose III in glycerol at high temperatures. The mechanical behavior of cellulose rests on the cellulose polymorph present, as well as on its crystalline and amorphous content. Nishino *et al.* (1995) measured by X-ray diffraction the elastic modulus of the crystalline regions of cellulose polymorphs in the direction parallel to the chain axis, determining values of 138 GPa and 88 GPa for Cellulose I and II, respectively (these two cellulose polymorphs are the ones relevant in ACCs) [15].

The strong network of interchain hydrogen bond interactions and dispersion forces in cellulose also has a significant effect on stabilizing the polymer bulk [16], and as a consequence cellulose is insoluble in water and in most common solvents. On the other hand, over the years cellulose has proved soluble in solvents or solvent systems such as N-methylmorpholine-N-oxide (NMMO), dimethylacetamide with lithium chloride (DMAc/LiCl), dimethyl sulfoxide with dinitrogen tetroxide (DMSO/N_2O_4), ionic liquids, trifluoroacetic acid (TFA) and alkaline systems such as sodium hydroxide-urea aqueous solutions (NaOH/Urea), among others [16–18]. These and other cellulose solvents which have been used in the preparation of all-cellulose composites will be further described in Section 4.1.

3.   Isolation of cellulose and nanocellulose from natural fibers

Although entire natural plant fibers have been sometimes used for the production of ACCs, these composites have been most frequently prepared from isolated celluloses. The cellulose content varies among plant fiber sources, but it is generally within the 40-80% interval. Table 1 summarizes the average cellulose content of the natural plant fibers (classified as bast, leaf, seed, straw and grass fibers [19]) that have been most commonly used in the production of ACCs either as the entire fibers or as cellulose extracted from them.

Table 1. Cellulose content of natural plant fibers that have been most frequently used in the preparation of ACCs (either as entire fibers or as cellulose extracted from them).

Type	Origin	Cellulose	References
Bast fibers	Flax	60-81[7]	[20,21,22]
	Hemp	70-78 [7]	[23]
	Ramie	68-76 [24]	[3,25,26,27]
	Jute	61-7 [24]	[28]
	Thespesia lampas	61[29]	[30]
Leaf fibers	Pineapple	70-82[31]	[32,33]
	Banana	62,5 [34]	[35]
Seed fibers	Cotton	89 [34]	[26,30,36,37,38,39,40,41,42,43]
Straw fibers	Rice	41-57 [31]	[44]
	Corn	39-42 [45]	[46,47]
	Canola	43 [48]	[49,50,51]
Grass fibers	Bagasse	55[31]	[52]
	Alfa	45 [34]	[53]
	African Napier	47 [54]	[38]

The structure of natural fibers is schematized in Fig. 1. Middle lamella (ML) is the outermost layer of cell wall and keeps the neighboring cells together. It is mainly conformed by pectins. The primary wall (P1), a generally flexible layer which is synthetized during cell growth, is mainly composed of poorly oriented crystalline cellulose, hemicellulose and pectin; and surrounds the secondary wall which is constituted by three layers (i.e. S1, S2 and S3). S2 is the thicker middle layer and because of its structure and cellulose content, it determines the mechanical properties of the entire fiber. There, cellulose chains are arranged in helically disposed microfibrils conforming an angle with the axis of the fiber. This is named the "microfibrillar angle" and determines the stiffness of the filament [55]. Cellulose microfibrils are present in primary cell walls as well, but there they are located at all angles resulting in a more disordered network.

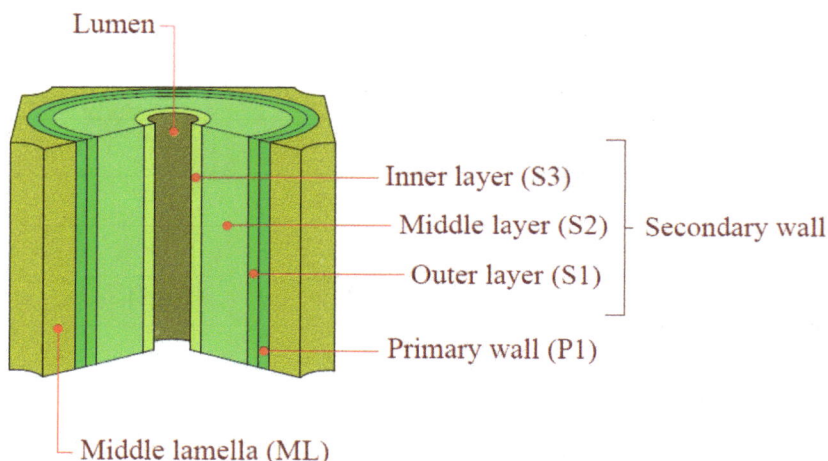

Figure 1. Schematic structure of a natural plant fiber. Adapted from John and Thomas, 2008 [55].

Cellulose isolation methods vary with the source material used, and they may have a significant impact on the final quality, structure and properties of the biopolymer. Purification methods are conducted to achieve total or partial removal of the components of the matrix in which cellulose is embedded; in the case of natural fibers, mainly lignin and hemicelluloses. During the entire process of isolation of cellulose from vegetal

Materials Research Forum LLC
https://doi.org/10.21741/9781644901854-1

biomass, the exposure of the S2 layer takes place due to a gradual peeling off of the P1 and S1 layers as a result of the combination of different treatments. These treatments can be physical, such as mechanical disruption conducted by milling, cutting, chopping, grinding, pulverizing, steam-explosion; chemical, like treatments with alkali, dilute acid, oxidizing agents, organic solvents; or a combination of them [34,56]. Among the approaches directed to hydrolyze hemicelluloses, treatments with alkali (KOH, NaOH) can be performed. The bleaching process (with H_2O_2, $NaClO_2$, CH_3COOH) is executed to dissolve lignin components and to get a purer product, and the concomitant whitening of the material is a good indicator of its removal. If the removal of pectins and waxy components is necessary, previous purification steps with hot water and extractions with different solvents or solvent systems can also be conducted. Some of them can be mixtures containing hexane/ethanol/water, benzene/methanol or toluene/ethanol [56,57]. An ideal isolation method of cellulose from natural plant fibers should avoid an excessive reduction of the degree of polymerization and crystallinity of cellulose, while being cost effective, minimizing the production of hazardous wastes and reducing the energy requirements.

Cellulose isolation methods can be followed by specific treatments devoted to produce cellulose nanomaterials, -frequently referred to as nanocelluloses-, which are those having at least one dimension in the nanoscale (i.e. 1-100 nm). Cellulose nanomaterials thus include both cellulose nanofibrils (CNF) (also called nanofibrillated cellulose (NFC) or simply cellulose nanofibers; typical length: up to 100 µm, typical cross-section dimension: 3-100 nm); and cellulose nanocrystals (CNC) (also called cellulose nanowhiskers (CNW) or nanocrystalline cellulose (NCC); typical length: from 100 nm to several µm, typical cross-section dimension: 3-50 nm) [58,59].

Production of nanofibrillated cellulose is generally performed by mechanical disintegration of cellulose slurries in high-pressure homogenizers or microfluidizers. Defibrillation steps involve high energy consumption and (often) equipment clogging due to an increase in the viscosity of the suspension, and so they are frequently combined with different pretreatments to reduce energy requirements and save time. The introduction of certain groups onto cellulose nanofibrils via chemical modifications (e.g. 2,2,6,6-tetramethylpiperidine-1-oxyl (TEMPO)-mediated oxidation, esterification, carboxymethylation) favor their separation because of electrostatic repulsion or steric impediment. Enzymatic partial hydrolysis with certain cellulases can also be employed in order to assist the isolation of CNF [34]. On the other hand, acid hydrolysis (typically 50–72% H_2SO_4) of cellulose sources combined with ultrasonication or homogenization steps, is commonly used to isolate CNC by preferential removal of the amorphous regions of cellulose microfibrils.

4. All-cellulose composites (ACCs)

In order to overcome some drawbacks of conventional polymer composites such as the already mentioned frequent incompatibility between matrix and reinforcement which leads to insufficient load transfer, as well as the difficulty for recycling, the concept of single-polymer composites was first presented by Capiati and Porter (1975) [60]. Authors focused on the significant difference in the melting temperature between the high-density polyethylene (HDPE) matrix and the HDPE reinforcement containing aligned and extended molecular chains, to obtain single-polymer composites by melt processing.

In SPCs the matrix and the reinforcement belong to the same polymer family, thus providing the resulting composites with high mechanical properties derived from the excellent compatibility between phases. The concept of single-polymer composites originally applied to HDPE was later used to develop composites with other polyolefins, and also with other polymers such as polyamides and polyesters, among others [61]. Nishino *et al.* (2004) first applied the concept of single-polymer composites to cellulosic materials, in this case via a "wet process", to prepare all-cellulose composites (ACCs) [3].

As it will be discussed later on, ACCs preparation requires the total or partial solubilization of cellulose, which will then be regenerated by use of proper non-solvents to conform the composite matrix. The final properties of ACCs are affected by a number of variables including the cellulosic material source, the preparation method and conditions chosen, the reinforcement content, the solvent system used and the regeneration process. The solvent systems most frequently employed in the preparation of ACCs are described in the following section.

4.1 Cellulose dissolution

Cellulose dissolution can be achieved by use of different solvents/solvent systems. Table 2 collects a summary of those that have been most frequently employed to dissolve cellulose for the purpose of ACCs preparation. Most of the solvents mentioned are not new since some of them have been used for years in the industrial processing of cellulosic fibers.

N-methylmorpholine-N-oxide (NMMO) is an organic compound which has traditionally been used to dissolve cellulose in textile industries for the production of regenerated cellulose under the name of Lyocell fiber. NMMO is recognized as one of the most successful commercial solvents for the non-derivatized dissolution of cellulose [62]. However, its use in the preparation of ACCs has not been as common as that of other solvent systems described below. Polymer dissolution in NMMO results from its ability to disrupt hydrogen bonding between cellulose chains and to establish new hydrogen bonds with the macromolecule. In this process water concentration has shown to play a key role, with recommended values between 4 and 17% [63].

On the other hand, many studies on ACCs have reported the use of DMAc/LiCl (the concentration of LiCl used in the solvent system is generally 8 wt%), which is a non-derivatizing solvent system recognized for its ability to dissolve high molecular weight cellulose with negligible chain degradation at moderate temperatures [4,16]. However, interactions involved in cellulose dissolution in this solvent system require that no cellulose bound water is present, for which solvent exchange of the cellulose or distillation of the DMAc/LiCl/cellulose system is often necessary [16]. Besides, dissolution of cellulose in DMAc/LiCl requires a "cellulose activation" pretreatment which is achieved by use of a polar medium that disrupts intermolecular hydrogen bonds, swells cellulose and opens its structure to promote access by the solvent at temperatures low enough to avoid significant polymer degradation [36,64]. Without activation steps (generally implying sequential immersion of cellulose in water, methanol/acetone/ethanol and finally DMAc), dissolution of the cellulosic material with DMAc/LiCl would demand very extended time internals (several months) regardless of cellulose crystallinity [4,65]. Even if DMAc/LiCl is still the most used solvent system for the preparation of ACCs [66], solvent exchange is a time and solvent demanding step, limiting its use at industrial scale [16].

DMSO in combination with ammonium fluorides or with N_2O_4 has also been used in solvent systems designed for the dissolution of cellulose for ACCs preparation. In particular, DMSO in combination with tetra-n-butylammonium fluoride (TBAF), is an efficient non-derivatizing solvent of cellulose which has been usually used at $60 - 80$ °C. The dissolution of the macromolecule is essentially influenced by the interaction of fluoride anions with the hydroxyl moieties of cellulose. Other DMSO/ammonium fluorides systems also dissolve cellulose, being crucial in this regard the solubility of the ammonium salt in DMSO [67]. On the other hand, DMSO or DMF with N_2O_4 are derivatizing solvent systems yielding cellulose nitrite as intermediate. The major disadvantage of the derivatizing solvents is the occurrence of side reactions during dissolution and sometimes the formation of undefined structures [18]. On the other hand, trifluoroacetic acid (TFA) is a non-aqueous derivatizing solvent for cellulose, which has been used in combination with trifluoroacetic anhydride (TFAA) to dissolve cellulose at 50 °C for the preparation of ACCs. TFA is a naturally occurring organic acid, recognized as biodegradable by microbial action and recyclable by distillation [68].

Other solvents that have been frequently used for the preparation of ACCs are ionic liquids (ILs), which are low-melting temperature salts that can dissolve cellulose without water. Both experimental and computational achievements have proved that the disruption of hydrogen bonds inside cellulose is the key factor in the dissolution process in which cations and anions of ILs act synergistically [69]. In the last years, ILs have been the focus of much research based on their characteristic very low vapor pressure, low flammability, high thermal stability and structural diversity. Besides, they can be recovered after cellulose dissolution, reducing the cost of their use [70]. The physical and chemical properties of ILs can be regulated by altering the structures of cations and anions for different purposes. The anion species of ILs is the major factor that determines the solubility of cellulose, but the choice of cation is also important for the solvent design [71].

On the other hand, recently several non-derivatizing solvent systems involving NaOH have been used to prepare ACCs. Sodium hydroxide solutions with different additives such as poly(ethylene glycol) (PEG), urea and thiourea have proved to dissolve cellulose more efficiently than the binary NaOH/H_2O system itself. The NaOH/Urea system is a promising non-toxic, non-flammable green solvent system for cellulose, although the need for low temperatures to dissolve cellulose is its main disadvantage. As pointed out by Hildebrandt (2018), the most accepted explanation for the role of NaOH and urea in cellulose solubilization is that hydroxide anions [OH^-] break hydrogen bonds between cellulose chains and that [Na^+] ions form a hydration layer and structure the water around the cellulose, preventing the rearrangement of cellulose molecules. Urea, as well as thiourea and polyethylene glycol, are thought to improve the penetration of [Na^+] and [OH^-] in the crystalline regions and collaborates with the formation of the hydration layer around the dissolved chains. The increase in cellulose solubility at low temperatures (below 0 °C) could be a consequence of the formation of a complex between cellulose, NaOH and urea due to reduced their thermal movement [72]. On the other hand, Lindmann et al. (2017) proposed that urea can collaborate with cellulose solubilization by its effect on the hydrophobic interactions among cellulose chains that lead to their stacking [73]. Restrictions on the maximum cellulose molecular weight to be dissolved, the relatively low cellulose concentration which may be achieved, and the low temperatures required (the cooling of the system increases the cost of the process) have triggered the incorporation of certain additives to the NaOH/Urea solution to try to improve cellulose dissolution [74]. In this context, Yang et al. (2011) have shown that the incorporation of 0.5 wt% of ZnO to a NaOH/Urea (7/12 wt%) solution can increase the solubility of cellulose, with results depending on the polymer molecular weight [75]. On the other hand, NaOH/PEG aqueous solutions in a 9:1:90 ratio were prepared via a freezing-thawing process. Although low temperatures were still required for cellulose dissolution (i.e. freezing at -15 °C during 12 h) authors demonstrated that the obtained solution flowed at room temperature for at least 30 days [76]. The fact that NaOH, PEG and urea are all environmentally friendly and pose relatively low toxicity towards humans and animals makes these solvent systems

Materials Research Forum LLC
https://doi.org/10.21741/9781644901854-1

interesting for large scale applications, although solvent recovery and reuse is always necessary [16].

Table 2. Solvent systems frequently used to dissolve cellulose for the preparation of ACCs

Solvent system	Toxicity	Dissolution temperature generally used	Derivatizing (Yes/No)	References
NMMO	High (oral) Low-Moderate (skin and inhalation)	≈ 90-140 °C	No	[23,77,78]
DMAc/LiCl	High (skin and inhalation)	Room temperature	No	[3,25,79,80]
DMSO/TBAF	High (skin and inhalation)	≈ 60-80 °C	No	[17,67]
Ionic liquids	Low	≈ 80-110 °C	No	[81,82]
TFA-TFAA	High (inhalation)	≈ 50 °C	Yes	[68,83]
NaOH/Urea	Moderate (skin, eyes and inhalation)	≈ -12- 4 °C	No	[72,84,85]
NaOH/PEG	Moderate (skin, eyes and inhalation)	≈ -15 °C	No	[76]

Sustainable Natural Fiber Composites   Materials Research Forum LLC
Materials Research Foundations **122** (2022) 1-36   https://doi.org/10.21741/9781644901854-1

Besides cellulose solvents, preparation of ACCs also requires the use of proper non-solvents to regenerate the dissolved cellulose. Upon contact with the dissolved cellulose, non-solvents induce the precipitation (coagulation) of cellulose, with process kinetics depending on the relative diffusion velocities of the solvent from cellulose to the coagulation medium and of the non-solvent into cellulose. This exchange of solvent with non-solvent is generally believed to lead to the reformation of intra- and inter-hydrogen bonds previously broken during dissolution. The type of solvent and non-solvent used strongly influences the properties (morphological and mechanical) of the regenerated cellulose material [73,86,87]. Non-solvents commonly used in the preparation of ACCs are ethanol, methanol and water. In alkaline cellulose solutions acids are often used to neutralize and achieve a rapid cellulose separation, although this generally generates bubbles within the composites, increasing the number of pores in the structure and reducing their mechanical performance.

4.2 Processing routes and derived ACCs properties

During the last years ACCs have attracted great interest due to their outstanding properties, especially their mechanical strength. This property derives not only from the intrinsic mechanical properties of cellulose, but also from the already mentioned excellent interaction between matrix and reinforcement in ACCs which are both of the same chemical nature. ACCs preparation requires the total or partial solubilization of cellulose (Cellulose I) to later obtain it as regenerated cellulose (Cellulose II). The X-ray pattern of Cellulose II shows peaks at $12.0°$, $20.0°$ and $22.0°$ (2θ angle), whereas Cellulose I exhibits peaks at $14.5°$, $16.5°$ and $22.5°$ [88,89].

ACCs are mainly prepared by two methods: the partial dissolution method (or "one-step method") and the impregnation method (or "two-step method"). Both methods have been schematized in Fig. 2. In the current section both ACCs preparation methods will be described, and key process variables together with their reported influence on main composite properties will be depicted by collection of ACCs literature examples. Revision will be restricted to ACCs containing at least one cellulosic phase (matrix or reinforcement) derived from natural plant fibers.

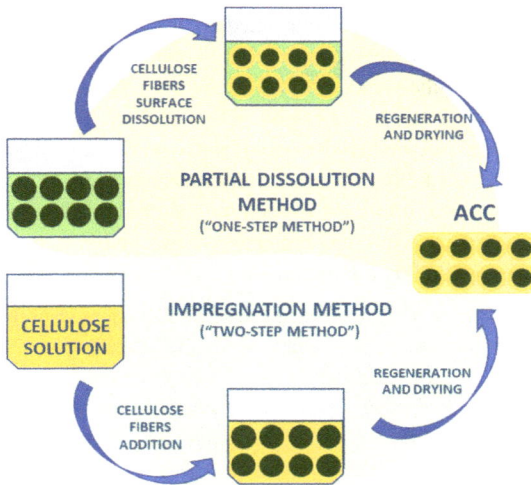

Figure 2. Diagram of ACCs production methods.

4.2.1 Partial dissolution method

In the partial dissolution method, also known as "selective dissolution method" or "one-step method", only the surface skin layer of the cellulosic fibers is dissolved; so that when regenerated it forms the matrix of the composite whereas the remaining undissolved cores of the fibers maintain their original structure and act as the reinforcement (Fig. 2) [90].

In this method, the cellulose dissolution system and conditions (the latter including dissolution time as the process variable most frequently assayed) play a very important role in the resulting ACCs properties, since they determine the extent of dissolution of the fibers. In this method the extent of dissolution has to be sufficiently high to guarantee proper composite consolidation quality, but low enough to allow the cellulosic fibers to retain their structure and intrinsic properties. Several examples of ACCs derived from natural plant fibers prepared by the partial dissolution method are listed in Table 3.

The effect of dissolution conditions, and more specifically the effect of dissolution time on the properties of ACCs based on cellulosic materials derived from natural plant fibers was described by Soykeabkaew *et al.* (2008). Authors dissolved aligned ramie fibers in DMAc/LiCl at different immersion times (1-12 h) and found that for an optimum immersion time of 2 h, a very high tensile strength and Young's modulus of the ACCs in the direction parallel to the fiber axis (longitudinal direction) could be obtained (460 MPa and 28 GPa, respectively). This was attributed to the concomitant effect of a strong interfacial interaction, a high fiber content and a high remaining fiber strength. That interaction was ascribed to the presence of strong hydrogen bonding within the cellulosic fiber, the matrix and the interphasial region. The latter was found to minimize stress concentrations and voids which otherwise could lead to crack initiation. On the other hand, dissolution times longer than 2 h were found to be detrimental to ACCs tensile properties due to the reduced remaining reinforcement volume fraction. In addition, authors observed that transparency increased with the immersion time, and that it required a perfect interphase with a perfect bonding between matrix and fiber [25].

Some years later, Arévalo *et al.* (2010) used the same solvent system to partially dissolve the surface of commercial cotton fibers and to study the effect of dissolution time (6-48 h) on the resulting ACCs properties. Authors found that intermediate dissolution times (i.e. 18 h) led to composites with the best overall mechanical performance in terms of stiffness, strength and ductility; which they attributed to an adequate balance between the extent of fiber surface dissolution (to provide sufficient interfacial adhesion to the composite) and the fraction of remaining fiber cores acting as reinforcement. Besides, authors also highlighted that heterogeneous dissolution processes which led to composite materials with regions with completely dissolved fibers (providing no reinforcement) and others with nearly undissolved fibers (providing no interfacial adhesion), resulted in reduced composites mechanical properties [36].

Also using DMAc/LiCl for the preparation of ACCs from pretreated (by steam explosion and alkaline treatment) pineapple leaf microfibers, Tanpichai & Witayakran (2015) observed that tensile strength and strain at break of ACCs significantly increased with dissolution time. The highest values (42.8 MPa and 28.7 %, respectively) were attained at 120 min (the longest time interval assayed), being the measured tensile strength value almost 28 times superior to that of pristine pineapple leaf fiber mats [32].

Adak & Mukhopadhyay (2016) also studied the effect of the dissolution time on the properties of jute based all-cellulose composite laminates obtained by contacting under pressure alkali-treated jute fabrics with the ionic liquid 1-butyl-3-methylimidazolium chloride ([EMIM][OAc]), followed by compression molding of a stack of five fabric layers. The alkali treatment performed on the fibers was aimed at partially removing the outer layer of non-cellulosic materials while retaining good tensile properties. Authors observed that the tensile modulus increased with increasing dissolution time from 2 h to 6 h. They attributed their results to the decrease in internal voids and the improvement in intra and inter-laminar adhesion for longer dissolution times. Further increase in dissolution time up to 8 h led to a slight reduction in stiffness [28].

Table 3. Some examples of the development of ACCs from natural plant fibers using the partial dissolution method.

Cellulosic material source	Solvent system	Activation pretreatment	Partial dissolution conditions assayed	References
As-received ramie fibers	DMAc/8 wt% LiCl	Succesive immersion in water, acetone and DMAc	Room temperature, 1-12 h	[25]
Commercial cotton pads	DMAc/8 wt% LiCl	Succesive immersion in water, acetone and DMAc	Room temperature, 6-48 h	[36]
Pretreated pineapple leaf fibers	DMAc/8 wt% LiCl	Succesive immersion in water, methanol, acetone and DMAc	Room temperature, 30-120 min	[32]
Pretreated commercial jute fabrics	[EMIM][OAc]	none	110 °C, 500 kPa, 2-8 h	[28]
Flax roving yarns	[EMIM][OAc]	none	Room temperature, 10-55 min + room temperature, 5 min, 80 bar	[21]
Commercial linen textiles	[BMIMAc]	none	110 °C, 60 min, 1.5 MPa + 110 °C, 20 min, 2.5 MPa	[20]
Commercial cotton fabrics	[BMIMCl]	none	100 °C, 30 min + 150 °C, 30 min, 5-15 MPa	[37]
Commercial hemp fabrics	[BMIMCl]	none	100 °C, 30 min + 100-130 °C, 10 min, 10 MPa	[91]

Chen *et al.* (2020) also prepared ACCs by the partial dissolution method and particularly studied the effect of dissolution time on their properties. In this case, authors unidirectionally aligned entire flax fibers and controlled their exposition to the [EMIM][OAc] ionic liquid during different times (15, 30, 45 and 60 min), followed by ionic liquid removal and hot pressing. Based on optical microscopy, X-rays diffraction studies and mechanical properties measurement as a function of dissolution time, this study clearly highlighted that the performance of the ACCs depended on a compromise between the formation of enough matrix to glue fibers together and fill voids (ideally only by pectin, hemicellulose and low molecular weight and amorphous cellulose in the outer layers, primary wall and S1 layers of flax fiber), and the non-dissolution of the Cellulose I crystalline domains in the S2 layer, as well as the restriction of the formation of mechanically weaker Cellulose II. The mechanical properties of these unidirectionally oriented ACCs obtained were higher than those of many "conventional" polymer composites reinforced with flax fibers. The best tensile properties values were observed for the composite obtained after 45 min dissolution, resulting in a tensile strength of 151.3 MPa and a Young's modulus of 10.1 GPa [21].

Besides dissolution time, which has been the process variable most frequently used to control ACCs properties, other key factors that have been reported to affect their performance are the cellulose source [20], the pressure applied during dissolution [37], and the hot-pressing temperature [91].

4.2.2 Impregnation method

The impregnation method (or two-step method) involves the dissolution of the cellulosic material in a solvent system followed by regeneration steps conducted in the presence of the undissolved cellulosic reinforcement (Fig. 2). Literature review on ACCs derived from plant fibers produced by this method illustrated that there are some key variables that highly influence their resulting properties, such as pretreatments applied to the reinforcing fibers, the amount of reinforcement used and the concentration of cellulose in the solution to be regenerated to conform the matrix. Another variable which has also proved to impact on ACCs properties is the length of the reinforcement. Some examples of ACCs prepared by the impregnation method from different cellulosic materials derived from natural plant fibers are listed in Table 4.

The influence of the pretreatment applied to the reinforcement on the effect of impregnation time and derived ACCs properties was studied by Nishino and coworkers (2004). Authors completely dissolved Kraft pulp fibers from coniferous trees (3 wt%) in an 8 wt% DMAc/LiCl solution and then regenerated the cellulose in the presence of unidirectionally aligned ramie fibers under a reduced pressure, attaining a fiber volume fraction of 80 %. Given the solvent system used (Section 4.1) pretreatments involving the successive immersion of the Kraft pulp fibers in water, acetone and DMAc (24 h each at 25 °C), were necessary for the successful dissolution of the cellulose to be used to conform the regenerated matrix. Aiming to allow partial swelling or dissolution of the fiber surface

Sustainable Natural Fiber Composites   Materials Research Forum LLC
Materials Research Foundations **122** (2022) 1-36   https://doi.org/10.21741/9781644901854-1

and thus stimulate interdiffusion of cellulose molecules across the interface, ramie fibers were also pretreated previous to their impregnation with the cellulose solution during 24 and 72 h. When the pretreated fibers were in contact with cellulose solution for a total impregnation time of 24 h, the surface of the pretreated fibers was partially dissolved into the solvent contributing to the interdiffusion of cellulose molecules across the fiber/matrix interface, whereas the crystal system of the remained core part of the fiber was maintained as Cellulose I. However, for this impregnation time, the mechanical properties of the ACCs with un-pretreated and pretreated fibers were similar. Further increase in the impregnation time (72 h) of pretreated fibers was observed to decrease the composite tensile strength and Young's modulus, which was attributed to their overdissolving during impregnation, thus hindering the high mechanical performance of the incorporated fibers [3].

Wei *et al*. (2016) used raw, alkali-treated (4% NaOH, 60 °C) and activated (water, ethanol and DMAc) straw cellulose fibers to prepare three different ACCs. The matrix in all cases was commercial MCC dissolved in DMAc/LiCl. After characterization authors found that, although the three types of fibers exhibited similar crystalline structures (i.e. Cellulose I), they showed differences in their surface morphologies. The highest tensile strength value of 650.2 MPa was reported for the ACC containing the activated fibers as reinforcement, which was attributed to the highest interfacial adhesion between the straw cellulose fibers and the regenerated MCC [92].

Tanpichai and Witayakran (2017) prepared ACCs from MCC dissolved in DMAc/LiCl and pineapple leaf fibers pretreated either by alkaline treatment or by steam explosion+alkaline treatment. Authors reported lower Young's modulus and tensile strength values in ACCs containing alkali-pretreated fibers, whereas the corresponding values were higher when fibers were additionally pretreated by steam explosion which was ascribed to the larger surface area of the fibers interacting with the cellulose matrix. The fiber width and amounts of the matrix filling in pores in a mat were found to dominate the mechanical properties of the prepared ACCs [33].

The content of reinforcement is also a frequently studied variable in the production of ACCs by the impregnation method. Yang *et al*. (2010) obtained all-cellulose films containing different contents of pretreated ramie fibers (from 5 to 25 wt.%) and analyzed the effect of fiber content on several films properties. The matrix was regenerated from an aqueous solution of cellulose from cotton linter pulps dissolved in a NaOH/Urea, whereas ramie fibers used as reinforcement were pretreated with 10 wt.% NaOH at 60°C for 4 h in order to remove lignin and other surface organic compounds. The increase in the ramie fibers content from 0 to 25 wt.% led to a sharp decrease (from 86.9% to 16.2%) in the films optical transmittance, which was attributed to the introduction of the fibers on the interface structure leading to optical scattering and refraction, especially when the ramie fiber content was high. Regarding the mechanical properties, authors observed that when the pretreated ramie fibers content was increased up to 15 wt%, tensile strength increased from 98.4 to 124.3 MPa; which was explained in terms of the intrinsic excellent strength of the pretreated fibers and strong interactions with the cellulose matrix through hydrogen bonds. On the other hand, Young's modulus was found to increase with fiber content in all the interval assayed [26].

Different reinforcement contents (i.e. 17-75 wt%) were also assayed by Labidi *et al.* (2019) during the preparation of ACCs in which both, matrix and reinforcing fibers, were obtained from alfa (*Stipa tenacissima* L) fibers. The matrix was prepared from previously acid hydrolyzed-alfa pulps dissolved in 8 wt% NaOH aqueous solution. Authors observed that the adhesion between the reinforcing fibers and the matrix was excellent. However, at high concentrations of reinforcing fibers there was "not enough" matrix to "glue" fibers together, leading to the presence of voids which was reflected by the progressive decrease in the composite density. As a result, Young's modulus decreased and yield stress showed a maximum as a function of reinforcement concentration [53].

More recently, Kumar and co-workers (2018) developed ACCs using a matrix made of cotton linters pulp dissolved in 8 wt% LiOH/15 wt% urea aqueous solution. Cellulose isolated from native African Napier grass fibers were incorporated at different contents (from 5 to 25 wt%) as reinforcement. Authors observed that the thermal stability of the ACCs was slightly higher than that of the cellulose matrix and that it increased with the reinforcement content. However, the tensile strength values of the ACCs produced were lower than those of the cellulose matrix [38]. Similar findings were also reported by Ashok *et al.* (2015) for ACCs based on cotton linters as matrix and *Thespesia Lampas* fibers as reinforcement [30].

On the other hand, Qin *et al.* (2008) explored the effect of the cellulose content of the solution prepared to generate the matrix on the ACCs mechanical properties. Unidirectionally aligned pretreated ramie fibers were impregnated by cellulose solutions (from 1 to 7% wt/v) made of ramie fibers dissolved in DMAc/LiCl after pretreatment by successive immersion in water/acetone/DMAc (24 h at 25 °C each). The best fiber wetting condition and related mechanical properties were found in the ACCs based on 4 % wt/v cellulose concentration in solution, and a total fiber volume fraction of 85%. Higher amounts of cellulose increased the viscosity of the matrix solution and prevented the reinforcement from wetting. Authors also applied an additional post-treatment with NaOH to the ACCs which resulted in an improvement in tensile strength. They attributed these results to the penetration of the alkaline solution through the composites and to the fiber swelling, successfully filling the internal voids and cracks [27].

Table 4. Some examples of the development of ACCs from natural plant fibers using the impregnation method.

Cellulosic matrix	Cellulosic reinforce-ment	Solvent system	Activation pretreatment applied to the cellulose used to form the matrix	Dissolution conditions	Reinfor-cement content	References
Kraft pulp from coniferous trees	Raw and pretreated refined ramie fibers	DMAc/ 8 wt% LiCl	Succesive immersion in water, acetone and DMAc	n.a.	80% volume fraction	[3]
Commercial MCC	Raw and pretreated straw cellulose fibers	DMAc/ 5 wt% LiCl	Succesive immersion in water, ethanol and DMAc	0 °C, 24 h	50 wt%	[92]
Commercial MCC	Pretreated pineapple leaf fibers	DMAc/ 8 wt% LiCl	Succesive immersion in water, methanol, acetone and DMAc	Room temperature, 24 h	3 wt%	[33]
Cotton linters pulp	Pretreated ramie fibers	7 wt% NaOH/ 12 wt % Urea	none	-12 °C, 2 min with stirring	5-25 wt%	[26]
Alfa stems pulp	Pretreated alfa stems fibers	8 wt% NaOH	Mild milling and sulphuric acid hydrolisis (pretreatment devoted to reduce de polymerization degree and enhace dissolution)	-7 °C, 2 h, 300 rpm	17-75 wt%	[53]
Cotton linters pulp	Pretreated African Napier grass fibers	8 wt% LiOH/15 wt% Urea	none	−12.5 °C with stirring	5-25 wt%	[38]

Cotton linters pulp	Pretreated *Thespesia Lampas* fibers	7 wt% NaOH/ 12 wt % Urea	none	−12 °C with stirring	1-5 wt%	[30]
Ramie fibers	As-received ramie fibers	DMAc/ 8 wt% LiCl	Succesive immersion in water, acetone and DMAc	n.a.	85-95% volume fraction	[27]
Hemp fibers	Pretreated hemp fibers	NMMO	none	135-140 °C, 12-15 min, 200-300 rpm	40 wt%	[23]

Finally, the effect of the length of the reinforcing fibers on ACCs performance was analyzed by Ouajai and Shanks (2009). NaOH pretreated-hemp fibers were introduced into a 12% w/v cellulose NMMO solution prepared using previously ground 200 μm hemp fibers. Authors kept the amount of the reinforcing fibers constant for every composite, but their length was varied (i.e. 45, 100 and 500 μm). Longer fibers were shown to be associated with larger voids and worse mechanical properties of the resulting ACCs [23].

Fig. 3 summarizes the best tensile properties obtained for ACCs produced using natural plant fibers (as matrix, as reinforcement or both). ACCs prepared by the partial dissolution method and by the impregnation method have been both included. As it is shown, different formulations, processing methods and conditions used have led to an extensive range of reported mechanical properties.

Materials Research Forum LLC

https://doi.org/10.21741/9781644901854-1

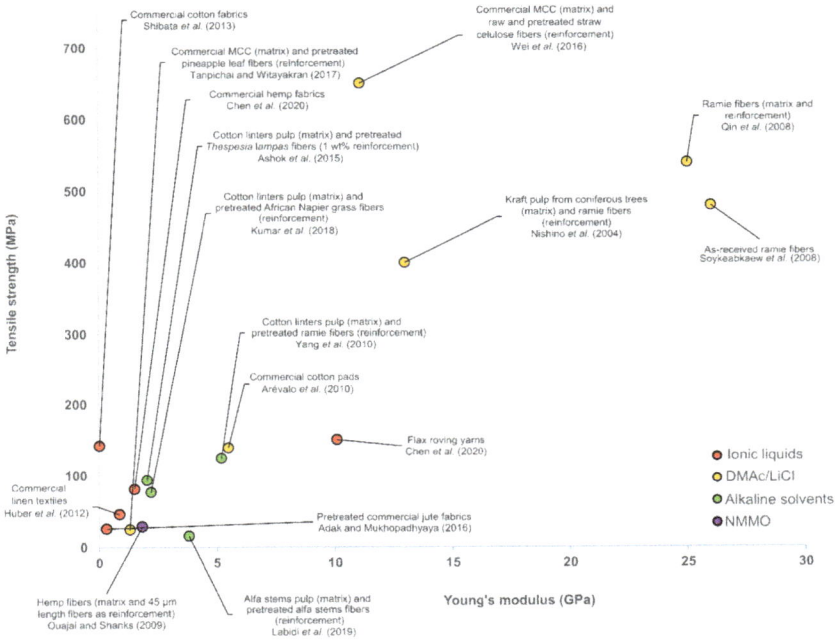

Figure 3. Young's modulus and tensile strength values for ACCs prepared from natural plant fibers.

5.   All-cellulose nanocomposites (ACNCs)

Besides the attractive properties of cellulose already described, over the past decades cellulose nanomaterials (i.e. cellulose nanofibrils (CNF) and cellulose nanocrystals (CNC)), have drawn especial attention due to their nano-sized scale and consequent large surface area, and to their high tensile strength and stiffness. These characteristics turn CNF and CNC into perfect candidates for the reinforcement of nanocomposites, although -as in the case of micrometric cellulose reinforcements-, their compatibility with common commercial hydrophobic biodegradable polymeric matrices is an issue which needs to be addressed. In this context, all-cellulose nanocomposites (ACNCs), and particularly those which include nanocellulosic materials derived from plant fibers, again appear as a very attractive option for a number of applications.

Sustainable Natural Fiber Composites   Materials Research Forum LLC
Materials Research Foundations **122** (2022) 1-36   https://doi.org/10.21741/9781644901854-1

ACNCs are often recognized for their high transparency resulting from the aforementioned nano-sized dimension of the reinforcement that prevents light scattering [93,94]. Nevertheless, this property strongly depends on the reinforcement content, since high loadings tend to form non-nanosized aggregates increasing light scattering and reducing the light transmittance. Optical transparency of ACNCs also relays on a good interfacial adhesion between matrix and reinforcement [4,93], which results in less light scattering at the composite interface/interphase [5,27]. ACNCs may also have distinguishing gas and water barrier properties as a result of their high specific surface area, high aspect ratio and the capacity of cellulose nanomaterials to form strong networks through hydrogen-bonding and to create a tortuous path length for molecules diffusion. Water vapor permeability of ACNCs has been reported to increase with dissolution time since gas transfer is enhanced in the less ordered composite regions [52]. ACNCs have also resulted in significantly improved mechanical properties when compared to their micrometer-scale counterparts. Yousefi *et al.* (2011) developed ACCs and ACNCs from microfiber and grinding-based nanofiber of canola straw, respectively, and found that ACNCs showed significantly higher tensile strength, stiffness and ductility values than ACCs [49].

Similarly to ACCs, ACNCs can be obtained by both, partial dissolution and impregnation methods. In the first approach, the solvent system employed and the dissolution conditions (especially dissolution time which is gain the most frequently studied variable) play a very important role, since -as already described for ACCs- these factors determine not only the possibility to adequately dissolve the surface of the cellulosic nano-reinforcements but also to retain their original structure and derived properties. Cellulose crystallinity loss due to excessively long dissolution times may affect the tensile strength and stiffness of the resulting ACNCs as well as their thermal stability, barrier capacity, etc. [49,52,95]. On the other hand, too short dissolution times would lead to a poor capacity to form a weld layer which is expected to act as the matrix by encapsulating the remaining undissolved nanofibers, filling the voids, and keeping adjacent nanofibers together [90].

The key effect of dissolution time (5-120 min) on ACNCs properties prepared by the partial dissolution method was illustrated by Ghaderi *et al.* (2014), who developed ACNCs based on sugarcane bagasse cellulose pulp nanosheets using as solvent system a solution 8 wt% DMAc/LiCl. Authors measured the thermal stability of the composites prepared using dissolution times of 10, 60 and 120 min and reported a decreasing trend attributed to reduced crystallinity and smaller cellulose crystallite size. Besides, water vapor permeability of the nanocomposites increased with dissolution time, which was associated with less ordered cellulose chains conforming the matrix and thus favoring the transfer of water vapor through the material. In reference to mechanical properties, the highest values of strength and stiffness were found for a dissolution time of 10 minutes which was

Sustainable Natural Fiber Composites   Materials Research Forum LLC
Materials Research Foundations **122** (2022) 1-36   https://doi.org/10.21741/9781644901854-1

ascribed to an optimum ratio of reinforcement and matrix phases for cementing adjacent nanofibers [52]. In a similar approach, Yousefi and co-workers (2010) obtained ACNCs by partially dissolving NFC sheets at different times (5-60 min) in DMAc/LiCl. Authors found that at a dissolution time of 10 min the ACNCs strength and stiffness reached the highest values (156 MPa and 13.2 GPa, respectively) [95]. Using the same solvent system, Yousefi *et al.* (2011) later reported a similar trend for ACNCs prepared from grinding-based nanofibers of canola straw [49]. On the other hand, when using as cellulose solvent the ionic liquid 1-butyl-3-methylimidazolium chloride (BMIMCl), the same group could dissolve microfibrillated cellulose sheets without previous defibrillation steps, obtaining ACNCs with better mechanical, optical and air barrier properties than the nanofibrillated sheet counterpart [50]. Besides dissolution time, ACNCs properties have also shown to be significantly affected by the concentration of cellulose in the solvent system [93].

ACNCs can also be developed through the impregnation method. In these composites, the cellulosic material is first completely dissolved in a proper solvent system, and then this solution is mixed with the nanosized cellulose reinforcement followed by regeneration. As already illustrated for ACCs, in this method a determining factor of the resulting ACNCs properties is the nano-reinforcement content. Qi *et al.* (2009) employed this method to develop ACNCs from cotton linters pulp dissolved in a NaOH/Urea and cellulose nanowhiskers obtained by acid hydrolysis of cotton linters with higher polymerization degree. Authors evaluated the effect of the nanofiller content (0-20 wt%) on chosen ACNCs properties, with higher amounts resulting in an increase in the Cellulose I content and intermediate values (i.e. 10 wt%) leading to optimum thermal stability, Young's modulus and tensile strength. The presence of agglomerates at higher filler contents also had a negative impact on the optical transparency of ACNCs [39]. Bian *et al.* (2020) also studied the effect of nano-reinforcement content (0-3 wt%) in ACNCs prepared with corn stalk cellulose nanofibrils obtained by TEMPO oxidation and a matrix of the same cellulosic source dissolved in DMAc/LiCl. Results showed that the tensile properties increased with the nanofibrils content [46]. Other variables that have also shown to significantly affect the performance of ACNCs (i.e. tensile properties, oxygen-barrier properties, thermal expansion coefficient and optical properties) obtained by the impregnation method are the nano-reinforcement characteristics resulting from the cellulose source and nanocellulose isolation methods used that may alter its surface groups (e.g. TEMPO-oxidation mediated isolation) [41,96], the cellulosic matrix source [46], the reinforcement orientation [41,42] and the regeneration conditions [43].

The best mechanical properties of selected ACNCs reported in the literature produced by using cellulosic materials derived from natural plant fibers (as matrix, as nano-reinforcement or both) are listed in Table 5. Examples of ACNCs obtained by both preparation methods have been included.

Table 5. Optimum mechanical properties of ACNCs derived from natural plant fibers.

Partial dissolution method (One-step method)							
Cellulosic material source	**Solvent system**	**Optimum dissolution time**	**Cellulosic materials transversal dimension (nm)**	**Young´s modulus (GPa)**	**Tensile strength (MPa)**	**Strain at break (%)**	**References**
Sugarcane bagasse pulp "nanosheets"	DMAc/ 8 wt% LiCl	10 min	39	12.8	140	8	[52]
Commercial microfibrillated cellulose "NFC sheets"	DMAc/ 8 wt% LiCl	10 min	42	13.2	156	12	[95]
Canola straw pulp "nanofiber sheets"	DMAc/ 8 wt% LiCl	10 min	32	15.2	164	9.9	[49]
Canola straw pulp "microfiber sheets"	[BMIMCl]	8 h	53	20	208	9.8	[50]
Commercial MCC	DMAc/ 8 wt% LiCl	12 h	-	13.1	242.8	8.6	[93]
Impregnation method (Two-step method)							
Cellulosic matrix/ reinfor-cement	**Solvent system**	**Optimum reinforcement content (wt%)**	**Reinforcement transversal dimension (nm)**	**Young´s modulus (GPa)**	**Tensile strength (MPa)**	**Strain at break (%)**	**References**
Cotton linters pulp (DP 500)/ CNW from hydrolysis of cotton linters pulp (DP 1985)	7 wt% NaOH/ 12 wt % Urea	10	21	5.1	124	~6	[39]
Corn stalk	DMAc/	3	20-50	1.5	52.6	n.a.	[46]

cellulose/ corn stalk TEMPO-oxidized cellulose nanofibrils	6 wt% LiCl						
Cotton linters pulp/ TEMPO-oxidized cellulose nanofibrils from bleached Kraft pulp	7 wt% NaOH/ 12 wt% Urea	1	2.7 (width)	6.2	167	10	[96]
Cotton linters pulp (DP 1196)/ TEMPO-oxidized cellulose nanofibrils from bleached Kraft pulp	DMAc/ 8 wt% LiCl	30	~3 (width)	13.9	317.3	4.4	[41]
Commercial MCC/ CNW from hydrolysis of tunicates pulp	DMAc/ 8 wt% LiCl	15% (v/v)	72.8 (aspect ratio)	12.5	175.6	4.1	[42]
Cotton linters pulp (DP 330)/ *in-situ* regenerated nanocellulose fibers	7 wt% NaOH/ 12 wt% Urea	n.a.	10-30	~4.8	135	18	[43]

Sustainable Natural Fiber Composites   Materials Research Forum LLC
Materials Research Foundations **122** (2022) 1-36   https://doi.org/10.21741/9781644901854-1

Conclusions and future trends

Composite materials with attractive properties to be used in different applications such as biomedical, electronic, automotive and aerospace industries are currently the subject of an increasing demand. However, attaining a proper interaction between matrix and reinforcement is a key and sometimes difficult issue in the field of composites; particularly in those involving natural fibers and commercially available biodegradable non-polar polymeric matrices. Incompatibility issues have been frequently assessed by physical or chemical modification of fibers/nanofibers surfaces; but more recently the concept of single polymer composites in which chemical similarity between matrix and reinforcement leads to strong interfacial adhesion, has merit special attention.

In this context, all-cellulose composites in which both matrix and reinforcement are cellulose-based, are of particular interest. In the current chapter, wet processing methods (i.e. partial dissolution of cellulosic material + regeneration, and complete dissolution of cellulosic material + regeneration in the presence of undissolved cellulosic fibers) and resulting properties of ACCs and ACNCs have been described. In all cases, attention has been centered on reports on ACCs and ACNCs involving cellulose-based materials derived from natural plant fibers, being ramie, cotton, flax and canola straw the cellulose fibers which have been mostly used.

The literature revision performed illustrated the key role of the cellulosic materials source, the solvent system, the dissolution conditions, the reinforcement content and alignment, and the regeneration process used on the overall ACCs and ACNCs performance; especially in terms of their resulting mechanical properties. In the case of composites prepared by the partial dissolution method, dissolution extent is the key factor since the amount of cellulose dissolved and regenerated should be enough to fuse the cellulosic fibers, while avoiding a detrimental decrease in crystallinity and fibers/nanofibers intrinsic strength. In reference to the impregnation method, reinforcement content plays a very important role. With respect to solvent systems, the most used one for the preparation of ACCs and ACNCs from natural plant fibers has shown to be DMAc/LiCl; although ionic liquids and NaOH/urea-based systems appear as greener alternatives with high potential.

High chemical compatibility between matrix and reinforcement, and appropriate processing conditions that allow the creation of a strong interface while guaranteeing that the excellent reinforcement/nano-reinforcement intrinsic mechanical properties are kept, have led to outstanding tensile properties of ACCs and ACNCs, which currently appear as promising biobased and biodegradable alternatives in the field of sustainable bio-composites. Among them, all-cellulose composites and nanocomposites derived from natural plant fibers deserve special attention in view of their huge availability, wide variety and relative low cost. Interesting ACCs and ACNCs applications in the field of corrugated board products and cellulose-based food packaging have recently been proposed, and a number of other uses which would benefit from their mechanical properties and fully bio-based and bio-degradable character are envisaged.

However, there are still some issues to assess for triggering large-scale production of ACCs and ACNCs, such as finding suitable cellulose solvents (i.e. in terms of cost, toxicity, flammability, cellulose dissolution extent as a function of molecular weight, operation conditions required (e.g. temperature), reciclability), extending the research activities on the dissolution evolution of low-cost natural fibers, and limiting ACCs moisture sensitivity which negatively impacts on the mechanical properties and dimensional stability of the composites.

In this context, derivatization of cellulosic materials opens a less explored but also very attractive opportunity in the field. As reviewed in this chapter, ACCs preparation relies on cellulose dissolution since due to the extensive intramolecular and intermolecular hydrogen bonds in native cellulose, this macromolecule cannot be melt processed. However, proper derivatization of cellulosic materials (e.g. benzylation, esterification, oxypropylation, etc.) may confer them/their surface thermoplastic behavior and allow derivatized all-cellulose composites (DACCs) production through processing methods similar to those originally proposed for SPCs (i.e. hot compaction). Derivatization aimed at converting the surface of the cellulose-based material into a thermoset polymer which can be later cross-linked during hot pressing is also of high interest for DACCs production. Although since they have been the focus of much less research their thorough description was beyond the scope of this chapter, DACCs derived from natural plant fibers do appear as interesting and versatile alternatives to continue exploring. When doing so, proper manipulation of sensitive process variables to tailor the extent of derivatization of cellulosic substrates is of special importance since, together with the nature of the modifying agent used, they will both play a key role in the resulting DACCs properties.

References

[1]   Fortune Business Insights, Cellulose: Global market analysis, insights and forecast, 2019-2026, (2020).

[2]   J.A. Ávila Ramírez, P. Cerrutti, C. Bernal, M.I. Errea, M.L. Foresti, Nanocomposites based on poly(lactic acid) and bacterial cellulose acetylated by an α-hydroxyacid catalyzed route, J. Polym. Environ., 27 (2019) 510–520. https://doi.org/10.1007/s10924-019-01367-5

[3]   T. Nishino, I. Matsuda, K. Hirao, All-cellulose composite, Macromolecules, 37 (2004) 7683–7687. https://doi.org/10.1021/ma049300h

[4]   T. Huber, J. Müssig, O. Curnow, S. Pang, S. Bickerton, M.P. Staiger, A critical review of all-cellulose composites, J. Mater. Sci., 47 (2012) 1171–1186. https://doi.org/10.1007/s10853-011-5774-3

[5]   S. Mukhopadhyay, B. Adak, Single-Polymer Composites, Chapman and Hall/CRC, Milton, 2018. https://doi.org/10.1201/9781351272247

[6]   K. Houston, M.R. Tucker, J. Chowdhury, N. Shirley, A. Little, The plant cell wall: A complex and dynamic structure as revealed by the responses of genes under stress conditions, Front. Plant Sci., 7 (2016) 1–18. https://doi.org/10.3389/fpls.2016.00984

[7]   A. Komuraiah, N.S. Kumar, B.D. Prasad, Chemical composition of natural fibers and its influence on their mechanical properties, Mech. Compos. Mater., 50 (2014) 359–376. https://doi.org/10.1007/s11029-014-9422-2

[8]   B. Madsen, E.K. Gamstedt, Wood versus plant fibers: similarities and differences in composite applications, Adv. Mater. Sci. Eng., 2013 (2013). https://doi.org/10.1155/2013/564346

[9]   H. Shaghaleh, X. Xu, S. Wang, Current progress in production of biopolymeric materials based on cellulose, cellulose nanofibers, and cellulose derivatives, RSC Adv., 8 (2018) 825–842. https://doi.org/10.1039/C7RA11157F

[10]   R.J. Moon, A. Martini, J. Nairn, J. Simonsen, J. Youngblood, Cellulose nanomaterials review: structure, properties and nanocomposites, Chem. Soc. Rev., 40 (2011) 3941–3994. https://doi.org/10.1039/c0cs00108b

[11]   A. Chami Khazraji, S. Robert, Interaction effects between cellulose and water in nanocrystalline and amorphous regions: a novel approach using molecular modeling, J. Nanomater., 2013 (2013). https://doi.org/10.1155/2013/409676

[12]   T. Kondo, Hydrogen bonds in cellulose and cellulose derivatives, in: S. Dimitriu (Ed.), Polysaccharides Struct. Divers. Funct. Versatility, 2nd Edition, CRC Press, Boca Raton, 2004: pp. 69–98. https://doi.org/10.1201/9781420030822.ch3

[13]   T. Heinze, Cellulose: Structure and Properties, in: O.J. Rojas (Ed.), Cellul. Chem. Prop. Fibers, Nanocelluloses Adv. Mater., Springer International Publishing, Cham, 2016: pp. 1–52. https://doi.org/10.1007/12_2015_319

[14]   L.C. Da Sousa, J. Humpula, V. Balan, B.E. Dale, S.P.S. Chundawat, Impact of ammonia pretreatment conditions on the Cellulose III allomorph ultrastructure and its

enzymatic digestibility, ACS Sustain. Chem. Eng., 7 (2019) 14411–14424. https://doi.org/10.1021/acssuschemeng.9b00606

[15]   T. Nishino, K. Takano, K. Nakamae, Elastic modulus of the crystalline regions of cellulose polymorphs, J. Polym. Sci. Part B Polym. Phys., 33 (1995) 1647–1651. https://doi.org/10.1002/polb.1995.090331110

[16]   C. Olsson, G. Westman, Direct dissolution of cellulose: background, means and applications, Cellul. - Fundam. Asp., (2013). https://doi.org/10.5772/52144

[17]   T. Liebert, Cellulose solvents – remarkable history, bright future, in: T.F. Liebert, T.J. Heinze, K.J. Edgar (Eds.), Cellulose solvents: for analysis, shaping and chemical modification, ACS Symposium Series, Washington, Volume 1033, 2010, pp. 3–54. https://doi.org/10.1021/bk-2010-1033.ch001

[18]   T. Heinze, A. Koschella, Solvents applied in the field of cellulose chemistry: a mini review, Polímeros, 15 (2005) 84–90. https://doi.org/10.1590/S0104-14282005000200005

[19]   S. Kalia, B.S. Kaith, I. Kaur, Pretreatments of natural fibers and their application as reinforcing material in polymer composites-A review, Polym. Eng. Sci., 49 (2009) 1253–1272. https://doi.org/10.1002/pen.21328

[20]   T. Huber, S. Pang, M.P. Staiger, All-cellulose composite laminates, Compos. Part A Appl. Sci. Manuf., 43 (2012) 1738–1745. https://doi.org/10.1016/j.compositesa.2012.04.017

[21]   F. Chen, D. Sawada, M. Hummel, H. Sixta, T. Budtova, Unidirectional all-cellulose composites from flax via controlled impregnation with ionic liquid, Polymers, 12 (2020) 1010. https://doi.org/10.3390/polym12051010

[22]   W. Gindl-Altmutter, J. Keckes, J. Plackner, F. Liebner, K. Englund, M.-P. Laborie, All-cellulose composites prepared from flax and lyocell fibres compared to epoxy–matrix composites, Compos. Sci. Technol., 72 (2012) 1304–1309. https://doi.org/10.1016/j.compscitech.2012.05.011

[23]   S. Ouajai, R.A. Shanks, Preparation, structure and mechanical properties of all-hemp cellulose biocomposites, Compos. Sci. Technol., 69 (2009) 2119–2126. https://doi.org/10.1016/j.compscitech.2009.05.005

[24]   L. Brinchi, F. Cotana, E. Fortunati, J.M. Kenny, Production of nanocrystalline cellulose from lignocellulosic biomass: Technology and applications, Carbohydr. Polym., 94 (2013) 154–169. https://doi.org/10.1016/j.carbpol.2013.01.033

[25]   N. Soykeabkaew, N. Arimoto, T. Nishino, T. Peijs, All-cellulose composites by surface selective dissolution of aligned ligno-cellulosic fibres, Compos. Sci. Technol., 68 (2008) 2201–2207. https://doi.org/10.1016/j.compscitech.2008.03.023

[26]   Q. Yang, A. Lue, L. Zhang, Reinforcement of ramie fibers on regenerated cellulose films, Compos. Sci. Technol., 70 (2010) 2319–2324. https://doi.org/10.1016/j.compscitech.2010.09.012

[27]   C. Qin, N. Soykeabkaew, N. Xiuyuan, T. Peijs, The effect of fibre volume fraction and mercerization on the properties of all-cellulose composites, Carbohydr. Polym., 71 (2008) 458–467. https://doi.org/10.1016/j.carbpol.2007.06.019

[28]   B. Adak, S. Mukhopadhyay, Jute based all-cellulose composite laminates, Indian J. Fibre Text. Res., 41 (2016) 380–384.

[29]   K.O. Reddy, B. Ashok, K.R.N. Reddy, Y.E. Feng, J. Zhang, A.V. Rajulu, Extraction and characterization of novel lignocellulosic fibers from *Thespesia Lampas* plant, Int. J. Polym. Anal. Charact., 19 (2014) 48–61. https://doi.org/10.1080/1023666X.2014.854520

[30]   B. Ashok, K.O. Reddy, K. Madhukar, J. Cai, L. Zhang, A.V. Rajulu, Properties of cellulose/*Thespesia Lampas* short fibers bio-composite films, Carbohydr. Polym., 127 (2015) 110–115. https://doi.org/10.1016/j.carbpol.2015.03.054

[31]   A.K. Mohanty, M. Misra, G. Hinrichsen, Biofibres, biodegradable polymers and biocomposites: An overview, Macromol. Mater. Eng., 276–277 (2000) 1–24. https://doi.org/10.1002/(SICI)1439-2054(20000301)276:1<1::AID-MAME1>3.0.CO;2-W

[32]   S. Tanpichai, S. Witayakran, Mechanical properties of all-cellulose composites made from pineapple leaf microfibers, Key Eng. Mater., 659 (2015) 453–457. https://doi.org/10.4028/www.scientific.net/KEM.659.453

[33]   S. Tanpichai, S. Witayakran, All-cellulose composite laminates prepared from pineapple leaf fibers treated with steam explosion and alkaline treatment, J. Reinf. Plast. Compos., 36 (2017) 1146–1155. https://doi.org/10.1177/0731684417704923

[34]   H. V Lee, S.B.A. Hamid, S.K. Zain, Conversion of lignocellulosic biomass to nanocellulose: structure and chemical process, Sci. World J., 2014 (2014) 631013. https://doi.org/10.1155/2014/631013

[35]   C. Álvarez, Ú. Montoya, P. Gañán, Materiales compuestos elaborados mediante la disolución parcial de celulosa obtenida a partir de residuos de la agroindustria platanera, Rev. Latinoam. Metal. y Mater., S1 (2009) 1213–1217.

Sustainable Natural Fiber Composites   Materials Research Forum LLC
Materials Research Foundations **122** (2022) 1-36   https://doi.org/10.21741/9781644901854-1

[36] R. Arévalo, O. Picot, R.M. Wilson, N. Soykeabkaew, T. Peijs, All-cellulose composites by partial dissolution of cotton fibres, J. Biobased Mater. Bioenergy, 4 (2010) 129–138. https://doi.org/10.1166/jbmb.2010.1077

[37] M. Shibata, N. Teramoto, T. Nakamura, Y. Saitoh, All-cellulose and all-wood composites by partial dissolution of cotton fabric and wood in ionic liquid, Carbohydr. Polym., 98 (2013) 1532–1539. https://doi.org/10.1016/j.carbpol.2013.07.062

[38] T. Senthil Muthu Kumar, N. Rajini, K. Obi Reddy, A. Varada Rajulu, S. Siengchin, N. Ayrilmis, All-cellulose composite films with cellulose matrix and Napier grass cellulose fibril fillers, Int. J. Biol. Macromol., 112 (2018) 1310–1315. https://doi.org/10.1016/j.ijbiomac.2018.01.167

[39] H. Qi, J. Cai, L. Zhang, S. Kuga, Properties of films composed of cellulose nanowhiskers and a cellulose matrix regenerated from alkali/urea solution, Biomacromolecules, 10 (2009) 1597–1602. https://doi.org/10.1021/bm9001975

[40] Q. Yang, T. Saito, L. Berglund, A. Isogai, Cellulose nanofibrils efficiently improve mechanical, thermal and oxygen-barrier properties of all-cellulose composites by nano- reinforcement mechanism and nanofibril-induced crystallization, Nanoscale, 7 (2015) 17957–17963. https://doi.org/10.1039/C5NR05511C

[41] S. Fujisawa, E. Togawa, N. Hayashi, Orientation control of cellulose nanofibrils in all-cellulose composites and mechanical properties of the films, J. Wood Sci., 62 (2016) 174–180. https://doi.org/10.1007/s10086-015-1533-4

[42] T. Pullawan, A.N. Wilkinson, S.J. Eichhorn, Influence of magnetic field alignment of cellulose whiskers on the mechanics of all-cellulose nanocomposites, Biomacromolecules, 13 (2012) 2528–2536. https://doi.org/10.1021/bm300746r

[43] Q. Cheng, D. Ye, W. Yang, S. Zhang, H. Chen, C. Chang, L. Zhang, Construction of transparent cellulose-based nanocomposite papers and potential application in flexible solar cells, ACS Sustain. Chem. Eng., 6 (2018) 8040–8047. https://doi.org/10.1021/acssuschemeng.8b01599

[44] Q. Zhao, R.C.M. Yam, B. Zhang, Y. Yang, X. Cheng, R.K.Y. Li, Novel all-cellulose ecocomposites prepared in ionic liquids, Cellulose, 16 (2009) 217–226. https://doi.org/10.1007/s10570-008-9251-3

[45] S.K. Ramamoorthy, M. Skrifvars, A. Persson, A review of natural fibers used in biocomposites: Plant, animal and regenerated cellulose fibers, Polym. Rev., 55 (2015) 107–162. https://doi.org/10.1080/15583724.2014.971124

[46]   H. Bian, P. Tu, J.Y. Chen, Fabrication of all-cellulose nanocomposites from corn stalk, J. Sci. Food Agric., 100 (2020) 4390–4399. https://doi.org/10.1002/jsfa.10476

[47]   L. Wu, F. Tian, J. Sun, On the use of cellulose nanowhisker as reinforcement in all-cellulose composite membrane from corn stalk, J. Appl. Polym. Sci., 138 (2021) 1–12. https://doi.org/10.1002/app.50206

[48]   M. Ahmadi, A.J. Latibari, M.M. Faezipour, S. Hedjazi, Neutral sulfite semi-chemical pulping of rapeseed residues, Turkish J. Agric. For., 34 (2010) 11–16.

[49]   H. Yousefi, M. Faezipour, T. Nishino, A. Shakeri, G. Ebrahimi, All-cellulose composite and nanocomposite made from partially dissolved micro- and nanofibers of canola straw, Polym. J., 43 (2011) 559–564. https://doi.org/10.1038/pj.2011.31

[50]   H. Yousefi, T. Nishino, M. Faezipour, G. Ebrahimi, A. Shakeri, Direct fabrication of all-cellulose nanocomposite from cellulose microfibers using ionic liquid-based nanowelding, Biomacromolecules, 12 (2011) 4080–4085. https://doi.org/10.1021/bm201147a

[51]   H. Yousefi, M. Mashkour, R. Yousefi, Direct solvent nanowelding of cellulose fibers to make all-cellulose nanocomposite, Cellulose, 22 (2015) 1189-1200. https://doi.org/10.1007/s10570-015-0579-1

[52]   M. Ghaderi, M. Mousavi, H. Yousefi, M. Labbafi, All-cellulose nanocomposite film made from bagasse cellulose nanofibers for food packaging application, Carbohydr. Polym., 104 (2014) 59–65. https://doi.org/10.1016/j.carbpol.2014.01.013

[53]   K. Labidi, O. Korhonen, M. Zrida, A.H. Hamzaoui, T. Budtova, All-cellulose composites from alfa and wood fibers, Ind. Crops Prod., 127 (2019) 135–141. https://doi.org/10.1016/j.indcrop.2018.10.055

[54]   V.P. Kommula, K.O. Reddy, M. Shukla, T. Marwala, E.V.S. Reddy, A.V. Rajulu, Extraction, modification, and characterization of natural ligno-cellulosic fiber strands from napier grass, Int. J. Polym. Anal. Charact., 21 (2016) 18–28. https://doi.org/10.1080/1023666X.2015.1089650

[55]   M. John, S. Thomas, Biofibres and biocomposites, Carbohydr. Polym., 71 (2008) 343–364. https://doi.org/10.1016/j.carbpol.2007.05.040

[56]   L.K. Kian, N. Saba, M. Jawaid, M.T.H. Sultan, A review on processing techniques of bast fibers nanocellulose and its polylactic acid (PLA) nanocomposites, Int. J. Biol. Macromol., 121 (2019) 1314–1328. https://doi.org/10.1016/j.ijbiomac.2018.09.040

[57]   A. Sharma, M. Thakur, M. Bhattacharya, T. Mandal, S. Goswami, Commercial application of cellulose nano-composites – A review, Biotechnol. Reports, 21 (2019) e00316. https://doi.org/10.1016/j.btre.2019.e00316

[58]   H. Charreau, E. Cavallo, M. L. Foresti, Patents involving nanocellulose: analysis of their evolution since 2010, Carbohydr. Polym., 237 (2020) 116039. https://doi.org/10.1016/j.carbpol.2020.116039

[59]   ISO/TS 20477:2017(E) (2017). Nanotechnologies – Standard terms and their definition for cellulose nanomaterial.

[60]   N.J. Capiati, R.S. Porter, The concept of one polymer composites modelled with high density polyethylene, J. Mater. Sci., 10 (1975) 1671–1677. https://doi.org/10.1007/BF00554928

[61]   J. Karger-Kocsis, T. Bárány, Single-polymer composites (SPCs): Status and future trends, Compos. Sci. Technol., 92 (2014) 77–94. https://doi.org/10.1016/j.compscitech.2013.12.006

[62]   B.D. Rabideau, A.E. Ismail, Effect of water content in n-methylmorpholine n-oxide/cellulose solutions on thermodynamics, structure, and hydrogen bonding, J. Phys. Chem. B, 119 (2015) 15014–15022. https://doi.org/10.1021/acs.jpcb.5b07500

[63]   T. Rosenau, A. Potthast, H. Sixta, P. Kosma, The chemistry of side reactions and byproduct formation in the system NMMO/cellulose (Lyocell process), Prog. Polym. Sci., 26 (2001) 1763–1837. https://doi.org/10.1016/S0079-6700(01)00023-5

[64]   A.F. Turbak, A. El-Kafrawy, J. Fred W. Snyder, A.B. Auerbach, Solvent system for cellulose, U.S. Patent 4,302,252. (1981).

[65]   U. Henniges, S. Schiehser, T. Rosenau, A. Potthast, Cellulose solubility: dissolution and analysis of "problematic" cellulose pulps in the solvent system DMAc/LiCl, in: T.F. Liebert, T.J. Heinze, K.J. Edgar (Eds.), Cellulose solvents: for analysis, shaping and chemical modification, ACS Symposium Series, Washington, Volume 1033, 2010, pp. 165-177. https://doi.org/10.1021/bk-2010-1033.ch009

[66]   J.M. Spörl, F. Batti, M.P. Vocht, R. Raab, A. Müller, F. Hermanutz, M.R. Buchmeiser, Ionic liquid approach toward manufacture and full recycling of all-cellulose composites, Macromol. Mater. Eng., 303 (2018) 1–8. https://doi.org/10.1002/mame.201700335

[67]   T. Heinze, S. Köhler, Dimethyl sulfoxide and ammonium fluorides-novel cellulose solvents, in: T.F. Liebert, T.J. Heinze, K.J. Edgar (Eds.), Cellulose solvents: for

analysis, shaping and chemical modification, ACS Symposium Series, Washington, Volume 1033, 2010, pp. 103-118. https://doi.org/10.1021/bk-2010-1033.ch005

[68]   S. Guzman-Puyol, L. Ceseracciu, G. Tedeschi, S. Marras, A. Scarpellini, J.J. Benítez, A. Athanassiou, J.A. Heredia-Guerrero, Transparent and robust all-cellulose nanocomposite packaging materials prepared in a mixture of trifluoroacetic acid and trifluoroacetic anhydride, Nanomater., 9 (2019). https://doi.org/10.3390/nano9030368

[69]   Y. Li, J. Wang, X. Liu, S. Zhang, Towards a molecular understanding of cellulose dissolution in ionic liquids: anion/cation effect, synergistic mechanism and physicochemical aspects, Chem. Sci., 9 (2018) 4027–4043. https://doi.org/10.1039/C7SC05392D

[70]   O.A. El Seoud, A. Koschella, L.C. Fidale, S. Dorn, T. Heinze, Applications of ionic liquids in carbohydrate chemistry: a window of opportunities, Biomacromolecules, 8 (2007) 2629–2647. https://doi.org/10.1021/bm070062i

[71]   Y. Fukaya, K. Hayashi, S.S. Kim, H. Ohno, Design of polar ionic liquids to solubilize cellulose without heating, in: T.F. Liebert, T.J. Heinze, K.J. Edgar (Eds.), Cellulose solvents: for analysis, shaping and chemical modification, ACS Symposium Series, Washington, Volume 1033, 2010, pp. 55-66. https://doi.org/10.1021/bk-2010-1033.ch002

[72]   N. Hildebrandt, Paper-based composites via the partial dissolution route with NaOH/urea, Dissertation thesis, University of Oulu, Finland, 2018.

[73]   B. Lindman, B. Medronho, L. Alves, C. Costa, H. Edlund, M. Norgren, The relevance of structural features of cellulose and its interactions to dissolution, regeneration, gelation and plasticization phenomena, Phys. Chem. Chem. Phys., 19 (2017) 23704–23718. https://doi.org/10.1039/C7CP02409F

[74]   A. Lue, L. Zhang, Advances in aqueous cellulose solvents, in: T.F. Liebert, T.J. Heinze, K.J. Edgar (Eds.), Cellulose solvents: for analysis, shaping and chemical modification, ACS Symposium Series, Washington, Volume 1033, 2010, pp. 67-89. https://doi.org/10.1021/bk-2010-1033.ch003

[75]   Q. Yang, H. Qi, A. Lue, K. Hu, G. Cheng, L. Zhang, Role of sodium zincate on cellulose dissolution in NaOH/urea aqueous solution at low temperature, Carbohydr. Polym., 83 (2011) 1185–1191. https://doi.org/10.1016/j.carbpol.2010.09.020

[76]   L. Yan, Z. Gao, Dissolving of cellulose in PEG/NaOH aqueous solution, Cellulose, 15 (2008) 789-796. https://doi.org/10.1007/s10570-008-9233-5

[77]   D.L. Johnson, Process for strengthening swellable fibrous material with an amine oxide and the resulting material, U. S. Patent 3,447,956. (1966)

[78]   A.J. Sayyed, N.A. Deshmukh, D. V. Pinjari, A critical review of manufacturing processes used in regenerated cellulosic fibres: viscose, cellulose acetate, cuprammonium, LiCl/DMAc, ionic liquids, and NMMO based lyocell, Cellulose, 26 (2019) 2913–2940. https://doi.org/10.1007/s10570-019-02318-y

[79]   C. McCormick, Novel cellulose solutions, U.S. Patent 4,278,790. (1980)

[80]   T.R. Dawsey, C.L. McCormick, The lithium chloride/dimethylacetamide solvent for cellulose: a literature review, J. Macromol. Sci. Part C, 30 (1990) 405–440. https://doi.org/10.1080/07366579008050914

[81]   A. Pinkert, K.N. Marsh, S. Pang, M.P. Staiger, Ionic liquids and their interaction with cellulose, Chem. Rev., 109 (2009) 6712–6728. https://doi.org/10.1021/cr9001947

[82]   S. Zhu, Y. Wu, Q. Chen, Z. Yu, C. Wang, S. Jin, Y. Ding, G. Wu, Dissolution of cellulose with ionic liquids and its application: a mini-review, Green Chem., 8 (2006) 325-327. https://doi.org/10.1039/b601395c

[83]   M. Hasegawa, A. Isogai, F. Onabe, M. Usuda, Dissolving states of cellulose and chitosan in trifluoroacetic acid, J. Appl. Polym. Sci., 45 (1992) 1857–1863. https://doi.org/10.1002/app.1992.070451020

[84]   A. Isogai, R.H. Atalla, Dissolution of cellulose in aqueous NaOH solutions, Cellulose, 5 (1998) 309–319. https://doi.org/10.1023/A:1009272632367

[85]   H. Qi, C. Chang, L. Zhang, Effects of temperature and molecular weight on dissolution of cellulose in NaOH/urea aqueous solution, Cellulose, 15 (2008) 779–787. https://doi.org/10.1007/s10570-008-9230-8

[86]   N. Isobe, S. Kimura, M. Wada, S. Kuga, Mechanism of cellulose gelation from aqueous alkali-urea solution, Carbohydr. Polym., 89 (2012) 1298–1300. https://doi.org/10.1016/j.carbpol.2012.03.023

[87]   H. Leipner, S. Fischer, E. Brendler, W. Voigt, Structural changes of cellulose dissolved in molten salt hydrates, Macromol. Chem. Phys., 201 (2000) 2041–2049. https://doi.org/10.1002/1521-3935(20001001)201:15<2041::AID-MACP2041>3.0.CO;2-E

[88]   J. Gong, J. Li, J. Xu, Z. Xiang, L. Mo, Research on cellulose nanocrystals produced from cellulose sources with various polymorphs, RSC Adv., 7 (2017) 33486–33493. https://doi.org/10.1039/C7RA06222B

[89]   E.N. Johnson Ford, S.K. Mendon, S.F. Thames, J.W. Rawlins, X-ray diffraction of cotton treated with neutralized vegetable oil-based macromolecular crosslinkers, J. Eng. Fiber. Fabr., 5 (2010) 10–20. https://doi.org/10.1177/155892501000500102

[90]   T. Nishino, T. Peijs, All -cellulose Composites, in: K. Oksman, A. Mathew, A. Bismarck, O. Rojas, M. Sain (Eds.), Handb. Green Mater., World Scientific Publishing, Singapore, 2014: pp. 201–216. https://doi.org/10.1142/9789814566469_0028

[91]   K. Chen, W. Xu, Y. Ding, P. Xue, P. Sheng, H. Qiao, J. He, Hemp-based all-cellulose composites through ionic liquid promoted controllable dissolution and structural control, Carbohydr. Polym., 235 (2020) 116027. https://doi.org/10.1016/j.carbpol.2020.116027

[92]   X. Wei, W. Wei, Y. Cui, T. Lu, M. Jiang, Z. Zhou, Y. Wang, All-cellulose composites with ultra-high mechanical properties prepared through using straw cellulose fiber, RSC Adv., 6 (2016) 93428–93435. https://doi.org/10.1039/C6RA20533J

[93]   W. Gindl, J. Keckes, All-cellulose nanocomposite, Polymer, 46 (2005) 10221–10225. https://doi.org/10.1016/j.polymer.2005.08.040

[94]   M. Nogi, K. Handa, A.N. Nakagaito, H. Yano, Optically transparent bionanofiber composites with low sensitivity to refractive index of the polymer matrix, Appl. Phys. Lett., 87 (2005) 1–3. https://doi.org/10.1063/1.2146056

[95]   H. Yousefi, T. Nishino, M. Faezipour, G. Ebrahimi, A. Shakeri, S. Morimune, All-cellulose nanocomposite made from nanofibrillated cellulose, Adv. Compos. Lett., 19 (2010) 190-195. https://doi.org/10.1177/096369351001900602

[96]   Q. Yang, T. Saito, L.A. Berglund, A. Isogai, Cellulose nanofibrils improve the properties of all-cellulose composites by the nano-reinforcement mechanism and nanofibril-induced crystallization, Nanoscale, 7 (2015) 17957–17963. https://doi.org/10.1039/C5NR05511C

Sustainable Natural Fiber Composites
Materials Research Foundations **122** (2022) 37-76

Materials Research Forum LLC
https://doi.org/10.21741/9781644901854-2

Chapter 2

Innovations in Natural Fibre Reinforced Composites for Improved Performance

N. Gokarneshan[1*], PG Anandhakrishnan[1], R. Sasirekha[1], Obulam Vidya Sagar[1], D. Haritha[1], and K.M. Pachiyappan[2]

[1]Department of Fashion Design and Arts,   Hindustan Institute of Technology and Science, Chennai India

[2]Department of Costume design and fashion, PSG College of arts and science, Coimbatore

India

* advaitcbe@rediffmail.com

Abstract

Natural fibre reinforced composites have become the new choice owing to their versatility in varied end use applications, economical and eco considerations as well. Efforts have been directed towards replacing them with their synthetic counter parts which prove to be more expensive. Mechanical and tribological characteristics of natural fibre reinforced composites have gained focus. Different types of natural fibre composites have been chemically treated to improve certain performance aspects. Also, properties related to energy absorption and water absorption have been evaluated. Studies have been carried out on fibres that include coconut sheath, flax seed fibre, jute, and areca sheath fibre. The ballistic performance, abrasive behaviour, ecological aspects, etc. have been given due consideration.

Keywords

Reinforced Composites, Cocos Nucifera, Jute, Properties, Applications, Fiber

Contents

Innovations in Natural Fibre Reinforced Composites for Improved Performance 37

1.   Introduction...39

2. **Epoxy composites reinforced with hybrid Kevlar VR /Cocos nucifera sheath** **40**

2.1 Influence of stacking sequence on the energy absorption of laminated composites..42

2.2 Influence of stacking sequence on the ballistic limit of laminated composites..44

2.3 Influence of hybridization on the ballistic performance of laminated composites..45

2.4 Influence of areal density on the ballistic performance of laminated composites..46

2.5 Absorption of specific energy...46

2.6 Studies on failure 47

3. **Flaxseed fiber bundle-reinforced polybutylene succinate composites48**

3.1 Studies on composition of flaxseed fiber bundles...........................50

3.2 Assessment of flaxseed fibre bundles by SEM 50

3.3 Tensile property of flaxseed fiber bundles 50

3.4 Tensile and flexural strength of flaxseed/PBS composites 51

3.5 Assessment of flaxseed/PBS composites using SEM 52

4. **Application of surface treatments on jute fibre reinforced polypropylene composites** **54**

4.1 Study of morphology 54

4.2 Studies using fourier transform infrared spectroscopy....................55

4.3 Studies on X-ray diffraction 56

4.4 Study of surface energy 56

4.5 Studies of tensile properties of single fiber 57

4.6 Study of tensile strength 57

4.7 Study of bending strength..58

5. **Abrasive behaviour of areca sheath fibre reinforced polyvinyl alcohol composites**...**58**

5.1 Tensile strength of composites 59

5.2 Abrasive wear behavior 59

5.3 Influence of load on wear behavior 59

5.4 Influence of sliding distance on wear behavior...............................60

5.5 Analysis of wear surface texture 60

6. **Optimization of surface treatment in jute fibre reinforced composites60**

6.1 Comparison between properties of untreated and alkali
treated jute fibres 61

6.2 Comparison between mechanical properties 62

6.3 Study of tensile strength of untreated and alkali treated jute
fibres by Weibull analysis 63

Conclusion...63

References 65

1. Introduction

Owing to their strength, stiffness and low density fiber reinforced polymer matrix composites have gained more acceptance in comparison with conventional metallic structures, in a number of end uses in engineering. In the case of end uses relating to ballistics like bullet proof helmets, vest, and other armor systems high-performance aramid fibers find broad application as a reinforcement in the polymer matrix Personal body armor could be classified into soft armor and hard armor. Soft body armor contains multiple layers of fabrics up to 50 layers [1]. Whereas, hard body armors normally comprise of three layered structures. The upper layer comprising of harder ceramic layer blocks the projectile and the centre layer comprising of lighter and flexible Kevlar VR fabric-based composites absorbs the kinetic energy of the projectile and ceramic fragments. The third ductile metallic layer acts as a final barrier to stop the projectile which penetrates into the ceramic and composite layers [2-4].

Ever since time immemorial flax has been cultivated as two varieties across the globe. One is fibre flax and the other is seed flax. Fiber flax or also called common flax is chiefly cultivated to get lengthy, soft and lustrous fibers (long-line fiber) and short fibers (tow). Long-line fibers have been used to produce high-quality, expensive and very fine linen fabrics used in fine clothing, tablecloths, bed sheets, drapes, and various high-end domestic goods [5]. Ropes and twines are made from tow. Moreover, flax fibers have also continuously been focused in numerous researches and used for various well-established industrial applications such as geotextiles; soil-associated permeable fabrics with characteristics like filtration, reinforcement, protection, drainage, insulation and absorbency [6-8].

Sustainable Natural Fiber Composites   Materials Research Forum LLC
Materials Research Foundations **122** (2022) 37-76   https://doi.org/10.21741/9781644901854-2

Owing to their eco-friendly, bio-degradable and sustainable properties, natural fibre reinforced polymer composite materials have attracted attention during recent years. Hence, these composites are associated with various application areas. An example tribological applications such as bearings, gears, and so on, in which liquid lubricants cannot be used due to various constraints. When considering tribological application, natural fibre polymer composites go through different kinds of wears. Of these, abrasive wear is assumes highest importance. Such abrasive wear relates to the loss of material which takes place due to sliding of one body over the other.

So far a number of researchers have turned their focus towards investigation of the mechanical and tribological characteristics of different natural fibre/polymer composites. In relation to tribological characteristics, specific work has been carried out on abrasive wear behavior of natural fibre/polymer composites. The wear and frictional behaviour of a new epoxy composite based on treated betel nut fibres, subjected to three-body abrasion with various abrasive particle sizes and sliding velocities have been investigated [9]. It is observed that the abrasive wear of the composite is based on the size of abrasive particles and sliding velocity. The wear and frictional characteristics of oil palm fibre reinforced polyester composite have been investigated [10]. It is found that the presence of oil palm fibre in the polyester improves the wear property 3-4 times than that of neat polyester. The natural fibres such as jute, flax, hemp, sisal, coir, have been used as reinforcement in composites amongst which the jute fibre based composites had found to exhibit moderate

tensile and flexural properties [11-13]. In order to enhance the interfacial adhesion with the hydrophobic matrices the natural bast fibres (hydrophilic) like jute, kenaf, hemp, flax, ramie can be treated with chemicals. In the past, the interfacial adhesion has been improved using alkali treatment, acetylation and silinisation methods [12-18].

2. Epoxy composites reinforced with hybrid Kevlar VR /Cocos nucifera sheath

Shock waves are created by ballistic impact and can result in serious trauma injuries to the soldiers. Aramid fiber-based protection system that are presently being used in defence industries are using offer affordable acceptable range of protection to the soldiers. But, disposal of Kevlar VR releases a great deal of carbon dioxide into the atmosphere, which is the major source of global warming. Hence, it is imperative to find an alternate material to Kevlar VR fabric [19,20]. Hybrid ballistic panels comprise of two or more high performance synthetic fibers. The merit of individual layers are combined in the case of hybrid synthetic/cellulosic fiber-based laminated composites combine the advantage of individual layers. Replacing KevlarVR fabric in the body armors with an eco-friendly light weight material, together with an improved kinetic energy absorption and dissipation have become an interesting approach to enhance the body armors ballistic performance [21].

Materials Research Forum LLC
https://doi.org/10.21741/9781644901854-2

Fibers which possess higher ballistic resistance and lower density are the most promising alternative to KevlarVR fabric [21]. The prospect of using eco friendly natural fibres have been recently studied by Brazilian (Malva, rami, curaua, bagasse, mallow, jute, and bamboo) as a substitute material to KevlarVR fabric in the ballistic composites. They have concluded that natural fibers can act as a potential reinforcement in the ballistic composites [22-28]. Jambari, Yahya, Abdullah, and Jawaid (2017) investigated the energy absorption and ballistic limit of hybrid kenaf/KevlarVR fabric reinforced polymer composites with different fiber wt.% and concluded that replacement of KevlarVR fabric with kenaf fiber reinforced epoxy composites declined the energy absorption and ballistic limit of the laminated composites [29]. The effect of adding more kenaf layers with the KevlarVR fabric, instead of replacing KevlarVR fabric, has been studied [30].

The ballistic limit and energy absorption have been enhanced by the surplus layers of kenaf. Studies have been done on the ballistic performance of twaron/coconut shell powder reinforced epoxy composites, and they reported that the laminates with outer tawron layers and inner coconut shell powder layers exhibited higher ballistic limit and energy absorption [31]. Investigation has been conducted on the ballistic behavior of hybrid KevlarVR /ramie fiber reinforced polyester composites for personal body armor applications and concluded that the projectile geometry (shape and size) plays a vital role in the energy absorption, ballistic limit, and life of the body armor [32]. On the other hand, ballistic performance of body armor also depends upon the fiber loading, weaving nature or architecture, panel thickness, and the layering sequence [33]. It has been found that the fibers exhibited different failure modes depending upon their layering pattern and location in the thickness direction, when it is subjected to ballistic impact [34].

In the event of ballistic impact, the upper layer of the ballistic panel shows shear failure whereas the lower layer fails by means of tensile mode of failure. Thus, while hybridizing various materials, optimized layering sequence of various laminae gets the merit of individual constituents and thus enhances the ballistic performance of the composite panel. Chen et al. investigated the effect of woven and longitudinally oriented fabrics on the ballistic performance of hybrid composite panels, and they have concluded that panels with woven fabric at the top layer exhibited higher shear resistant against ballistic impact [34]. Whereas, owing to greater tensile strength and modulus "longitudinal" or "unidirectional" fibers can be utilized on the backside of the composite panel.

It has been reported that woven kenaf/KevlarVR -based laminates showed improvement in ballistic performance than nonwoven kenaf/KevlarVR reinforced epoxy composites [35]. The research has been directed towards assessment of the influence of hybridizing naturally woven innovative *Cocos nucifera* sheath (CS) with KevlarVR fabric/epoxy composites on the ballistic performance of laminated composites. Naturally woven CS has been selected

Sustainable Natural Fiber Composites
Materials Research Foundations **122** (2022) 37-76

Materials Research Forum LLC
https://doi.org/10.21741/9781644901854-2

for hybridization owing to its special naturally woven architecture, low cellulose content (21%), allowable mechanical properties, economy, and availability.

Low cellulose content declines the hydrophilic nature of the natural fiber and forms a rough and hard surface which is suitable to absorb impact energy [36]. A question may arise as to how a relatively weak CS can compare with a high performance KevlarVR fabric in the ballistic composites. It has been proved that the ballistic performance of the CS/epoxy composite is not only based on the strength but also on its ability to absorb and dissipate the kinetic energy of the projectile rapidly away from the area of impact.

Assessment has been done with regard to energy absorption, ballistic limit, and specific energy absorption for the ballistic performance of various layering patterns with 9 and 12 layers of KevlarVR and CS reinforced epoxy composites. The theoretical considerations for performing the ballistic impact test are as follows

a) The velocity at the points of commencement and impact are unchanged, and

b) The loss of projectile energy is proportional to the energy absorption of the samples.

The impact velocity, residual velocity, impact energy, and residual energy of various laminates have been compared. Impact velocity is the velocity at which the projectile hits the specimen with a predefined pressure and the corresponding kinetic energy is termed as impact energy. The velocity at which the projectile emerges from the sample, following impact is termed residual velocity and residual energy is its corresponding kinetic energy. The impact velocity of all the laminates has been kept at 300–320 m/s.

2.1 Influence of stacking sequence on the energy absorption of laminated composites

The main purpose of using Kevlar VR fabric reinforced polymer composites in the personal body armor is to absorb and dissipate the impact energy of the projectile. Hence energy absorption plays a vital role in the ballistic material [37]. The percentage of change in energy absorption has been calculated by an equation.

The energy absorption of all laminated composites has been compared. On comparison of the energy absorption of five different stacking sequences having 9 layers, the laminate having 9 layers of Kevlar exhibit lowest energy absorption even though all the 9 layers were high performance Kevlar VR fabric.

Hybrid laminates with 7 layers of Kevlar and 2 layers of *Cocos Nicifera* sheath show 6.9% higher absorption of energy in comparison with laminated 9 layered Kevlar composites due to the replacement of 2 Kevlar VR layers with *Cosos Nucifera* Sheath. Likewise, the laminates with 5 layers Kevlar and 4 layers *Cosos Nucifera* Sheath, and 9S3 and 9S4 hybrid

Materials Research Forum LLC
https://doi.org/10.21741/9781644901854-2

laminates having 5 layers of Kevlar and 4 layers of *Cosos Nucifera* Sheath, and 2 layers Kevlar and 7 layers *Cosos Nucifera* Sheath possess 21.7% and 28% higher energy absorption in comparison with pure KevlarVR /epoxy composites with 9 layers of kevlar). *Cocos Nucifera* sheath reinforced epoxy composites having laminates of 9 layers *Cosos Nucifera* Sheath show greater absorption of energy among the laminated composites (with 9 layers) and it has shown 30% greater absorption of energy in comparison with KevlarVR fabric reinforced epoxy composites having 9 layers of kevlar.

Similar trend has been observed upon comparison with the energy absorption of the laminated composites having 12 layers. The energy absorption with regard to laminated composite having 12 layers of kevlar has been found to be lower. Whereas the hybrid laminates having 9 layers of Kevlar and 3 layers of *Cocos Nucifera* sheath shows 3.2% greater energy absorption in comparison with laminates having 12 layers of kevlar. Hybrid laminates having 6 layers of Kevlar and 6 layers of *Cocos Nucifera* sheath and 3 layers of Kevlar and 9 layers of *Cocos Nucifera* sheath show 11.2% and 18.6% greater energy absorption in comparison with KevlarVR fabric reinforced epoxy composites with 12 layers of kevlar. Among the laminated composites (with 12 layers), CS reinforced epoxy composites having 12 layers of *Cocos Nucifera* show 20% greater energy absorption in comparison with KevlarVR /epoxy composites having 12 layers of kevlar. While increasing the no of layers, laminated composites can absorb more energy [38]. It can be validated with the following results. Energy absorption in the case of laminate with 12 layers of kevlar has been found to be 11.3% greater in comparison with laminated composites having 9 layers of kevlar. Likewise, laminated composites having 12 layers of *Cocos Nucifera* sheath have shown 2.53% greater energy absorption comparison with laminates having 9 layers of *Cocos Nucifera* sheath. Hybrid composites having 9 layers of Kevlar and 3 layers of *Cocos Nucifera* sheath, 6 layers of Kevlar and 6 layers of *Cocos Nucifera* sheath and 3 layers of Kevlar and 9 layers if *Cocos Nucifera* sheath exhhibit 7.5%, 1.7%, and 3% higher energy absorption in comparison with laminated composites having 7 layers of Kevlar and 2 layers of *Cocos Nucifera* sheath, 5 layers of Kevlar and 4 layers of *Cocos Nucifera* sheath and 2 layers of Kevlar and 7 layers of *Cocos Nucifera* sheath, respectively. Energy absorption of the composite is the most important factor in the armor system since it protects the wearer from blunt trauma [39]. Natural fibers comprise of cellulose which is the cause of their hydrophilic nature. Each anhydro-D-glucose element of cellulose contains three alcohol hydroxyls. These hydroxyls form hydrogen bonding in between the cellulose macromolecules and with the hydroxyl groups which is present in the air [40]. Low cellulose content of the *Cocos Nucifera* sheath result in a rough fiber surface and it forms a hydrogen bond with the adjacent cellulose and hydrophobic polymer matrix rather than the atmospheric air molecules. On the other hand greater lignin content considerably enhances the bonding of distinct cells of hard natural fiber. This nature of the *Cocos Nucifera* sheath is considered crucial for greater energy absorption in comparison than KevlarVR /epoxy composites. Thus, in the work considered herein it has been identified that naturally woven novel *Cocos Nucifera* sheath and hybrid KevlarVR / *Cocos Nucifera* sheath reinforced epoxy composites have greater energy absorption in comparison with KevlarVR /epoxy composites.

One way ANOVA has been used for statistical analysis of the findings using Minitab software. The ANOVA test results of energy absorption (Eabs) of the laminated composites has been determined. The variance of "Eabs" has been decomposed into two categories such as between the groups (BG) and within the groups (WG). F value is the ratio between the mean square (BG) to the mean square (WG). The P-value of the F-test is less than 0.05 which rejects the null hypothesis. Thus, it has been concluded that there is a statistically significant difference between the mean energy absorption of the laminated composites with 95% confidence level. The normal probability plot of energy absorption has been determined. The data points in the normal probability plots are almost nearer to the normalization line. It has been found that there is a minimal deviation from the normalization line. In general, the data points follow a relatively straight line for energy absorption which confirms the goodness of fit of the model in the ANOVA.

2.2 Influence of stacking sequence on the ballistic limit of laminated composites

Ballistic limit or limit velocity can be termed as the velocity above which a specific material is not able to endure the impact of the projectiles or it results in complete penetration of the material. The velocity of all the test laminates have been compared. Of the 9 layered laminates considered, the laminates having 9 layers of *Cocos Nucifera* sheath have shown greater ballistic limit. In comparison with laminates having 9 layers of kevlar, hybrid composites having 7 layers of Kevlar and 2 layers of *Cocos Nucifera* sheath, 5 layers of Kevlar and 4 layers of *Cocos Nucifera* sheath and 2 layers of Kevlar and 7 layers of *Cocos Nucifera* sheath indicate 3.1%, 11.01%, and 12.9% greater ballistic limit, respectively. The limit velocity of laminated composite having 9 layers of *Cocos Nucifera* sheath has been found to be 13.86% greater in comparison with laminates having 9 layers of kevlar. In the case of 12 layered laminates similar trend has been observed. Of the 12 layered laminates, laminates having 12 layers of *Cocos Nucifera* sheath have shown maximum limit velocity, whereas laminate having 12 layers of Kevlar have minimum limit velocity. The limit velocity of hybrid laminates having 9 layers of Kevlar and 3 layers of *Cocos Nucifera* sheath is found to be 1.24% greater in comparison with laminates having 12 layers of Kevlar owing to the replacement of 3 KevlarVR layers with *Cocos Nucifera* sheath. Owing to the replacement 6 and 9 layers of Kevlar VR with *Cocos Nucifera* sheath the limit velocity of hybrid laminates with 6 layers of Kevlar and 6 layers of *Cocos Nucifera* sheath and 3 layers of Kevlar and 9 layers of *Cocos Nucifera* sheath has been 5.6% and 8.8% greater than KevlarVR fabric reinforced epoxy composites (12layers of kevlar). Moreover, composites with *Cocos Nucifera* sheath (12layers of *Cocos Nucifera* sheath) have 9.6% greater ballistic limit than composites with KevlarVR fabric (12 layers of kevlar). More number of layers enhance the ballistic limit of the laminated composites. The ballistic limit of laminates having 12 layers of Kevlar has been 5.3% greater in comparison with laminates having 9 layers of Kevlar. On the other hand laminated composites with 12 layers of *Cocos Nucifera* sheath have shown 1.4% greater ballistic limit in comparison with laminates having 9 layers of *Cocos Nucifera* sheath. Hybrid composites having 6 layers of Kevlar and 6 layers of *Cocos Nucifera* sheath have shown slightly greater ballistic limit than laminates 5 layers of Kevlar and 4 layers of *Cocos Nucifera* sheath. Laminates having 9 layers of Kevlar and 3 layers of *Cocos Nucifera* sheath and 3 layers of Kevlar and 9 layers of *Cocos Nucifera* sheath exhibit 3.3%, 1.5% greater ballistic limit than to laminated composites with 2 layers of Kevlar and 7 layers of *Cocos Nucifera* sheath and 7 layers of Kevlar and 2 layers of *Cocos Nucifera* sheath, respectively.

The fabric design plays a vital role in the ballistic resistance of the composites [41]. Deflection can be reduced if the weaving nature of the fabric is too tight. Loose weaving nature of the fabric allows more penetration rate [42]. The projectile can be resisted by naturally woven randomly interlaced dense architecture of the *Cocos Nucifera* sheath. There is a strong mechanical interlocking with the adjacent lamina and matrix that render the structure more rigid and enhances the limit velocity by dense natural weaving style of the *Cocos Nucifera* sheath. The ANOVA test results of ballistic limit (V50) of the laminated composites have been determined. The variance of V50 has been divided into two categories like between the groups (BG) and within the groups (WG). F-value is the ratio between the mean square (BG) to the mean square (WG). The P-value of the F-test is less than 0.05 which rejects the null hypothesis. Thus, it has been concluded that there is a statistically significant difference between the mean ballistic limit of the laminated composites with 95% confidence level. Normal probability plot of ballistic limit has been determined. The data points in the normal probability plot are almost nearer to the normalization line which confirms the goodness of fit of the model in the ANOVA.

2.3   Influence of hybridization on the ballistic performance of laminated composites

The influence of hybridizing KevlarVR and *Cocos Nucifera* sheath on the energy absorption behavior of the laminated composites has been studied. The influence of hybridizing Kevlar on the energy absorption of CS/epoxy composites has been compared.

In comparison with the hybrid composites the energy absorption of CS/epoxy composites has been greater. The energy absorption has decreased due to replacement of *Cocos Nucifera* sheath with KevlarVR fabric in the epoxy composites (S5 > S4 > S3 > S2). The influence of hybridizing *Cocos Nucifera* sheath on the energy absorption of Kevlar/epoxy composites has been compared. The absorption of energy shows improvement in the epoxy composites in which KevlarVR fabric has been replaced with *Cocos Nucifera* sheath (S1 < S2 < S3 < S4).

The Study has been carried out regarding the influence of KevlarVR /CS hybridization on the ballistic limit. The ballistic limit of hybrid and *Cocos Nucifera* sheath reinforced epoxy composites has been compared. Study has shown that replacement of *Cocos Nucifera* sheat with KevlarVR fabric reduced the ballistic limit [43]. The ballistic limit of hybrid and KevlarVR fabric reinforced epoxy composites has been compared. The KevlarVR fabric reinforced epoxy composites has been found show greater ballistic limit in comparison with hybrid composites. It has been reported that hybridizing KevlarVR with kenaf fabric declined the energy absorption and ballistic limit. However, the work has proved that hybridizing KevlarVR with *Cocos Nucifera* sheath improved ballistic performance in comparison with KevlarVR /epoxy composites.

2.4   Influence of areal density on the ballistic performance of laminated composites

Energy absorption and ballistic limit of the laminated composites have been given in relation to areal density. There has been enhancement in energy absorption and the ballistic limit of the laminates with a rise in areal density in the case of 9 as well as 12 layered laminated composites. The energy absorption and ballistic limit of the hybrid and *Cocos Nucifera* sheath reinforced epoxy composites is found to be better than KevlarVR fabric reinforced epoxy composites [43]. The difference in density of *Cocos Nucifera* sheath in comparison with Kevlar VR fabric has resulted in greater areal density of laminates having 9 layers as well as 12 layers of *Cocos Nucifera* sheath and 12 layers. Thus from the study it has been found that areal density is crucial with regard to the ballistic performance of the composites.

2.5   Absorption of specific energy

The absorption of energy has been normalized in relation to areal density because of slight variation in thickness and density difference of individual constituents of the laminated composites. A mathematical relation has been used to calculate the specific energy absorption.

Sustainable Natural Fiber Composites   Materials Research Forum LLC
Materials Research Foundations **122** (2022) 37-76   https://doi.org/10.21741/9781644901854-2

The absorption of specific energy of laminates having 9 layers and 12 layers of *Cocos Nucifera* sheath have been determined. In the case of Kevlar VR fabric reinforced expoxy composites with laminates having 9 layers Kevlar VR showed lower specific energy of absorption among the nine layered laminates. On the other hand, there has been slightly greater specific energy absorption in the case of the hybrid and *Cocos Nucifera* sheath based laminates. Similar trend has been found with the 12 layered laminates [43]. In comparison with hybrid and pure Kevlar VR fabric reinforced epoxy composites the laminates having 12 layers of *Cocos Nucifera* sheath have greater specific energy absorption among the 12 layered laminates.

2.6   Studies on failure

The outcome of the ballistic tests have been validated by visual inspection of the failed samples. Shear failure has been observed on the impact surface of the KevlarVR and hybrid laminates whereas the bottom layer exhibits tensile mode of failure. By means of shear plugging, coupled with delamination and tensile fracture, KevlarVR and hybrid panels have been failed. Another research reports similar observation has been reported. Whereas, *Cocos Nucifera* sheath based laminates have been failed by means of bulging, coupled with delamination and fiber fracture [43].

When considering the ballistic performance of the laminated composites, energy dissipation mechanism has a key function. The schematic of damage profile and wave propagation of the failed samples have been determined. The projectile penetrates into the laminates due to its kinetic energy that acts over a small area. The kinetic energy with regard to the laminates having 9 and 12 layers of Kevlar VR in the KevlarVR /epoxy composite samples has been dissipated in the radial direction. The kinetic energy of the projectile is quickly dissipated from the area of impact through replacing of KevlarVR with *Cocos Nucifera* sheath. This has been clearly observed in the case of laminates having 3 layers of Kevlar and 9 layers of *Cocos Nucifera* sheath. Further, based on the study it has been found that the *Cocos Nucifera* sheath follows a unique kinetic energy dissipation profile rather than a regular radial kinetic energy dissipation as can be observed with laminates having 2 layers of Kevlar and 7 layers of *Cocos Nucifera* sheath and laminates having 3 layers of Kevlar and 9 layers of *Cocos Nucifera* sheath.

Thus, from the study it has been observed that the combined radial wave propagation caused by KevlarVR fabric and diagonal energy dissipation because of *Cocos Nucifera* sheath contributes for the enhanced ballistic performance of the hybrid laminates in comparison with KevlarVR fabric reinforced epoxy composites. In the case of laminates having 9 layers as well as 12 layers of *Cocos Nucifera* sheath show somewhat improved

Sustainable Natural Fiber Composites   Materials Research Forum LLC
Materials Research Foundations **122** (2022) 37-76   https://doi.org/10.21741/9781644901854-2

because of the rapid diagonal dissipation of kinetic energy of the projectile away from the impact zone, and could be clearly observed in the back side of the laminate samples.

3.   Flaxseed fiber bundle-reinforced polybutylene succinate composites

The seed flax, mostly termed as flaxseed, oil seed or linseed is an important agricultural crop extensively grown globally, for the production of seed and its oil derivatives [44,45]. According to Food and Agriculture Organization of the United Nations Statistics (FAOSTAT), in 2016, world's total cultivation and production of flaxseed crop was over 2.7 million hectare and 2.9 million tonnes, however, average annual production of flaxseed in million tonnes during the period 1994–2016 [46]. Globally, seeds and its derivatives have more nutritional and industrial applications including flaxseed oil, meal and fertilizer [47]. Flaxseed products have been used for human and animal consumption as well as in industrial applications such as paints, varnishes, linoleum, printing inks and soaps. It is used in resins, paints, printing inks, varnishes, and linoleum for flooring, whereas there is limited flaxseed straw usage for pulp or specialty papers, e.g. currency and tea bags [48]. But, owing to the rise in the activity of modern agricultural sector, a great deal of residual straws pose serious issue related to this agricultural crop. In some areas, flax straw has become an environmental problem due to lack of alternative uses [49]. Flaxseed straws are mostly burnt as they require comparatively much longer degradation time than many other agricultural biomass [50]. In the world, every year this by-product accounts for about millions of metric tons of straws which are burnt. Thus the most abundant lignocellulose fiber resource is not only wasted but also causes a serious challenge to the environmental.

The present effect of pollution can be avoided by use of such lignocellulose fibers from flaxseed straws. It can also offer advantage of a double income to the poor farmers and producers of value-added product, like bio-composites sector. Shift in trends due to novel materials, processing techniques, applications and higher costs has also drawn attention towards the multi-purpose flaxseed plants to establish a long-term strategy to benefit from the tremendous potential of this currently underutilized natural plant fiber source[51]. Foulk et al. reported that short staple market potential in the USA was prevailing, which indicated the trend of research and development under process for short staple and uniform fiber production from fiber-flax as well as flaxseed plants [52]. The huge quantities of unutilized flaxseed fibers could offer an optional source of abundant, economical and saleable fibers being shorter but having comparable mechanical properties, having no need for extensive processing time, techniques, costly equipment or facilities.

Cellulosic fibers are associated with non-cellulosic components such as hemicellulose, pectin, lignin and waxes, etc. [53 -55]. The interaction between fibers was reduced by the removal of hemicellulose [56]. Alkali treatment has been simplest traditional method for cellulosic fibers to remove impurities, but fiber quality depended upon the alkali concentration, temperature and time [58, 59   17-19]; however, decrease in the surface impurities resulted in better mechanical properties [60   20]. The enzymatic methods for retting and surface modification of flaxseed straws reported limitations in parameters and processability [61   21]. Lamb et al. [62   22] carried out flaxseed fiber individualization using mechanical separation and chemo-mechanical combined treatments, which resulted in very short, non-spinnable and damaged flaxseed fibers, indicating higher alkali concentration significantly increased fiber damage and weight loss. Hence, in the investigation concerned, fiber individualization has not been tried. But, flaxseed fiber bundles having least chemical treatment and processing have been chosen.

Various renowned automobile manufacturers, e.g. Mercedes, Audi, BMW, Ford, Opel, etc. have already established commercially using natural fiber composites for interiors, panelling and door linings, such as wood fibers as rear enclosure of seat backrest; cotton fibers as sound proofing material; coconut fibers for interior trim and seat cushioning [63,64   23,24]. Flax/sisal polyurethane-reinforced mat is used for weight reduction door trims; soya-based natural fibers as form filling seats; cellulose-based cargo floor tray for improved noise reduction; kenaf fiber/polypropylene-based boards; hemp fiber/polyester hybrid composites for lightweight lotus designed seats; and sisal for the carpet in Eco Elise due to higher toughness and abrasion resistance [65   23–25]. Many researches emphasized the importance of innovative materials [66,67   26,27]. In these decades, environmental protection and government policies in various countries require environmental friendly composites to alleviate the environmental pollution [68,69   28,29]. Natural fiber-reinforced composites with biodegradable polymer matrices called 'green composites' have been the centre of attention for most researchers [30]. Due to advantages of natural fibers of low density, biodegradability and renewability over the manmade fibers [70,71 30,31], researchers proposed that natural fiber-reinforced polymer composites can be the replacement for manmade fibers composite materials and even further incorporating biodegradable polymer matrices [72-76   32–36].

Polybutylene succinate (PBS) is a white crystalline biodegradable thermoplastic resin, which has drawn considerable attention as a polymer matrix for green composites due to its exceptional mechanical properties and processability [77   37]. The flax/PBS composites also reported to have enhanced thermal properties due to charring of matrix. In addition, PBS is a biodegradable polymer formed by polymerization of butanediol and succinic acid and also available from bio-based renewable resources [78].

In the investigation considered herein, flaxseed fiber bundles have been utilized in place of individualized fibres as reinforcement and composites have been designed with PBS matrix by hot compression technique. In order to improve the interfacial bonding, flaxseed fiber

Materials Research Forum LLC
https://doi.org/10.21741/9781644901854-2

bundles have been treated with 10 g/L and 20 g/L NaOH solutions, prior to be composed with PBS resin. Also, the chemical composition and mechanical properties of the flaxseed fiber bundles have been studied. Tests have been carried out on the tensile and flexural characteristics of the flaxseed fiber bundles/PBS composites. SEM studies have been done on the fractured cross sections.

3.1   Studies on composition of flaxseed fiber bundles

Studies on chemical constituents has been carried out on test samples for each constituent. The findings have been considerable since the analyzed composition exhibit cellulose 40.11%, hemi-cellulose 28.27%, lignin 15.08%, pectin 6.3%, wax 3.1%, and dissolvable substances including ash 7.14% [79]. They have been observed to be totally different in comparison with the common flax fiber composition. With regard to the flaxseed fibers, the wax has been nearly two times, lignin three times (almost equal to jute), pectin two times, dissolvable substances and hemicellulose are twice to the respective component values present in common flax fibers. On the other hand the cellulose shows 40–50% less than that present in common flax fibers. It shows that in the absence of strong chemical treatment the flaxseed fiber bundles have been difficult to degum. This can result in fiber damage and serious environmental issue. Hence, the application of the pre-degum flaxseed fiber bundle has been a suitable option.

3.2   Assessment of flaxseed fibre bundles by SEM

The untreated fiber bundles have been totally covered with gum and cementing materials as revealed by SEM. On the other hand surface gum has been partly removed. However, there has been no removal of the inter-fiber gums in 10 g/L NaOH treated fibers and surface has been obsesrved to be comparatively rough. However, the fibers have not been seperated and have still remained intact in bundles. But, surface of the 20 g/L alkali treated fibers have been relatively clean and rough subsequent to removal of the superficial gums and cementing materials. However, the fibers have not been individualized and still adhered in smaller bundles. It has been considered that 10 g/L NaOH was not sufficient. But, the 20 g/L NaOH treatment has been relatively better since enhanced surface roughness and greater surface contact area of fiber has been noticed, and is considered crucial for improvement in interfacial properties and mechanical interlocking between the fiber and matrix.

3.3   Tensile property of flaxseed fiber bundles

This study aimed at the use of flaxseed fiber bundles instead of individualized fibers. However, the literature review revealed that the dynamic failure of flax fiber being pulled in tension, was caused due to combined effect of spiral angle, hierarchical fiber pull-out, and crack bridging [80]. The fractographic method of the failure of flax fiber uncovered the complex structure of flax fiber such as the arrangement of mesofibrils and microfibrils, which supported the observations of Thuault et al. [81]. Findings have revealed that untreated fiber bundles exhibit greater strength of 1.14 cN/dTex, whereas 10 g/L and 20 g/L NaOH treated had 0.88 cN/dTex and 0.85 cN/dTex, respectively. As the variation in tensile properties has been within acceptable limit and thus untreated as well as treated fiber bundles can be used as reinforcements in composites. But, treated fibers exhibit better surface morphology with comparable tensile property and can be anticipated to show better interfacial properties in composites. Also, it has been found that there is a direct relationship between the caustic soda concentration and the weight loss. The loss in weight calculated are respectively 22.2% and 45.2% in the case of 10 g/L and 20 g/L treated with caustic soda, which shows that rise in the concentration further can cause damage to the macromolecular chains of cellulose. The effect of caustic soda in aqueous solution prevents does not permit binding between cellulose molecules and ultimately weakens the binding force of cellulose molecules. This swelling mechanism of the aqueous mixture is that sodium hydroxide hydrates destroy the intermolecular and intramolecular hydrogen bonds in cellulose.

In order to assess the influences of the pressure and temperature the bundle fiber stress–strain curves under the temperature from 110°C to 150°C and 1.5MPa pressure has been determined. It has been found that the flaxseed fiber bundles have been sensitive to high temperature because of low decomposition temperature of cellulose based materials. An abrupt reduction of 41% has been noticed in the fiber strength by the rise in temperature to 150°C. The untreated fibre bundles as well as those compressed at 1.5MPa have been compared at room temperature. There has been reduction in strength by 7.9% due to rise in pressure, which shows a limited negative effect on the fiber bundle strength. This valuable data offers the means for the choice of suitable low melting temperature matrix, guidelines for choice technique and process temperatures.

3.4   Tensile and flexural strength of flaxseed/PBS composites

It has been found that there is reduction in tensile strength of alkali treated flaxseed fiber bundle/PBS composites from 78.20MPa to 53.85MPa for 10 g/L NaOH-treated composites and 54.29MPa for 20 g/L NaOH-treated composites, revealing about 30% reduction. It has been presumed that with the decrease in the tensile strength during alkali treatment, and hence the fiber bundles used in respective composites also bear similar phenomenon. Moreover, it is found that the wax of the surface has been removed by the alkali. However, the flax seed fibres have been damaged, thereby reducing stiffness of treated fiber composites. It also appears prospective with regard to the usefulness of untreated fiber bundles to be used in composite materials where strength is required. Normally, the caustic-treated fiber bundle composites have shown the acceptable range of the tensile properties that related to the use in civil industry. Also, the caustic treated fiber bundle composites have shown a lower modulus and greater elongation at break, which is due to the rotation and alignment of partially degummed fiber bundles during tensile process.

The flexural strength of alkali-treated flaxseed fiber bundle/PBS composites have been determined. There has been considerable rise in the flexural strength of alkali treated flaxseed fiber bundle/PBS composites in comparison with the composites with the untreated fiber bundles (26.70 MPa), which approximated to 43% for 10 g/L caustic treated (38.26 MPa) and 84% for 20 g/L NaOH treated (49.16 MPa) composite samples. The interfacial properties of the composites have been directly reflected by flexural properties. Hence, alkali treatment, in particular 20 g/L caustic treated composites reveal enhanced fiber-matrix interface and interlocking. It appears prospective with regard to the usefulness of treated fiber bundles in composite materials applications requiring interfacial properties that are superior.

3.5   Assessment of flaxseed/PBS composites using SEM

Using different magnifications the tensile fractured surfaces of the untreated as well as treated composite samples have been seen. The photographs of SEM reveal that a superficial coating is formed on the fibre bundles by the PBS resin. But, the impurities continue to remain intact on the surface of fiber bundles. Also, the adhesion between the untreated fibre composites and matrix has been found to be weak as revealed by the evident separation between flaxseed bundles and PBS resin. There has been superficial adhesion of PBS to the 10 g/L caustic-treated fibers bundles with formation of a tight layer resin on the surface of fiber bundle, showing the enhancement in interaction of the fiber and matrix. Also, the loose structure of flaxseed fiber bundles could affect the fiber alignment and cause reduction in the tensile strength of the fiber bundles and their reinforced composites. It is found that the 20 g/L NaOH-treated samples showed considerable enhancements in the morphology fiber surface, fiber–matrix interface, adherence to the matrix and a little inter-fiber penetration.

Factors like fiber aspect ratio, alignment, and fibre strength have a major contribution towards the tensile strength of fiber-reinforced composite. On the other hand, its flexural strength is considerably influenced by the interfacial bonding between fiber and matrix. The flaxseed fibre bundles exhibit relative reduction in tensile strength subsequent to caustic treatment. However, they show better interfacial bonding with resin. The cross section morphologies of untreated fibre reinforced composites have been studied using

Sustainable Natural Fiber Composites   Materials Research Forum LLC
Materials Research Foundations **122** (2022) 37-76   https://doi.org/10.21741/9781644901854-2

SEM images. It has been found that there is severe delamination between fiber bundles, which shows poor fiber-matrix interfacial bonding. Also, some parts visibly show that there has been no proper envelopment of fibre bundles by matrix.

Such gaps and spaces resulted in the poor interface between the fiber–matrix and thus fiber pull out and delamination of the untreated composite layers have been noticed at the fractured points. There has been poor interfacial bonding and restricted resin penetration into assembly of fibre bundle due to the smooth and slippery wax on the untreated flaxseed fiber bundle surface. It is found that there is an evident delamination in the cross-sectional images of 10 g/L caustic-treated fiber bundle reinforced composites. However, relatively good dispersion and penetration of PBS resin have shown an improved interaction between the fiber and matrix. The surface roughness, interfacial properties and interlocking were considerably improved with the chemical treatment, as reported by various studies [82 19,20, 32, 43]. Most of the wax and impurities present in the flax seed bundles have been removed by alkali treatment. It lead to improvement in property and loose bundle structure, which enabled to strengthen the adhesion and penetration of the PBS resin. Composites treated with caustic soda at 20 g/L concentration exhibit better-embedded fibers within the matrix layers, leading to improved fiber surface morphology and interaction with the matrix. Owing to improved surface roughness and fibre wetting the composite samples so treated had more reactive sites and fiber surface area available for better interlocking and adherence with the matrix. In brief it can be stated that surface modification of flaxseed fiber bundles by caustic treatment considerably eliminated the impurities and reduced the moisture absorption. Thus, it lead to rise in the surface roughness, wettability and interfacial bond strength with matrix, and thereby lead to better flexural properties of bio-composites. Because of less micro voids and fibre-PBS deboning in the interface area the structural integrity of PBS composites has been improved. It is found that the gaps have been visible in untreated composites showing weak adhesion. But, there has been decrease subsequent to caustic treatment at concentration of 10 g/L and nearly disappeared subsequent to treatment of caustic soda at concentration of 20 g/L, exhibiting improved compatibility in composites. Hence, it is found that fibres have been pulled out easily from interfacial region with poor compatibility in untreated composites. On the other hand alkali-treated fiber bundles exhibit better adhesion with PBS matrix by effectively dispersing and transferring stress. This lead to overall deformation of composites. On the whole, the flaxseed fiber bundles, untreated as well as alkali treated, show good potential acceptability as reinforcements in green composites or other thermoplastic composites. The technique used for processing sample and the comparable mechanical properties of flaxseed fiber bundle composites assert the end uses in different civil, automobile and industrial areas.

Sustainable Natural Fiber Composites   Materials Research Forum LLC
Materials Research Foundations **122** (2022) 37-76   https://doi.org/10.21741/9781644901854-2

4. Application of surface treatments on jute fibre reinforced polypropylene composites

Many approaches had been attempted for surface treatments such as grafting treatment and low temperature plasma treatment [83-85]. Alkali treatment was universally accepted as an efficient method to improve the interface compatibility of natural fiber-reinforced composites [86,87]. Cai et al. [88] confirmed that abaca fiber with the mild alkali treatment (5 wt.% sodium hydroxide solution for 2 h) exhibited the increased crystallinity, tensile strength and interfacial shear strength of the reinforced composites. Ferreira et al. [89] found that the tensile strength and frictional shear strength of sisal fibers with the alkali treatment were improved by 32% and 78%, respectively, which was attributed to the removal of non-cellulosic materials and impurities. Yan et al. [90] reported that the coir fiber possessed a much cleaner and rougher fiber surface after the alkali treatment. In the case of alkali treated fibre reinforced composites there has been a rise in tensile and bending strength, respectively, in comparison with the control fiber-reinforced composites.

Sinha and Rout [91] used 5 wt.% sodium hydroxide solution to treat jute fibers for 4 h, the flexural strength of composites was improved by 9.5% compared with the control fiber-reinforced composites. It is found that interfacial compatibility of natural fiber-reinforced composites can be improved to a certain level by alkali treatment, while the anticipated outcome has been inaccessible. Currently, there is scantly literature available with regard to the combination of two kinds of surface treatments in enhancement of the interfacial compatibility of natural fiber-reinforced composite.

The jute fibers have been treated with alkali, ethylenediamine/alkali and acid/alkali, respectively, for enhancement of the interfacial compatibility of jute fibers reinforced polypropylene composite. Comparison has been made between the influence of ethylenediamine/alkali, acid/alkali treatment and alkali treatments by means of studies using various investigation techniques that include electron microscopy (SEM), atomic force microscopy (AFM), Fourier transform infrared (FT-IR), X-ray diffraction (XRD), surface energy measurement. Also, investigation has been carried out with regard to the effect of the three types of surface treatments on the interfacial compatibility of jute fiber-reinforced composites.

4.1 Study of morphology

Evidently, there has been a change in the surface morphology of jute fibers subsequent to surface treatments. Waxy substance is found to be present on the surface of control fibers. It has been found that there have been few narrow grooves that are distributed in the longitudinal direction of fibre in comparison with the control fibers. It is possible that such surface treatments are able to remove the waxy substance and thereby cause rise in the surface roughness of jute fibers. The hydroxyl groups in the waxy substance could react with alkali, which helped the removal of waxy substance from the fiber surface. But, some residual waxy substance is found to remain on the surface of alkali treated fiber. which indicates that the alkali treatment has not been so effective in the removal of the waxy substance from jute fiber [92]. The alkali treatment has been combined with ethylenediamine and acid treatment so as to enhance the influence of alkali treatment. When treated with ethylenediamine, amino groups (–NH2) penetrated into the waxy substance swelling the hydrogen bonds between molecular chains, and proved advantageous for the alkali solution to penetrate into the waxy substance. The acidic hydrolysis of acid treatment could decrease polymerization degree of waxy substance, which was beneficial to remove the waxy substance [93]. The waxy substance has been cleaned out by the acid/alkali treatment, and the surface grooves of jute fibers are made visible. Hence, the acid treatment in combination with the alkali treatment prove to be the most effective technique for removal of the waxy substance.

The morphology and surface roughness of jute fibres has been studied using AFM. The control fibres have been found to have a relatively smooth surface of 38.9 mm due to the waxy substance covering on the surface grooves of jute fibers. There is appearance of irregular grooves on the surface of jute fibres due to the removal of waxy substance. The surface roughness of the alkali treated jute fibers is found to be 120 nm, which was 2.08 times greater in comparison with that of the control one. The improved effect of surface treatments have caused the surface roughness of acid/alkali treated fibers to get enhanced by 7.02 times, which supposes that the waxy substance can be effectively removed through acid/alkali treatment.

4.2   Studies using fourier transform infrared spectroscopy

The infrared spectra of jute fibers having various surface treatment are presented have been obtained. The bending of the C-H bond of cellulose is related to the vibration peak at 898 cm $^{-1}$ of jute fibers. The appearance of this peak reveals that the cellulose structure of jute fibers remain unaffected by surface treatments. The vibration peak at 1735 cm^{-1} was corresponding to the C=O stretching of carboxyl and acetyl groups in hemicellulose [94,95]. The disappearance of this peak is caused by the removal of waxy substance. Subsequent to surface treatments there is reduction in the intensity of band at 1640 cm $^{-1}$ assigned to bonded water molecules in hemicellulose, particularly for the acid/alkali treatment. It can be established that surface treatments removes the waxy substance from the surface of jute fibers. The characteristic absorption peak at 1246 cm^{-1} related to the acetyl group present in the hemicellulose of waxy substance was decreased with the surface

Sustainable Natural Fiber Composites   Materials Research Forum LLC
Materials Research Foundations **122** (2022) 37-76   https://doi.org/10.21741/9781644901854-2

treatments, this mainly contributed to the removal of waxy substance, which agreed with the report of Basak et al. [96].

4.3   Studies on X-ray diffraction

Powder XRD patterns have been obtained for jute fibers having various treatments are shown. The peaks at 17° and 22.5° were indexed to the planes (1001) and (002) of cellulose, respectively [97,98]. The XRD peaks at 22.5° appeared in the XRD pattern of treated jute fibers, which indicate that there is no change in macromolecular chain structure of jute fiber in the process of surface treatments. This finding is consistent with the study of infrared spectrum. Interestingly, in comparison with the control fibres the diffraction peak intensity of jute fibers with surface treatments has been strengthened. The XRD peak at 17° was shown, particularly the acid/alkali treatment. The crystalline and amorphous regions were involved in cellulose. The surface of fibres has been entirely by amorphous waxy substance. The removal of amorphous waxy substance may increase the ratio of cellulose after the surface treatments, which could contribute to the increase of diffraction peak intensity. This result showed agreement with the previous reports. There has been considerable enhancement in the diffraction peak intensity of jute fibres treated with acid/alkali due to the striking influence on the removal of waxy substance.

4.4   Study of surface energy

The contact angle and surface energy of jute fibers having various surface treatments have been determined. The control fibres exhibit low contact angle and surface energy because of the presence of hydrophobic pectin in the waxy substance covering on the jute fibers. There has been a reduction in contact angle with distilled water and ethylene glycol of alkali treated from 75.13° to 64.65° and from 68.32° to 58.91° in comparison with the control fibres, respectively. There has been enhancement in surface energy of alkali treated fibers 33.67%. The findings can be explained with regard to two aspects:
(1) the elimination of weak boundary layer and the increased roughness of fibers improved the contact area with the test liquids;

(2) the hydrophobic pectin presence in the waxy substance was removed by surface treatments to increase the surface energy of fibers. There has been reduction in the contact angle from 75.13° to 59.85° and from 68.32° to 52.17°, respectively in the case of distilled water and ethylene glycol of acid/alkali treated fibers in comparison with the control fibers. There has been an enhancement in the surface energy of acid/alkali treated fibers by 45.58%, which indicates that the surface energy of jute fibers has been gonverned by the composition and the surface morphology of jute fibers.

4.5 Studies of tensile properties of single fiber

The tensile strength of jute fibers having various surface treatments have been determined. There has been a reduction in tensile strength and tensile modulus values in the case of alkali treated fibres in comparison with control fibers (5.30 % and 1.15%). Although the strength, stiffness and structure of jute fibers were mainly provided by the cellulose, the surface grooves of jute fibers which were regarded as the stress concentration area were revealed by the alkali treatment resulting in the reduction of tensile strength and the premature failure of fibers. Also, fibrillation results to some extent due to the removal of waxy substance between fibers, leading to lower stiffness of fibers. In the case of acid/alkali treated fibres there is a reduction in the values of tensile strength and tensile modulus in comparison with the control fibers to the extent of 13.41% and 4.70%, respectively. This implied that the waxy substance was more sensitive to the acidic hydrolysis of acid treatment. According to the previous research the acid used in surface treatments could affect the cellulose structure leading to shorter cellulose chains. Therefore, the acid dosage of acid treatment should be noticed, preventing the damage from excessive acid.

4.6 Study of tensile strength

The tensile strength of jute fiber-reinforced polypropylene composites having various surface treatments has been determined. The fracture morphology of jute fiber-reinforced PP composites have been observed. There has been a rise in the tensile strength of control fibre reinforced composite in comparison with pure polypropylene. It indicates that the jute fibers holds a great potential to reinforce PP matrix. There has been a rise in the tensile strength of alkali treated fibre reinforced composite (@18%) in comparison with the control fiber-reinforced composite. The control fiber-reinforced composites exhibit the lowest tensile strength, which indicate that the weak interface between the control fibers and PP matrix restricted the enhancement of tensile performance. The control fibers have been pulled out from the matrix. There has been a rise in the surface energy and surface roughness of jute fibres by the removal of waxy substance on the surface of the fibre. It resulted in a good surface energy and stronger interfacial bonding between the fibers and PP matrix.

There has been an evident improvement in the interfacial compatibility between jute fibers and PP matrix in comparison with the alkali treated fibers. The composites exhibit a flat fracture surface particularly in the case of the acid/alkali treatment, which points out to an improved interfacial adhesion. Even though there has been a slight decrease in the tensile strength of jute fibres by the acid/alkali treatment, the acid/alkali treated fiber-reinforced composite achieved the maximum tensile strength among other fiber reinforced composites. There has been a rise in the tensile strength of acid/alkali treated fibre reinforced composite in comparison with the control fiber-reinforced composite. The exposed surface grooves could increase the effect of interlocking and the secondary forces (dispersion, induced dipole, dipole–dipole and H-bonds, etc.) between acid/alkali treated jute fibers and PP matrix, which was in favor of forming a solid interface of composites and transferring the stress [99].

4.7   Study of bending strength

The bending strength of jute fiber-reinforced PP composites has been determined. There has been a rise in bending strength of control fibre reinforced composite to about 30% in comparison with the pure polypropylene. The finding has been consistent with the conclusion of the tensile test. In the case of all the surface treated fibres, the bending strength of composites have been anticipated to show more improvement than in the case of control fiber-reinforced composite. The property of interface influenced the fracture mode of the fiber-reinforced composite.

The stress is effectively transferred by the solid interface and promote the crack propagation across the fibers up to the breaking of fibers weak point, which improved the fibre reinforcement. Also, the damage of a solid interface is required for consumption of more energy, contributing to the increase of bending strength.

But, the presence of interfacial defects caused the interface debonding, instead of the crack propagation [100]. Hence, the improved roughness and surface energy of jute fibers have proved advantageous to create a solid interface with PP matrix. The bending strength of acid/alkali treated fiber-reinforced composite (49.93MPa) has been enhanced by 22.53% in comparison with the control fiber-reinforced composites,

5.   Abrasive behaviour of areca sheath fibre reinforced polyvinyl alcohol composites

A number of research works have been reported on abrasive wear characteristics of various natural fibres reinforced in polymers. These include works on coir3, cotton4, kenaf5, jute6, bamboo7, 8, surgarcane9 and wild cane grass [101-108]. As far as areca sheath fibre reinforced polymer composite is concerned, some works have been done on evaluating mechanical and biodegradable properties rather than on tribological properties [109-114]. The objective of the research work discussed herein is thus to investigate the abrasive wear behaviour of chemically modified areca sheath (AS) fibre reinforced polyvinyl alcohol (PVA) composite.

Sustainable Natural Fiber Composites   Materials Research Forum LLC
Materials Research Foundations **122** (2022) 37-76   https://doi.org/10.21741/9781644901854-2

5.1   Tensile strength of composites

The tensile strengths of PVA/AS composites at various fibre loadings have been plotted against various wt% of benzyl chloride treated fibres. As pointed out by Shalwan and Yousif in their work, the tensile strength of PVA/AS composites first increases with fibre loading and after reaching a maximum value, it decreases [115]. The optimum fibre loading, at which maximum tensile strength of the composite is achieved, has been obtained by regression analysis by fitting a 2nd degree polynomial curve (R value 0.982) to the data points. The optimum weight percentage of fibre obtained from the plot is 27 wt%, at which the tensile strength is 46.52 ± 1.02 MPa.

5.2   Abrasive wear behavior

For the investigation of the tribological behavior of PVA/AS composites, the composite samples have been prepared adopting the same procedure but taking various fibre wt% (0, 10, 20, 27, 30 and 40%). Study has been done with regard to the influence of load and sliding distance on specific wear rate.

5.3   Influence of load on wear behavior

Investigation has been carried out regarding the impact of load on specific wear rate (W_s), friction coefficient (μ) and weight loss (ΔW) of PVA/AS composite at different fibre loadings and sliding speed of 0.471 m/s. In order to avoid replication the comparative results for sliding speeds of 0.392 and 0.549 m/s have not been reported herein. The results have been found to be similar. As the fibre content and load rises there is a reduction in specific rate of wear. However, it is further observed that the least wear exists at optimum fibre loading (27 wt%), where tensile strength indicates the optimum value. It arises from the improved interfacial bonding between fibre and matrix, which plays an important role during the process of wear process. Similar results are also obtained while studying the abrasive wear behavior of composites made by reinforcing locally available fibres such as rice straw, elephant grass and vakka, into unsaturated polyester resin material [116]. The friction co-efficient exhibits the similar outcomes as with specific wear rate. It is found to reduce with the rise in load since there in rise in load enhances the surface roughness and causes rise in wear debris. This could attribute for reduction in friction coefficient. The findings of researchers also support the above result [78]. As the load rises, there is deeper penetration of abrasive particles inside the softer matrix, by greater removal of material, thus representing it as loss of weight (ΔW) of the composites. At fibre content up to optimum loading of fibre there is reduction in weight loss (27 wt %), and subsequent rise again. It arises from the fact that with rise in content of fibre, there is rise in adhesion between matrix-fibre, which decreases the material evacuation. The trend progresses upto

27 wt% of fibre content, which is considered as the ideal value of AS fibre in PVA matrix that renders wear resistance significant. Moreover, it is seen that as the load in each fibre loading rises, there is rise in loss of material. The highest wear resistance (least wear rate) occurs at optimum fibre loading (27 wt %) which justifies the novelty of above evidence.

5.4 Influence of sliding distance on wear behavior

It is found that with the continual reduction of the specific wear with the increment of sliding distance. Tensile strength of PVA/AS composites at various fibre loading causes very less abrasion of material with longer sliding distance. It is observed that the specific wear rate is in the scope of to m3/Nm [117]. Similar to earlier findings, the specific wear rate also exhibits least value at 27 wt% (optimum loading) of fibre loading, due to maximum tensile strength of the composite and good fibre-matrix bonding. Also, it is seen that there is reduction in specific wear rate up to the ideal loading of fibre after which there is rise in the value.

5.5 Analysis of wear surface texture

The SEM images of PVA/AS composites (27 wt% fibre loading) have been obtained at specified load and related different sliding velocities respectively. The distance of sliding has been maintained constant at a particular value. The micrographs reveal that there is rise in fibre debonding with rise in speeds of sliding [117]. The majority of the wear is attributed to debonding, matrix damage, presence of cracks and fibre pull-out. Also, the appearance of furrow and cutting action justifies the abrasive wear mechanism. The damage caused to the fibre is pronounced at greater speeds in comparison with lower rates.

6. Optimization of surface treatment in jute fibre reinforced composites

Alkali treatment is amongst the widely used chemical treatment for the surface modification of natural fibres [118]. It has been shown that alkali treatment of natural fibre increases surface roughness leading to better mechanical bonding and simultaneously, it also increased the amount of cellulose exposed on the fibre surface, resulting in enhancing the number of possible reaction sites. In particular, jute fibres are complex heterogeneous lignocellulosic polymers that comprise of cellulose, hemi-cellulose and lignin. The fibres are hydrophilic owing to the presence of many hydroxyl groups and thus, they are cannot match with hydrophobic matrices including polypropylene. Alkali treatment of jute fibres eliminates the hydroxyl groups and improves the compatibility between jute fibres and PP matrix by reducing the hydrophilicity of the fibres [119].

The alkali treatment causes the removal of lignin and hemicellulose that leads to less dense and less rigid interfibrillar regions [120]. Thus, the fibrils can rearrange themselves better

Sustainable Natural Fiber Composites   Materials Research Forum LLC
Materials Research Foundations **122** (2022) 37-76   https://doi.org/10.21741/9781644901854-2

in the event of the application of tensile load leading to better load sharing. The removal of non-cellulosic materials including lignin, hemi-cellulose and pectin has resulted in the weight loss, reduction in fibre diameter and enhancement in tensile properties [121]. The increment in tensile properties of the alkali treated jute fibres has also been correlated with the molecular orientation of fibres [122].

But, in order to improve the mechanical properties the amount of alkali has to be optimised.

In general, the alkali treatment of cellulosic fibres is well known for decades in the textile industry, that is, mercerization process [123-125]. In this process, the related parameters including concentration of alkaline treatment, temperature, time, tension of the material and additives, play a significant role in determining the structure and properties of fibres. The alkali treatment of natural fibres not only altered their crystallinity but also led to a reduction in the spiral angle, that is closer to fibre axis in molecular orientation [126]. Nevertheless, it has been reported a considerable increase of 120% in tensile strength of jute fibres after treating with 25 wt.% NaOH at ambient temperature over 20 min duration. But, the usage of high concentrations of alkaline solution is not viable on a commercial scale. Recently, investigation has been carried on the alkali treatment of jute fibres under ambient, elevated temperatures and high pressure steaming conditions [127]. The investigation considered herein justifies the treatment of jute fibres with 4 wt.% caustic soda under ambient conditions and has lead to maximum tensile strength.

More recently, Roy et al. have treated the jute fibres with 0.5 wt.% NaOH for 24 hours duration that has yielded maximum tensile strength and even higher tensile strength of jute fibres than that reported by Saha et al. [128]. Nevertheless, it has been clearly demonstrated that the alkali treatment of jute fibres with low concentration of alkali under minimum duration of time (30 min) and ambient conditions can alter the surface characteristics of jute fibres resulting in the significant improvement in tensile properties. But, it is hard to directly compare between different experiments owing to variation in experimental conditions.

In order to study the tensile properties of untreated jute as well as those treated with alkali under similar experimental conditions two-parameter Weibull distribution has been used. The aim of the investigation is to optimize the alkali treatment processes of jute fibres using various concentrations of caustic soda at room temperature by means of Weibul Analysis.

6.1 Comparison between properties of untreated and alkali treated jute fibres

The physical and mechanical properties of untreated and alkali treated jute fibres have been determined. The typical stress–strain curves of untreated and alkali treated jute fibres have been obtained. The mechanical characteristics (modulus strength) of the jute fibre are

Materials Research Forum LLC

https://doi.org/10.21741/9781644901854-2

determined by its cross sectional area, which is considered an important geometrical property. Hence, it is critical to determine the fibre cross-sectional area of jute fibres at the pre and post stages of alkali treatment. Jute fibre have been compared for its cross sections before and after alkali treatment.

SEM has been used with certain magnification to obtain the cross-sections of the jute fibres. The average cross-sectional areas of selected number of jute fibres have then been determined by using freely available image analysis tool. Also, assuming the jute fibres to have circular cross section, additional calculations of cross-sectional area of jute fibres have been done. In the past, many researchers have considered the cross-section of jute fibres to be circular in nature [129-131]. The cross-sectional areas of jute fibres before and after alkali treatment have been compared by using SEM analysis and circular fibre assumption. Based on the SEM images it is observed that there is a decrease in the cross-sectional area of jute fibres by at specified concentrations of caustic soda.

The findings clearly show that the improvement in tensile properties of jute fibres subsequent to alkali treatment can be considerably overestimated under the assumption that they are circular in cross-section. Hence, further study of tensile properties of jute fibres has been conducted depending on the actual measurements of cross-sectional areas through SEM images.

6.2   Comparison between mechanical properties

It has been found that in comparison with untreated jute fibres the tensile strength of jute fibres treated with 0.5, 4 and 25 wt. % caustic soda have increased nearly by 59, 39 and 49% respectively. But, the increment of breaking extension (%) for the alkali treated jute fibres are found to be between 17–18%. Since, the alkali treatments of jute fibres with different concentration and time duration cause the interfibrillar region to be less rigid and less dense due to leaching out of hemicellulose resulting in rearrangement amongst fibrils towards the tensile loading direction. It is found that there is noticeable increment of the extension at break of the alkali treated jute fibres. The alkali treated jute fibres became more flexible owing to the removal of the surface impurities [132]. In the case of the untreated jute fibres, it is generally found that hemicellulose remains scattered in the interfibrillar region separating the cellulose chains from one another. The removal of hemicellulose by alkali treatment results in a closer packing of the cellulose chains due to the release of internal strain.

Eventually, the fibrils will be able to rearrange themselves in a more compact manner and leading to higher tensile strength in alkali treated jute fibres [133]. There is a reduction in the calculated bending rigidity (termed as the product of fibre modulus and its area moment of inertia) of jute fibres treated with 0.5, 4 and 25 wt.% caustic soda by 12.78, 23.20 and

Sustainable Natural Fiber Composites
Materials Research Foundations **122** (2022) 37-76

Materials Research Forum LLC
https://doi.org/10.21741/9781644901854-2

5.28% respectively as per the assumption that cross-sectional shape of jute fibres is unchanged. The decrease in bending rigidity of alkali treated jute fibres enhances the coherence of textile structures causing further improvement in their mechanical properties.

6.3 Study of tensile strength of untreated and alkali treated jute fibres by Weibull analysis

It has been observed that the shape parameter (Weibull modulus) of alkali treated jute fibres is lesser in comparison with that of the untreated jute fibres. A lower value of the shape parameter indicated higher scattering in tensile strengths [134]. It also implies that greater the value of shape parameters, lower is the variability in the tensile strength.

This may be attributed to the fact that the surface area of jute fibres was not uniformly treated with alkali as these fibres are occasionally found in bundles [135].

There is a very good agreement with regard to the tensile strength between Weibull and experimental average results. The findings clearly indicate that two-parameter Weibull distribution is very appropriate in prediction of the tensile strength of untreated and alkali treated jute fibres. Moreover, the probabilistic tensile strengths of untreated as well as alkali treated jute fibres have also been studied [98]. It is found that there is a decrease in differences in tensile strengths between untreated and alkali treated jute fibres with the rise in the probability of survival. Nevertheless, the tensile strengths at 95% confidence level of jute fibres treated with 0.5, 4 and 25 wt. % caustic soda have been nearly 37, 25 and 37% greater in comparison with that of the untreated jute fibres. Whereas, the improvement in the average tensile strengths of these alkali treated fibres using Weibull technique experimentally has been found to be 59, 39 and 49% respectively. The findings clearly show that probabilistic tensile strength of jute fibres is an effective method rather than presenting the average tensile strength. The findings can prove very beneficial for modelling the tensile properties of textile structures and their composites.

Conclusion

Various types of natural fibre reinforced composites have been discussed with regard to their properties. The hybrid KevlarVR /Cocos nucifera sheath reinforced epoxy composites have been assessed for ballistic performance. Multilayered hybrid and non hybrid laminates have been designed having various layering sequence following specified methods. The multi-layered laminates have been evaluated for energy absorption and ballistic limit. The studies have revealed that in comparison with hybrid KevlarVR /Cocos nucifera sheath the hybrid composites and CS/epoxy composite panels show greater energy absorption and ballistic limit. This can be attributed to the chemical composition of CS, architecture and unique shock wave dissipation mechanism. Thus the newly developed eco friendly material can offer a good substitute for Kevlar fabric in the ballistic composites. Innovative untreated and alkali-treated flaxseed fiber bundles reinforced polybutylene succinate (PBS) composites have been designed and studied. Greater tensile strength has been found in the case of untreated fibre bundles than treated ones. Treatment with alkali, caustic soda show significant improvement in the flexural properties of the flaxseed fiber bundles reinforced PBS composites. It could arise from better interfacial properties, interlocking and wetting of flaxseed fiber bundles with PBS matrix. In order to better exploit the use of flaxseed fiber bundles, the untreated fiber bundles can be used for the composites demanding greater tensile strength. But fibre bundles treated with caustic soda at specified concentration offer a better option of reinforcement for composites that require better interfacial properties. Short areca sheath (AS) fibre reinforced polyvinyl alcohol (PVA) composites modified by randomly oriented benzyl chloride have been investigated for tribological properties, particularly the abrasive behaviour. Also, the tribological behavior under multipass abrasion condition and the impacts of applied load & sliding distance on specific wear rate and weight loss have been studied. It is found that nearly 30% of the weight of fibre loading shows better resistance to wear similar to tensile strength. The wear surface at optimum fibre loading have been studied by use of SEM in order to understand the wear mechanism of the composite. The natural bast fibres such as jute, flax, kenaf, hemp, ramie are chemically modified for improving the interfacial adhesion with the hydrophobic matrices. Comparison has been made between untreated as well as alkali (caustic soda) treated jute fibres. Further, the cross-sectional areas of jute fibres before and after alkali treatment have been compared by SEM studies assuming circular fibre. The alkali treatment processes of jute fibres with different concentrations of NaOH at room temperature have been optimized. Two-parameter Weibull distribution is also applied to analyse the tensile properties of untreated and different alkali treated jute fibres. It has been observed that probabilistic tensile strength is an effective technique rather than presenting the average tensile strength. The study clearly demonstrates the jute fibre treated with 0.5 wt.% NaOH is more feasible and effective way to improve the mechanical properties of natural fibre reinforced composites. The chapter thus highlights newer types of natural fibre reinforced composites which offer as substitutes for their synthetic counterparts and also show improved properties. The studies indicate an innovative approach in the composites with improved properties and a march towards green revolution in textile composites.

References

[1] A.R. Hani, A. Roslan, J. Mariatti & M. Maziah. Body armor technology: a review of materials, construction techniques and enhancement of ballistic energy absorption. Paper presented at the Advanced Materials Research (2012). https://doi.org/10.4028/www.scientific.net/AMR.488-489.806

[2] F. de Oliveira Braga, L.T. Bolzan, F.S. da Luz, P.H.L.M. Lopes, E.P Jr. Lima, & S.N. Monteiro. High energy ballistic and fracture comparison between multilayered armor systems using non-woven curaua fabric composites and aramid laminates. Journal of Materials Research and Technology, (2000)6(4), 417-422. https://doi.org/10.1016/j.jmrt.2017.08.001

[3] L.A. Rohen, F.M. Margem, S.N. Monteiro, C.M.F. Vieira, B. Madeira de Araujo, & E.S. Lima. Ballistic efficiency of an individual epoxy composite reinforced with sisal fibers in multi-layered armor. Materials Research, (2015)18(Suppl2), 55-62. doi:10.1590/1516-1439.346314. https://doi.org/10.1590/1516-1439.346314

[4] I.K. Yilmazcoban, & S. Doner. Ballistic protection evaluation of sequencing the composite material sandwich panels for the reliable combination of armor layers. Acta Physica Polonica A, (2016)130(1), 343-346. https://doi.org/10.12693/APhysPolA.130.342

[5] H.G.M. Edwards, D.W. Farwell and D. Webster. FT Raman microscopy of untreated natural plant fibres. Spectrochim Acta Part A: Mol Biomol Spectrosc 1997; 53: 2383-2392. https://doi.org/10.1016/S1386-1425(97)00178-9

[6] K. Oksman, M. Skrifvars and J.F. Selin. Natural fibres as reinforcement in polylactic acid (PLA) composites. Compos Sci Technol (2003) 63: 1317-1324. https://doi.org/10.1016/S0266-3538(03)00103-9

[7] S. Goutianos, T. Peijs, B. Nystrom. Development of flax fibre based textile reinforcements for composite applications. Appl Compos Mater (2006)13: 199-215. https://doi.org/10.1007/s10443-006-9010-2

[8] A. Kulma, K. Skorkowska-Telichowska, K. Kostyn. New flax producing bioplastic fibers for medical purposes. Industrial Crops Product (2015) 68: 80-89. https://doi.org/10.1016/j.indcrop.2014.09.013

[9] B.F. Yousif, N. Umar & K.J. Wong, Materials Design, 31 (2010) 4514. https://doi.org/10.1016/j.matdes.2010.04.008

[10] B.F. Yousif & N.S,M El-Tayeb, Surface Rev Letters, 14(6) (2007) 1095. https://doi.org/10.1142/S0218625X07010561

[11] J.A. Khan, M.A. Khan & R. Islam. Mechanical, thermal and degradation properties of jute fabric-reinforced polypropylene composites: Effect of potassium permanganate

as oxidising agent. Polymer Composites, 34(5) (2013), 671-680. https://doi.org/10.1002/pc.22470

[12] A.K. Mohanty, M.A. Khan, & G. Hinrichsen. Surface modification of jute and its influence on performance of biodegradable jute-fabric/Biopol composites. Composites Science and Technology, 60(7) (2000), 1115-1124. https://doi.org/10.1016/S0266-3538(00)00012-9

[13] H.U. Zaman, M.A. Khan, & R.A. Khan.. Comparative experimental measurements of jute fiber/polypropylene and coir fiber/polypropylene composites as ionizing radiation. Polymer Composites, 33(7) (2012), 1077-1084. https://doi.org/10.1002/pc.22184

[14] X. Li, G.L. Tabil, & S. Panigrahi. Chemical treatments of natural fiber for use in natural fiber-reinforced composites: A review. Journal of Polymers and the Environment, 15(1) (2007), 25-33. https://doi.org/10.1007/s10924-006-0042-3

[15] L.Y. Mwaikambo & M.P. Ansell.. Chemical modification of hemp, sisal, jute, and kapok fibers by alkalization. Journal of Applied Polymer Science, 84(12) (2002), 2222-2234. https://doi.org/10.1002/app.10460

[16] D. Ray & B.K. Sarkar. Characterization of alkali-treated jute fibers for physical and mechanical properties. Journal of Applied Polymer Science, 80(7) (2001), 1013-1020. https://doi.org/10.1002/app.1184

[17] C.A. Hill, H.P.S. Khalil & M.D. Hale. A study of the potential of acetylation to improve the properties of plant fibres. Industrial Crops and Products, 8(1) (1998), 53-63. https://doi.org/10.1016/S0926-6690(97)10012-7

[18] M.J. Jhon, B. Francis, K.T. Varughese, & S. Thomas. Effect of chemical modification on properties of hybrid fibre biocomposites. Composites Part A: Applied Science and Engineering, 39(2008), 352-363. https://doi.org/10.1016/j.compositesa.2007.10.002

[19] P. Liu, & M.S. Strano. Toward ambient armor: Can new materials change longstanding concepts of projectile protection? Advanced Functional Materials, 26(6) (2016), 943-954. https://doi.org/10.1002/adfm.201503915

[20] P. Wambua, J. Ivens, & I. Verpoest. Natural fibres: Can they replace glass in fibre reinforced plastics? Composites Science and Technology, 63(9) (2003), 1259-1264. https://doi.org/10.1016/S0266-3538(03)00096-4

[21] Z. Benzait, & L. Trabzon. A review of recent research on materials used in polymer-matrix composites for body armor application. Journal of Composite Materials, 52(2018), 3241-3263. https://doi.org/10.1177/0021998318764002

[22] Y. Zhou, J. Hou, X. Gong, & D. Yang. Hybrid panels from woven KevlarVR and DyneemaVR fabrics against ballistic impact with wearing flexibility. The Journal of the Textile Institute, 1(2017), 8. https://doi.org/10.1080/00405000.2017.1398122

[23] R.B.D. Jr. Cruz, P. Lima, M. Edio, S. Neves, & L.H.L. Louro. Giant bamboo fiber reinforced epoxy composite in multi-layered ballistic armor. Materials Research, 18(2015) (Suppl2), 70-75. https://doi.org/10.1590/1516-1439.347514

[24] F.S.D. Jr. Luz, P.E. Lima, P., L.H.L. Louro & S.N. Monteiro. Ballistic test of multilayered armor with intermediate epoxy composite reinforced with jute fabric. Materials Research, 18(2015) (Suppl2), 170-177. https://doi.org/10.1590/1516-1439.358914

[25] S.N. Monteiro, L.H.L. Louro, W. Trindade, C.N. Elias, C.L. Ferreira, E. de Sousa Lima, E.P. Lima. Natural curaua fiber-reinforced composites in multi layered ballistic armor. Metallurgical and Materials Transactions A, 46(10) (2015), 4567-4577. https://doi.org/10.1007/s11661-015-3032-z

[26] S.N. Monteiro, V.S. Candido, F.O. Braga, L.T. Bolzan, R.P. Weber & J.W. Drelich, Sugarcane bagasse waste in composites for multilayered armor. European Polymer Journal, 78 (2016), 173-185. https://doi.org/10.1016/j.eurpolymj.2016.03.031

[27] S.N. Monteiro, T.L.Milanezi, L.H.L. Louro, E. P., Jr. Lima, F. O. Braga, , A.V. Gomes, & J.W. Drelich,. Novel ballistic ramie fabric composite competing with KevlarTM fabric in multi layered armor. Materials & Design, 96(2016), 263-269. https://doi.org/10.1016/j.matdes.2016.02.024

[28] L.F.C, Nascimento, L.H.L. Louro, S.N. Monteiro, A.V. Gomes, E.D. Jr. Lima, , & R. L. S. B. Marc¸al. Ballistic performance in multilayer armor with epoxy composite reinforced with Malva fibers. Paper presented at the Proceedings of the 3rd Pan American Materials Congress. Pp. 331-338. https://doi.org/10.1007/978-3-319-52132-9_33

[29] S. Jambari, M. Y. Yahya, M. R. Abdullah, & M. Jawaid, Woven Kenaf/Kevlar Hybrid Yarn as potential fiber reinforced for anti-ballistic composite material. Fibers and Polymers, 18(3) (2017), 563-568. https://doi.org/10.1007/s12221-017-6950-0

[30] Yahaya, R., Sapuan, S. M., Jawaid, M., Leman, Z., & Zainudin, E. S. (2016c). Investigating ballistic impact properties of woven kenaf-aramid hybrid composites. Fibers and Polymers, 17(2), 275-281. https://doi.org/10.1007/s12221-016-5678-6

[31] M. S..Risby, S. V. Wong, A. M. S. Hamouda, A. R.. Khairul, & M. Elsadig, Ballistic performance of coconut shell powder/twaron fabric against non-armour piercing projectiles. Defence Science Journal, 58(2) (2008), 248. https://doi.org/10.14429/dsj.58.1645

[32] Z.S. Radif, A. Aidy & K.Abdan, Development of a green combat armour from rame-Kevlar-polyester composite. Pertanika Journal of Science and Technology, 19(2) (2011), 339-348.

[33] I. M.De Rosa, H. N.Dhakal,, C.Santulli, F.Sarasini, & Z. Y. Zhang, Post-impact static and cyclic flexural characterisation of hemp fibre reinforced laminates. Composites Part B: Engineering, 43(3) (2012), 1382-1396. https://doi.org/10.1016/j.compositesb.2011.09.012

[34] X. Chen, F. Zhu, & G. Wells. An analytical model for ballistic impact on textile based body armour. Composites Part B: Engineering, 45(1) (2013), 1508-1514. https://doi.org/10.1016/j.compositesb.2012.08.005

[35] R. Yahaya, S. M. Sapuan, M. Jawaid, Z. Leman, & E. S Zainudin,. Effect of fibre orientations on the mechanical properties of kenaf-aramid hybrid composites for spall-liner application. Defence Technology, 12(1) (2016a), 52-58. https://doi.org/10.1016/j.dt.2015.08.005

[36] R.Yahaya , S. M. Sapuan, M. Jawaid, Z. Leman, & E. S. Zainudin, Effect of moisture absorption on mechanical properties of natural fibre hybrid composite. Paper presented at the Proceedings of the 13th International Conference on Environment, Ecosystems and Development (EED'15) (2016b).

[37] Yahaya, , Sapuan, S. M., Jawaid, M., Leman, Z., & Zainudin, E. S. (2016c). Investigating ballistic impact properties of woven kenaf-aramid hybrid composites. Fibers and Polymers, 17(2), 275-281. https://doi.org/10.1007/s12221-016-5678-6

[38] R.Yahaya, Sapuan, Jawaid, M., Leman, Z., & Zainudin, E. S. (2016d). Measurement of ballistic impact properties of woven kenaf-aramid hybrid composites. Measurement, 77, 335-343. https://doi.org/10.1016/j.measurement.2015.09.016

[39] Jambari, S., S. M.Yahya, M. R. Abdullah, & M. Jawaid, Woven Kenaf/Kevlar Hybrid Yarn as potential fiber reinforced for anti-ballistic composite material. Fibers and Polymers, 18(3) (2017), 563-568. https://doi.org/10.1007/s12221-017-6950-0

[40] A. K. Bledzki, S. Reihmane, & J. Gassan,. Properties and modification methods for vegetable fibers for natural fiber composites. Journal of Applied Polymer Science, 59(8) (1996), 1329-1336. https://doi.org/10.1002/(SICI)1097-4628(19960222)59:8<1329::AID-APP17>3.3.CO;2-5

[41] D.Zhang, Y.Sun, L.Chen, S. Zhang, & N.Pan. Influence of fabric structure and thickness on the ballistic impact behavior of ultrahigh molecular weight polyethylene composite laminate. Materials & Design (1980-2015), 54(2014), 315-322. https://doi.org/10.1016/j.matdes.2013.08.074

[42] P.V. Cavallaro.. Soft body armor: An overview of materials, manufacturing, testing, and ballistic impact dynamics: Naval Undersea Warfare Center Div Newport Ri. (2011). https://doi.org/10.21236/ADA549097

[43] J. Naveen, M. Jawaid, E. S. Zainudin, Mohamed T. H. Sultan & R. Yahaya: Evaluation of ballistic performance of hybrid Kevlar®/Cocosnucifera sheath reinforced epoxy composites, The Journal of The Textile Institute (2018). https://doi.org/10.1080/00405000.2018.1548801

[44] Singh KK, Mridula D, Rehal J, et al. Flaxseed: A potential source of food, feed and fiber. Crit Rev Food Sci 2011; 51: 210-222. https://doi.org/10.1080/10408390903537241

[45] K. Heller, Q.C. Sheng, F. Guan. A comparative study between Europe and China in crop management of two types of flax: Linseed and fibre flax. Industrial Crops Product 2015; 68: 24-31. https://doi.org/10.1016/j.indcrop.2014.07.010

[46] FAOSTAT. FAOSTAT for world linseed production 1994-2016: Food and Agriculture Organization of the United Nations, http://www.fao.org/faostat/en/?#data/QC/visualize (2018).

[47] D.T.E. Ehrensing. "FLAX" February 2008. Department of Crop and Soil Science, Oregon State University, https://catalog.extension.oregonstate.edu/sites/catalog/files/project/pdf/em8952.pdf (2008).

[48] K. Richardson. Flaxpage (2007). NDSU Libraries, Fargo, ND 58105, http://www.ag.ndsu.edu/agnic/flax/index.html

[49] A. Ulrich, R. Marleau, Soils and Crops Workshop Using agronomic practices to increase the per hectare yield of flax. In: 91st PAPTAC annual meeting; 17 February 2005. https://harvest.usask.ca/handle/10388/9472.

[50] W.S. Anthony. Separation of fiber from seed flax straw. Appl Eng Agri 2002; 18: 227-233. https://doi.org/10.13031/2013.7788

[51] W.S. Anthony. New technology to separate fiber and shive from seed flax straw. In: ASAE Annual International Meeting, paper no. 056060. St. Joseph, MI:American Society of Agricultural and Biological Engineers.(2005). www.asabe.org. https://doi.org/10.13031/2013.19565

[52] J.A. Foulk, D.E. Akin and R.B. Dodd. New low cost flax fibers for composites. SAE Technical Paper 2000-01-1133, (2000). https://doi.org/10.4271/2000-01-1133

[53] D.E. Akin, R.B. Dodd, W. Perkins. Spray enzymatic retting: A new method for processing flax fibers. Textile Res J 2000; 70: 486-494. https://doi.org/10.1177/004051750007000604

[54] I. Ansari, G. East and D. Johnson. Structure-property relationships in natural cellulosic fibres. Part I: Characterisation. J Textile Inst .90(1999): 469-480. https://doi.org/10.1080/00405000.1999.10750046

[55] K. Wong, X. Tao, C. Yuen. Effect of plasma and subsequent enzymatic treatments on linen fabrics. Coloration Technol. 116(2000): 208-214. https://doi.org/10.1111/j.1478-4408.2000.tb00040.x

[56] C. Wang. Exploitation on plant fiber reinforced thermoplastic composite-bamboo fiber and the properties of the composites. Dissertation, School of Textiles, Tianjin Polytechnic University, Tianjin, China. (2006).

[57] D. Ray, B.K. Sarkar, A. Rana. Effect of alkali treated jute fibres on composite properties. Bull Mater Sci 24(2001): 129-135. https://doi.org/10.1007/BF02710089

[58] M. Baiardo, G. Frisoni, M. Scandola. Surface chemical modification of natural cellulose fibers. J Appl Polym Sci. 83(2002): 38-45. https://doi.org/10.1002/app.2229

[59] A.M.M. Edeerozey, H.M. Akil, A.B. Azhar. Chemical modification of kenaf fibers. Mater Lett. 61(2007): 2023-2025. https://doi.org/10.1016/j.matlet.2006.08.006

[60] P. Maijala, M. Makinen, S. Galkin. Enzymatic modification of flaxseed fibers. J Agri Food Chem. 60(2012): 10903-10909. https://doi.org/10.1021/jf303965k

[61] P.R. Lamb and R.J. Denning. Flax-Cottonised fibre from linseed stalks: a report for the Rural Industries Research and Development Corporation. Report no. RIRDC, (2004). Australia: Kingston, Act.

[62] B.C. Suddell. Natural fibre composites in automotive applications in natural fibres in biopolymers & their biocomposites. CRC Press. (2005). https://doi.org/10.1201/9780203508206.ch7

[63] B. Suddel. Industrial fibres: recent and current developments. In: Proceeding of the symposium of natural fibres. Rome, Italy: Food and Agriculture Organization (FAO) and Common Fund for Commodities (CFC). 20(2008): 71-82

[64] N. Abilash and M. Sivapragash. Environmental benefits of eco-friendly natural fiber reinforced polymeric composite materials. IJAIEM. 2(2013): 53-59.

[65] L. Chen, F. Wu, Y. Li. Robust and elastic superhydrophobic breathable fibrous membrane with in situ grown hierarchical structures. J Membrane Sci. 547(2018): 93-98. https://doi.org/10.1016/j.memsci.2017.10.023

[66] L. Chen, Z. Hu, Z. Wu. POSS-bound ZnO nanowires as interphase for enhancing interfacial strength and hydrothermal aging resistance of PBO fiber/epoxy resin composites. Compos Part A: Appl Sci Manuf. 96(2017): 1-8. https://doi.org/10.1016/j.compositesa.2017.02.013

[67] A.N. Netravali and S. Chabba. Composites get greener. Mater Today. 6(2003): 22-29. https://doi.org/10.1016/S1369-7021(03)00427-9

[68] A. Mohanty, M. Misra and G. Hinrichsen. Biofibres, biodegradable polymers and biocomposites: An overview. Macromol Mater Eng. 276(2000): 1-24. https://doi.org/10.1002/(SICI)1439-2054(20000301)276:1<1::AID-MAME1>3.0.CO;2-W

[69] M.K. Hossain, M.R. Karim, M.R. Chowdhury. Comparative mechanical and thermal study of chemically treated and untreated single sugarcane fiber bundle. Indus Crops Product. 58(2014): 78-90. https://doi.org/10.1016/j.indcrop.2014.04.002

[70] J.A. Foulk, D.E. Akin, R.B. Dodd. Flax fiber: potential for a new crop in the Southeast. Trends in new crops and new uses. Alexandria, VA: ASHS Press, 2002: 361-370.

[71] N. Zafeiropoulos, D. Williams, C. Baillie.. Engineering and characterisation of the interface in flax fibre/polypropylene composite materials. Part I. Development and investigation of surface treatments. Compos Part A: Appl Sci Manuf. 33(2003): 1083-1093. https://doi.org/10.1016/S1359-835X(02)00082-9

[72] Gassan J. A study of fibre and interface parameters affecting the fatigue behaviour of natural fibre composites. Compos Part A: Appl Sci Manuf . 33(2002): 369-374. https://doi.org/10.1016/S1359-835X(01)00116-6

[73] S.V. Joshi, L. Drzal, A. Mohanty. Are natural fiber composites environmentally superior to glass fiber reinforced composites? Compos Part A: Appl Sci Manuf. 35(2004): 371-376. https://doi.org/10.1016/j.compositesa.2003.09.016

[74] S. Sam-Brew and G.D. Smith. Flax and Hemp fiber-reinforced particleboard. Industrial Crops Product. 77(2015): 940-948. https://doi.org/10.1016/j.indcrop.2015.09.079

[75] L. Chen,. Si L, F. Wu. Electrical and mechanical self-healing membrane using gold nanoparticles as localized "nano-heaters". J Mater Chem C. 4(2016): 10018-10025. https://doi.org/10.1039/C6TC03699F

[76] L. Liu, J Yu, L Cheng. Biodegradability of poly (butylene succinate)(PBS) composite reinforced with jute fibre. Polym Degrad Stabil. 94(2009): 90-94. https://doi.org/10.1016/j.polymdegradstab.2008.10.013

[77] G. Dorez, A. Taguet, L. Ferry. Thermal and fire behavior of natural fibers/PBS biocomposites. Polym Degrad Stabil. 98(2013): 87-95. https://doi.org/10.1016/j.polymdegradstab.2012.10.026

[78] Lee SH and Wang SQ. Biodegradable polymers/bamboo fiber bio composite with bio based coupling agent. Compos Part A-Appl Sci. 37(2006): 80-91. https://doi.org/10.1016/j.compositesa.2005.04.015

[79] W.A. Sayed, X. Fujun, Z. Yinnan and Q. Yiping. Fabrication and mechanical properties of flaxseed fiber bundle-reinforced polybutylene succinate composites, Journal of industrial textiles, 50(1)(2020): 98. https://doi.org/10.1177/1528083718821876

[80] S. Ahmed and A.C. Ulven. Dynamic in-situ observation on the failure mechanism of flax fiber through scanning electron microscopy. Fibers. 6(2018): 17. https://doi.org/10.3390/fib6010017

[81] A.B.D. Thuault, I. Hervas and M. Gomina. Investigation of the internal structure of flax fibre cell walls by transmission electron microscopy. Cellulose. 22(2015): 3521-3530. https://doi.org/10.1007/s10570-015-0744-6

[82] R.A. Shanks, A. Hodzic and D. Ridderhof. Composites of poly(lactic acid) with flax fibers modified by interstitial polymerization. J Appl Polym Sci. 99(2006): 2305-2313. https://doi.org/10.1002/app.22531

[83] A. Orue, A. Jauregi, U. Unsuain. The effect of alkaline and silane treatments on mechanical properties and breakage of sisal fibers and poly(lactic acid)/sisal fiber composites. Compos A Appl Sci. 84(2016): 186-195. https://doi.org/10.1016/j.compositesa.2016.01.021

[85] R. Sepe, F. Bollino, L. Boccarusso. Influence of chemical treatments on mechanical properties of hemp fiber reinforced composites. Compos B Eng. 133(2018): 210-217. https://doi.org/10.1016/j.compositesb.2017.09.030

[86] X. Liu and L. Cheng . Influence of plasma treatment on properties of ramie fiber and the reinforced composites. J dhes Sci Technol. 31(2016): 1723-1734. https://doi.org/10.1080/01694243.2016.1275095

[87] M. Cai, H. Takagi, A.N. Nakagaito. Influence of alkali treatment on internal microstructure and tensile properties of abaca fibers. Ind Crop Prod. 65(2015).: 27-35. https://doi.org/10.1016/j.indcrop.2014.11.048

[88] Z. Zhang, Y. Li and C. Chen. Synergic effects of cellulose nanocrystals and alkali on the mechanical properties of sisal fibers and their bonding properties with epoxy. Compos A Appl Sci. 101(2017): 480-489. https://doi.org/10.1016/j.compositesa.2017.06.025

[89] M. Cai, H. Takagi, A.N. Nakagaito. Effect of alkali treatment on interfacial bonding in abaca fiber-reinforced composites. Compos A Appl Sci. 90(2016): 589-597. https://doi.org/10.1016/j.compositesa.2016.08.025

[90] S.R. Ferreira, F.D.A Silva, P.R.L. Lima. Effect of fiber treatments on the sisal fiber properties and fiber-matrix bond in cement based systems. Constr Build Mater.101(2015): 730-740. https://doi.org/10.1016/j.conbuildmat.2015.10.120

[91] L. Yan, N. Chouw, L. Huang. Effect of alkali treatment on microstructure and mechanical properties of coir fibres, coir fibre reinforced-polymer composites and reinforced-cementitious composites. Constr Build Mater. 112(2016): 168-182. https://doi.org/10.1016/j.conbuildmat.2016.02.182

[92] E. Sinha and S.K. Rout. Influence of fibre-surface treatment on structural, thermal and mechanical properties of jute fibre and its composite. Bull Mater Sci.32(2009): 65-76. https://doi.org/10.1007/s12034-009-0010-3

[93] L. Xuan, J.H. Sen, H.C. Yi and Y.C. Hai. Improvement on the interfacial compatibility of jute fiber-reinforced polypropylene composites by different surface treatments, 49(7)(2020)906. https://doi.org/10.1177/1528083718801366

[94] M. Mariano, R. Cercena and V. Soldi.. Thermal characterization of cellulose nano crystals isolated from sisal fibers using acid hydrolysis. Ind Crop Prod. 94(2016): 454-462. https://doi.org/10.1016/j.indcrop.2016.09.011

[95] H. Wang, G. Xian and H. Li Grafting of nano-TiO2 onto flax fibers and the enhancement of the mechanical properties of the flax fiber and flax fiber/epoxy composite. Compos A: Appl Sci Manuf..76(2015): 172-180. https://doi.org/10.1016/j.compositesa.2015.05.027

[96] K. Roy, S.C. Debnath, A. Das. Exploring the synergistic effect of short jute fiber and nanoclay on the mechanical, dynamic mechanical and thermal properties of natural rubber composites. Polym Test. 67(2018): 487-493. https://doi.org/10.1016/j.polymertesting.2018.03.032

[97] R. Basak, P.L. Choudhury and K.M. Pandey. Effect of temperature variation on surface treatment of short jute fiber-reinforced epoxy composites. Mater Today Proc. 5(2015):1271-1277. https://doi.org/10.1016/j.matpr.2017.11.211

[98] S.P. Kundu, S. Chakraborty, S.B. Majumder. Effectiveness of the mild alkali and dilute polymer modification in controlling the durability of jute fibre in alkaline cement medium. Constr Build Mater.174(2018): 330-342. https://doi.org/10.1016/j.conbuildmat.2018.04.134

[99] J.W. Dormanns, J. Schuermann, J. Mü¨ssigl. Solvent infusion processing of all cellulose composite laminates using an aqueous NaOH/urea solvent system. Compos A: Appl Sci Manuf. 82(2015): 130-140. https://doi.org/10.1016/j.compositesa.2015.12.002

[100] Y. Li, C. Chen, J. Xu. Improved mechanical properties of carbon nanotubes-coated flax fiber reinforced composites. J Mater Sci. 50(2015): 1117-1128. https://doi.org/10.1007/s10853-014-8668-3

[101] H. Aireddy, J Metallurgy Mater Sci, 53 (2011) 139.

[102] S.A.R. Hashmi, U.K. Dwivedi U K & Chand N, Wear, 262 (2007) 1426. https://doi.org/10.1016/j.wear.2007.01.014

[103] N. Singh, B.F. Yousif & D. Rilling, Tribology Transactions, 54(5) (2011) 736. https://doi.org/10.1080/10402004.2011.597544

[104] A.K. Rana, B.C. Mitra & A.N. Banerjee, J Appl Polym Sci, 71 (1999) 531. https://doi.org/10.1002/(SICI)1097-4628(19990124)71:4<531::AID-APP2>3.0.CO;2-I

[105] J. Tong, Y. Ma, D. Chen, J. Sun & L. Ren, Wear, 259 (2005) 37. https://doi.org/10.1016/j.wear.2005.03.031

[106] N. Chand & U.K. Dwivedi, J Mat Processing Technol, 183 (2007) 155. https://doi.org/10.1016/j.jmatprotec.2006.09.036

[107] N.S.M El-Tayeb, Wear, 266 (2009) 220. https://doi.org/10.1016/j.wear.2008.06.018

[108] A.V. Prasad, K.B. Rao, K.M. Rao, K. Ramanaiah & S.P.K. Gudapati, Int J Polym Anal Character, 20 (2015) 541. https://doi.org/10.1080/1023666X.2015.1053335

[109] S. Chikkol, B. Bennehalli, M.G. Kenchappa & R. Patel, Bio Res, 3 (2010) 1846.

[110] N.K.S. Gowda, S. Anandan, D.T. Pal, N.C. Vallesha, S. Verma & K.T. Sampath, Indian Dairyman, 64 (2012) 58.

[111] N.H. Padmaraj, M. VijayKini, P.B. Raghuvir & S.B. Satish, Procedia Engg, 64 (2013) 966. https://doi.org/10.1016/j.proeng.2013.09.173

[112] K.M. Sunil, K.H.R. Shiva, S.G. Gopalakrishna & K.S. Rai, Int J Mech Eng Tech, 5 (2014) 55.

[113] K.M. Sunil, S.G. Gopalakrishna, M. Sujay & K.H.R. Shiva, Int J Adv Res Sci Eng, 4 (2015) 1647.

[114] M.R. Chethan , S.G. Gopala Krishna, R. Chennakeshava & M. Manjunath, Int J Adv Engg Tech Management Appl Sci, 3 (2016) 412.

[115] A. Shalwan & B.F. Yousif , Materials Design, 48 (2013) 14. https://doi.org/10.1016/j.matdes.2012.07.014

[116] P.A.V. Ratna, A.M. Durga A M, Rao K M, Ramanaiah K & Reddy B V, Proceedings, 18th International Conference on Advance Trends in Engineering Materials and their Applications (AES-ATEMA, Canada). (2014): 129.

[117] A.K. Mohanty, M. Misra, & L.T. Drzal, Surface modifications of natural fibers and performance of the resulting bio composites: An overview. Composite Interfaces, 8(5)(2001): 313-343. https://doi.org/10.1163/156855401753255422

[118] X.Y. Liu & G.C. Dai. Surface modification and micro mechanical properties of jute fibre mat reinforced polypropylene composites. Express Polymer Letters, 1(5)(2007): 299-307. https://doi.org/10.3144/expresspolymlett.2007.43

[119] K. Bledzki, & S.J. Gassan. Composites reinforced with cellulose based fibers. Progress in Polymer Science, 24(2)(1999): 221-274. https://doi.org/10.1016/S0079-6700(98)00018-5

[120] A. Rawal & M.M.A. Sayeed. Tailoring the structure and properties of jute blended nonwoven geotextiles via alkali treatments of jute fibres. Materials and Design, 53(2014): 701-705. https://doi.org/10.1016/j.matdes.2013.07.073

[121] S.J. Gassan, & K. Bledzki, Alkali treatment of jute fibres: Relationship between structures and mechanical properties. Journal of Applied Polymer Science, 71(4)(1999): 623-629. https://doi.org/10.1002/(SICI)1097-4628(19990124)71:4<623::AID-APP14>3.0.CO;2-K

[122] D.S. Varma, M. Varma, & I.K. Varma, Coir fibres Part 1: Effect of physical and chemical treatments on properties. Textile Research Journal, 54(12)(1984), 827-832. https://doi.org/10.1177/004051758405401206

[123] P. Saha,, S. Manna, S. Roy Chowdhury, R. Sen, D. Roy, & B. Adhikari, Enhancement of tensile strength of lignocellulosic jute fibres by alkali-steam treatment. Bioresource Technology, 101(9)(2010): 3182-3187. https://doi.org/10.1016/j.biortech.2009.12.010

[124] A. Roy, S. Chakraborty, S.P. Kundu, R..K. Basak, S.B. Majumder, & B. Adhikari. Improvement in mechanical properties of jute fibers through mild alkali treatment as demonstrated by utilisation of the Weibull distribution model. Bioresource Technology, 107(2012): 222-228. https://doi.org/10.1016/j.biortech.2011.11.073

[125] N. Defoirdt, S. Biswas, L. De Vriese, L. Ngoc Tran, L. Q., Acker, J. V. & Q. Ahsan, Verpoest, I. Assessment of the tensile properties of coir, bamboo and jute fibre. Composites Part A: Applied Science and Manufacturing, 41(2010): 558-595. https://doi.org/10.1016/j.compositesa.2010.01.005

[126] I.M. De Rosa, J.M. Kenny, D. Puglia, C. Santulli, & F. Sarasini. Morphological, thermal and mechanical characterization of okra (Abelmoschus esculentus) fibres as potential reinforcement in polymer composites. Composites Science and Technology, 70(1)(2010a): 116-122. https://doi.org/10.1016/j.compscitech.2009.09.013

[127] F.A. Silva, N. Chawla & R.D.T. Filho, Tensile behaviour of high performance natural (sisal) fibres. Composites Science and Technology, 68(15-16)(2008): 3438-3443. https://doi.org/10.1016/j.compscitech.2008.10.001

[128] H. Gu. Tensile behaviours of the coir fibre and related composites after NaOH treatment. Materials Letters, 30(2009): 3931-3934. https://doi.org/10.1016/j.matdes.2009.01.035

[129] D. Ray, & B.K. Sarkar. Characterization of alkali-treated jute fibers for physical and mechanical properties. Journal of Applied Polymer Science, 80(7)(2001): 1013-1020. https://doi.org/10.1002/app.1184

[130] M.W. Barsoum. Fundamentals of Ceramics, International ed. India: The McGraw Hill Companies (1997).

[131] A. Mukherjee, P.K. Ganguly, & D. Sur. Structural mechanics of jute: The effects of hemicellulose or lignin removal. Journal of the Textile Institute, 84(3)(2003): 348-353. https://doi.org/10.1080/00405009308658967

[132] C.A. Hill, H.P.S. Khalil, & M.D. Hale. A study of the potential of acetylation to improve the properties of plant fibres. Industrial Crops and Products, 8(1)(1998): 53-63. https://doi.org/10.1016/S0926-6690(97)10012-7

[133] M.J. Jhon, B. Francis, K.T. Varughese, & S. Thomas. Effect of chemical modification on properties of hybrid fibre biocomposites. Composites Part A: Applied Science and Engineering, 39(2008): 352-363. https://doi.org/10.1016/j.compositesa.2007.10.002

[134] K.L. Pickering, G.W. Beckermann, S.N. Alam, & N.J. Foreman. Optimising industrial hemp fibre for composites. Composites Part A: Applied Science and Manufacturing, 38(2)(2007), 461-468. https://doi.org/10.1016/j.compositesa.2006.02.020

[135] M.M. Alamgir Sayeed & P. Ayush P. Optimisation of the surface treatment of jute fibres for natural fibre reinforced polymer composites using Weibull analysis, The Journal of The Textile Institute (2019). https://doi.org/10.1080/00405000.2019.1610998

Sustainable Natural Fiber Composites   Materials Research Forum LLC
Materials Research Foundations **122** (2022) 77-95   https://doi.org/10.21741/9781644901854-3

Chapter 3

Next Generation Biodegradable Material for Dental Teeth and Denture base Material: Vegetable Peels as an Alternative Reinforcement in Polymethylmethacrylate (PMMA)

Anirudh Kohli[1], Arun Y Patil[1*], Mahesh Hombalmath[1]

[1] School of Mechanical Engineering, [1,2,3] Centre for Material Science, KLE Technological University, Hubballi, India

* patilarun7@gmail.com

Abstract

Today, oral diseases are the most quotidian and conventional issues of the 21st Century. Further investigations about the oral disorders stated that it comprises tooth decay, mouth sores, tooth erosion, tooth sensitivity, gum diseases, toothaches, and dental emergencies are predominant in today's generation. The further study identified that the improper bonding between the denture base and dental teeth ends up with costly repair along with mouth sores to provide relief from these problems. This effect is not age-restricted and can be observed in any age range but it is prevalent in middle age (35 years) to old age (<80 years) people in the world. Approximately one-third of the world's population is suffering because of untreated caries of natural teeth (32.7%). The optimal composition of PMMA, organic filler materials, zinc or titanium oxides etc. as denture base, dental tooth, clinical trials with observation, and experimental methodology is yet to be unleashed to reach its optimized state for this considered problem. A call for an alternative material for denture base or the dental tooth is still an unanswered question for a long period. Selection of appropriate reinforcement material encompassing the guidelines for liquid/powder ratio, avoid processes terrible for bond strength, wax-free model. Current work focuses on the investigation of mechanical properties with or without the reinforcement of new and novel biomaterials made of epoxy with carrot peel, epoxy with onion peel, lemon peel, and epoxy with potato peel. Comparative analysis of coupons for with/without the reinforcement (in the form of Short/Continuous fiber) was investigated. The Simulation work correlated with experimental work to explore the possibilities of new material in the arena of dental.

Sustainable Natural Fiber Composites
Materials Research Foundations **122** (2022) 77-95

Materials Research Forum LLC
https://doi.org/10.21741/9781644901854-3

Keywords

Dental Material, Potato Peel, Onion Peel, Lemon Peel, PMMA, Mechanical Properties

Contents

Next Generation Biodegradable Material for Dental Teeth and Denture base Material: Vegetable Peels as an Alternative Reinforcement in Polymethylmethacrylate (PMMA) 77

1.   Introductions 79

2.   Selected materials 80

3.   Methods for material fabrication 81

4.   Method for mechanical testing 82

4.1   Methods for material physical analysis 82

4.1.1 Archimedean density test..82

4.1.2 Water absorption test 83

4.1.3 Discoloring, saliva, and comfort test..83

4.2   Methods for mechanical analysis 83

4.2.1 Tension test...83

4.2.2 Flexural test 83

4.2.3 Rockwell hardness test 83

4.2.4 Scanning electron microscopy (SEM)..85

4.3   Methods for virtual analysis 85

4.3.1 Contact generation 85

4.3.2 Mesh formation..86

4.3.3 Boundary and Load related mechanical Conditions............................87

5.   Results and discussion 88

5.1   Flexural strength 88

5.2   Hardness...89

5.3   Archimedean density 90

6.   Virtual testing...90

6.1   Maximum principal stress 90

6.2   Total deformation 92

Acknowledgment..**93**

Reference...**93**

1. Introductions

Presently, the proper disposing and utilization of waste is the most pressing issue among the general public and material science researcher's community. The appropriate usage of waste materials is overdue for a long period. The waste that is considered here is the waste produced by domestic usage from your home kitchen etc. Commercial waste from restaurants, malls, etc., medical waste comprising used syringes, Gauss, medical wastes, etc. and finally plastic and E-waste [1,2]. The organic waste is considered here due to various applications, available abundantly, and showcasing of varied mechanical properties. The wastes that come under the vast umbrella of organic waste are vegetable waste, stems from trees, broken eggshells, rotten plants, various stems, shoots, roots, and leaves of plants [3,4,5].

Even carbon footprint produced is a hype among the general public and by the utilization of biodegradable denture base and material. The use of this organic material can tend to help to cater to fill the void in the dental industry for advancement in the field of denture base and material. Tending to cater to the denture industry towards more sustainable and efficient utilization of organic materials that were disposed of thinking to be waste.

India is the third-largest producer of organic waste in the world accounting for 5.37% of the world organic waste which goes underutilized in the garbage bins and which could be utilized like other usages of natural fiber embedded poly matrix which is used in applications varying from aerospace, automobile, wind, hydrothermal, engines, manufacturing etc. According to, the latest report of Material Forum 2012 it was concluded that there is an emerging trend towards bio-composites as more people are moving away from glass or carbon fiber-based matrix to natural fiber-based matrix due to less CO_2 production irrespective of the fact that glass fiber-based. These natural fiber-based materials are "carbon positive" that tend to absorb more carbon than what it produces.

Fatigue life and failure of denture teeth and denture base materials have not yielded any fruitful results for the dental community [1,2]. PMMA as denture base and dental teeth is early staining, noisy while mastication process, low impact strength, shelf life, unpredictable and inconsistent [3]. To date, there was no such attempt to achieve an optimal condition of PMMA for denture base, dental teeth, clinical trials, and process methodology. Fiber-based reinforcements tried with glass fiber [4-9], Saline glass fiber [10-12], Aramid

Sustainable Natural Fiber Composites   Materials Research Forum LLC
Materials Research Foundations **122** (2022) 77-95   https://doi.org/10.21741/9781644901854-3

[13-14], Polyethylene [15-16], Polypropylene [17], OPEFB [18], Vegetable fiber [19] in holding matrix PMMA to know the mechanical characterization like flexural strength, Tensile strength, Compressive strength, micro-hardness, and impact strength. Since ancient days bio materials are used in the human body for the betterment. The new entry proposed in this domain is potato peel with epoxy, onion peel with epoxy, lemon peel, and carrot peel with epoxy. Simulation using ANSYS and many other tools making life easier for researchers [23, 24].

2. Selected materials

Since ancient times bio materials have played a significant part in the medical domain. Still, in rural settings the use of bio materials in surgeries and operations conducted by doctors specializing in surgeries conducted by ayurvedic doctors is prevalent. During this study, testing on various materials was conducted to obtain desired mechanical properties.

Taking inspiration from our ancient studies, a test was conducted on the peels of various vegetables in varied proportions to obtain desired properties. Finally, 4 compositions of materials were selected which are potato peel with epoxy, lemon peel with epoxy, onion peel with epoxy, and carrot peel with epoxy.

The use of Nanomaterials and Bio-composite was studied to find the right material for the denture application as we have seen the use of bio-materials in bone repair, prosthetic limbs, automobiles, etc. We have seen many bio-materials namely pineapple leaf, sugarcane bagasse, and banana pseudo-stem for their thermal stability. One of the predominant bases for deciding is its strength and aspect ratio. Fiber dimension, defects, crystallization, visibility, dimension are the secondary factors that are needed to be kept in mind while going ahead with any specific natural fiber composite as well as their reinforcement type which is fibrous, particulate, etc.

By the means of an exhaustive literature survey, it was revealed there is very limited work done in denture base and material in the field of utilization of natural fiber as a material to substitute PMMA.

During our review mechanical properties of natural fiber had been reviewed to find the research gaps. After doing a literature survey it could be concluded that the utilization of vegetable waste could be used in the study. The composition is stated in Table 1.

Sustainable Natural Fiber Composites   Materials Research Forum LLC
Materials Research Foundations **122** (2022) 77-95   https://doi.org/10.21741/9781644901854-3

Table 3.1 *Composition of Different fibers*

SI. NO	PARAMETERS	UNIT	POTATO FIBER	ONION FIBER	CARROT FIBER	SWEET LIME	LEMON
1	MOISTURE	%	4.26	4.81	5.14	6.38	7.13
2	TOTAL ASH	%	5.34	6.5	7.61	3.70	6.23
3	CRUDE FIBER	%	33.82	32	34.26	32.66	33.15
4	CELLULOSE	%	42.60	41.54	42.68	41.58	41.80
5	SPECIFIC GRAVITY		1.075	0.847	0.751	0.725	0.781
6	LIGNIN	%	44.13	43.11	44.81	38.63	43.51
7	pH		6.9	6.6	6.82	7	6.8

3. Methods for material fabrication

For the fabrication of the polymer-based composite, we employed an open casting hand laying method. Six different samples of these composites were prepared with 10-30% composition by weight of the polymer holding matrix. This was cured for 1 complete day at STP condition. In this case the specimen was prepared taking into consideration the ASTM standards for various mechanical testing. Special care was taken to maintain structural and dimensional stability as well as uniformity thus preventing error to a great extend.

It is suggested that in case of any cellulose-based medium to high composition fibers, that it should undergo alkali treatment. This tends to enhance its mechanical characteristics such as its flexural. Impact, tensile, and absorption property. In this, each of the selected fiber was firstly treated with 5% Na-OH solution by volume for 2 hours after which it was followed by rinsing it from distilled water 3 times and then finally left to dry at 50°C for 1 hour. This process is carried out to remove moisture from the natural fiber and remove any unwanted particles. The fabricated material sample is shown in Figure 1.

Sustainable Natural Fiber Composites   Materials Research Forum LLC
Materials Research Foundations **122** (2022) 77-95   https://doi.org/10.21741/9781644901854-3

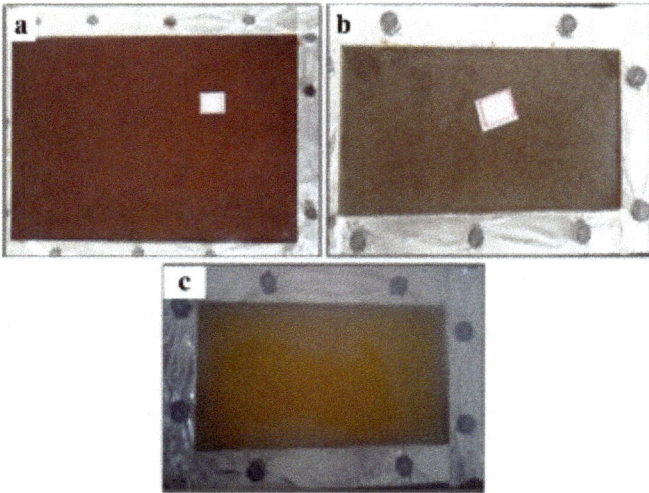

Fig. 1 *Mould of the composite material formed using vegetable peel fillers as a reinforcement with epoxy **a)** onion composite **b)** potato composite **c)** Carrot composite*

4.   Method for mechanical testing

4.1   Methods for material physical analysis

4.1.1   Archimedean density test

In this case the density test of the specimen is performed with regards to ASTM D792. As the name suggests this is based on the Archimedes Principle. The engineering density of each of the specimen is found out by the help of setup comprising of a cantilever along with a weight machine. The steps to perform the above mentioned test is as following:

For the first step we calibrate the weighing machine to zero using the Tare option by placing the cantilever on weighing machine and then calibrating the weight to zero.

- For the second step , we will place the specimen on the cantilever at STP conditions and record its weight when air (W_a).

- For the last step, we weights the same specimen after placing in water and taken weight in water.(W_w)
- Density of the above materials is found using the stated formula

$$\rho = \frac{W_a}{W_a - W_w}$$

(3.1)

Where p, the engineering density of different vegetable peel composite specimen (g/cm^3).

4.1.2 Water absorption test

The specimen used for this testing is formed according to the ASTM D5229 standards. To perform this experiment, we first take the specimens and record their weight. After this these specimens are placed in water and this weight is monitored regularly after 24 hours and the rate of absorption of water is noted, this is done for 11 days

4.1.3 Discoloring, saliva, and comfort test

Each of the specimens was exposed to all the oils, spices, and saliva composition to check the effect of it on a person's health. For the comfort of the person, the denture has to be a little moist and not completely in powder form so that it does not affect the health of the patient

4.2   Methods for mechanical analysis

4.2.1 Tension test

The tensile test is performed by utilizing the micro Universal Testing Machine (UTM) which was made in Enkaky Enterprise, Bangalore which had a capacity of 10 tonnes. For each specimen dimension was chosen according to ASTM D-3039 standards which were 250mm X 25mm X 3mm. To perform this test, we had taken the specimen gauge length to be 138 mm having a cross head speed of 3 mm/sec that was kept as constant. To perform this test, we apply a uni-axial singular point load at both the ends of the specimen.

4.2.2 Flexural test

To perform this test the specimen of dimensions 154Millimetre X 13Millimetre X 3Millimetre which was formed with reference to the ASTM D-7264 standards has to undergo a 3 point bend test in which the support span length is only 100 millimetre having a cross head speed of 3 millimetre/min as shown in the figure 2.

4.2.3 Rockwell hardness test

The specimen for this case was prepared according to ASTM D785, in this case, the indenter selected was a steel ball of diameter 1/16". Its basic concept is Rockwell hardness tester which is done utilizing Saroj Udyog Private Limited, Bangalore as shown in Figure 3. In this test the specimen which is going to be tested is applied with a load of 60Kg while it is lying horizontally on a horizontal surface, the load is applied for 15 seconds for each of the specimen to find the harness of each specimen individually. All the hardness values were recorded for reference purpose [25,26,27].

Fig 2. *Flexural testing setup*

Fig. 3 *Rockwell hardness testing setup for the specimen*

Sustainable Natural Fiber Composites   Materials Research Forum LLC
Materials Research Foundations **122** (2022) 77-95   https://doi.org/10.21741/9781644901854-3

4.2.4 Scanning electron microscopy (SEM)

In this test we utilize the ZEISS SEM to find the fracture in the tensile tested specimen having dimensions as 10 Millimetre X 10 Millimetre X 2.5 Millimetre that had an accelerating voltage of 2 kV .In this case to make the specimen to be electrically conducting we coat the specimen with a layer of Palladium . For the testing purpose we test the specimen of 10% by volume fraction as it was found that those show the best mechanical properties , to perform this test the specimen was placed vertically to view its surface under SEM.

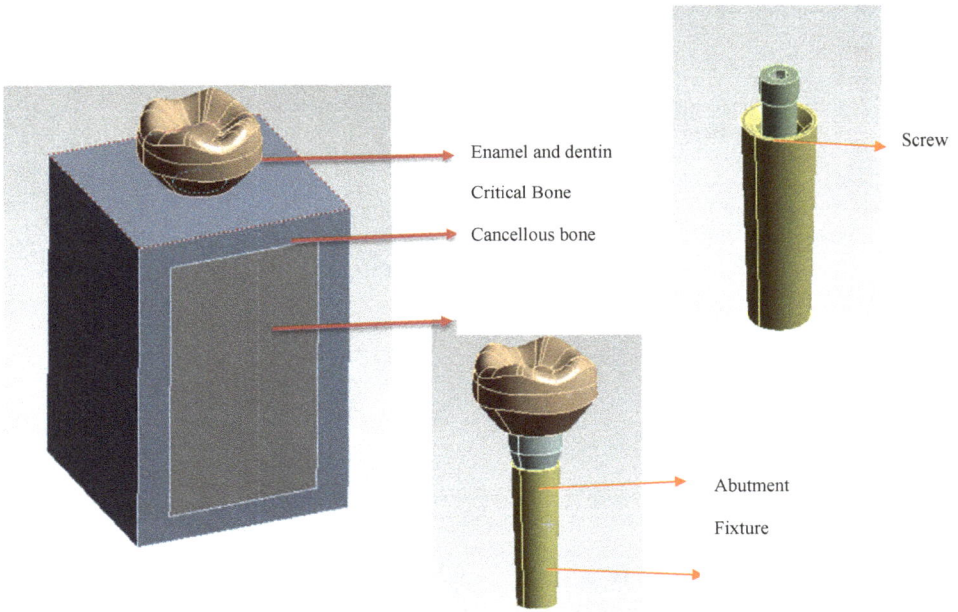

Enamel and dentin

Critical Bone

Cancellous bone

Screw

Abutment

Fixture

Fig. 4*. Parts of Denture of tooth*

4.3   Methods for virtual analysis

4.3.1   Contact generation

The complete assembly of a tooth implant comprises 6 parts which are shown in figure 4. To understand the intricacies of the implant we have designed the abutment implant fixture

Sustainable Natural Fiber Composites   Materials Research Forum LLC
Materials Research Foundations **122** (2022) 77-95   https://doi.org/10.21741/9781644901854-3

along with the screw. We followed it by forming bonded contact which acts like a permanent joint just like welding as shown in figure 5. This behavior is considered to be "asymmetric" thus leading to the selection of the Penalty Method rather than the Augmented Lagrange Method. This is done as in that case the contact stiffness is higher for Penalty method in comparison to the Augmented Lagrange method as well as it has higher accuracy thus making them apt for penetration related issues.

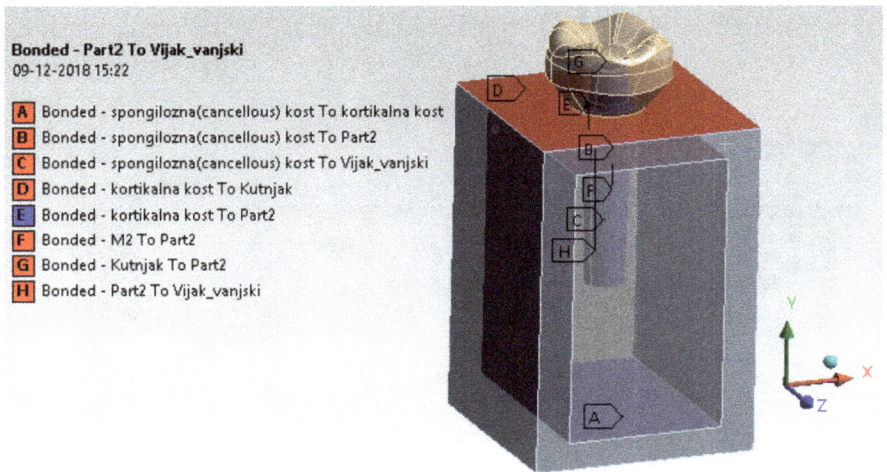

Fig. 5. *All the contacts of the Model*

4.3.2 Mesh formation

To prepare the mesh H and P type elements were utilized. The element size in the case of H -Type model varies from 0.25-1 mm. In the case of P-type method, 8 nodes elements were used to solve first-order elements whereas 20 node element was used to solve the model as shown in figure 6.

Sustainable Natural Fiber Composites   Materials Research Forum LLC
Materials Research Foundations **122** (2022) 77-95   https://doi.org/10.21741/9781644901854-3

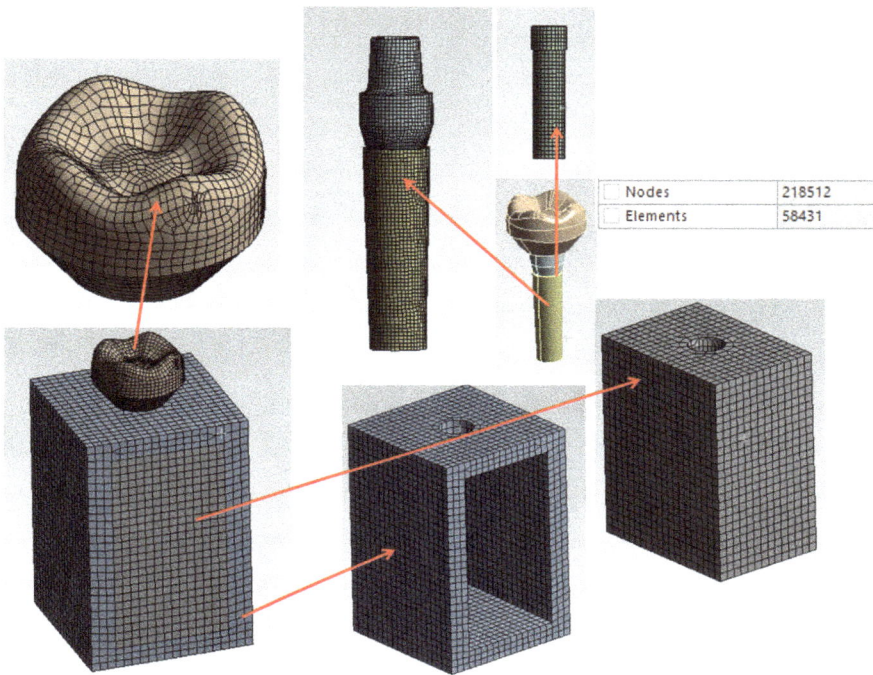

Nodes	218512
Elements	58431

Fig. 6. *Mesh generated of the model*

4.3.3 Boundary and Load related mechanical Conditions

All the boundary and mechanical load related conditions to solve the assigned model is as following:

- As specified in the figure 7(a) an axial mastication force applying normal to the unit cell tooth of force 250N is applied at the central junction

- As depicted in figure 7(b) we are taking the worst scenario of chewing into consideration by applying pressure at 5 different junctions with a total force of 800N ,which is a result of maximum mastication force .

- As palpable from figure 7 (c) a force of 225 N was applied at 45° angle to the longitudinal axis

• As visible from figure 7(d) the model is fixed beneath the denture

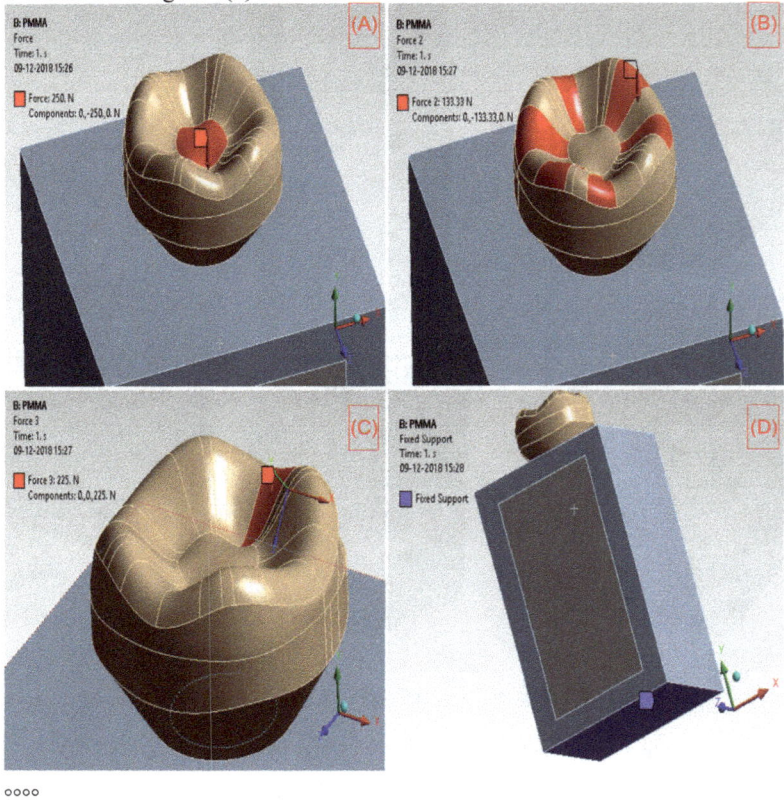

Fig. 7. *Boundary and mechanical load conditions with regards to the denture base unit*

5. Results and discussion

5.1 Flexural strength

We calculate the flexural strength of any composite material using the formula stated below:

$$\text{Flexural strength } (\sigma f) = 3FL/2bh^2$$

In table 2 the flexural strength of the natural fiber-reinforced composites is stated. All the composites developed with maximum flexural strength at a 10% volume fraction. Cross-head speed could be a significant factor in the decrease of flexural strength compared to literature values.

Table 3.2 Flexural strength

SERIAL NUMBER	Vegetable peel and Epoxy COMPOSITE	Volume fraction percentage (%)	Modulus of rupture of composite (MPa)
1	ONIONCOMPOSITE	10	36.06
2	POTATOCOMPOSITE	10	35
3	CARROT COMPOSITE	10	34.55
4	LEMON COMPOSITE	10	50.48
	SWEET LIME COMPOSITE	10	57.69
5	PURE EPOXY COMPOSITE	100	112

5.2   Hardness

Table 3 states data about the hardness of different composite compositions when subjected to a load of 60kN. Hardness is generally taken as the mean of 4 specimens of each specified volume. Sweet lime has the highest hardness of 55.4 HRB as compared to the HRB of the other natural fiber-reinforced composites. It is found that the hardness value of the sweet lime and epoxy reinforced composite ensures that sweet lime and epoxy composite has the highest tensile strength among the tested specimens and this is in agreement with the physical tested values of the tensile tested results.

Table 3. Rockwell Hardness Test

SERIAL NUMBER	Vegetable peel and epoxy Composite	Volume fraction (%)	Mechanical Load (kN)	Rockwell Hardness values obtained (HRB)
1	Onion Composite	10	60	50.75
2	Potato Composite	10	60	49.75
3	Carrot Composite	10	60	32.33
4	Lemon Composite	10	60	21.6
5	Sweet lime Composite	10	60	55.4
6	Pure Epoxy composite	100	60	81.5

5.3   Archimedean density

The density is measured using the Archimedes principle. The values are stated in Table 4

Table 4. *Density*

SI Number.	VEGETABLE PEEL AND EPOXY COMPOSITE	VOLUME FRACTION (%)	ARCHIMEDEAN DENSITY(g/cm^3)
1	Onion Composite	10	1.138
2	Potato Composite	10	1.153
3	Carrot Composite	10	1.121
4	Lemon Composite	10	1.192
5	Sweet lime Composite	10	1.206
6	Pure Epoxy Composite	100	1.3

6.   Virtual testing

6.1   Maximum principal stress

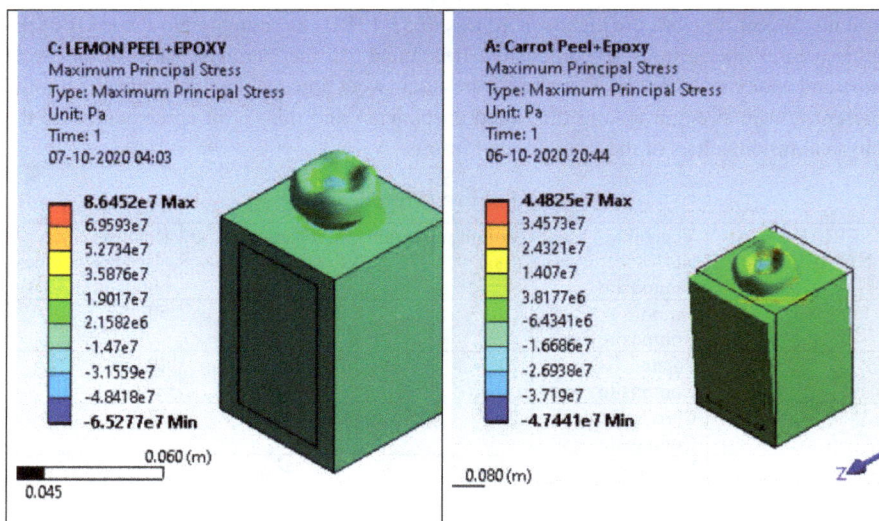

Sustainable Natural Fiber Composites   Materials Research Forum LLC
Materials Research Foundations **122** (2022) 77-95   https://doi.org/10.21741/9781644901854-3

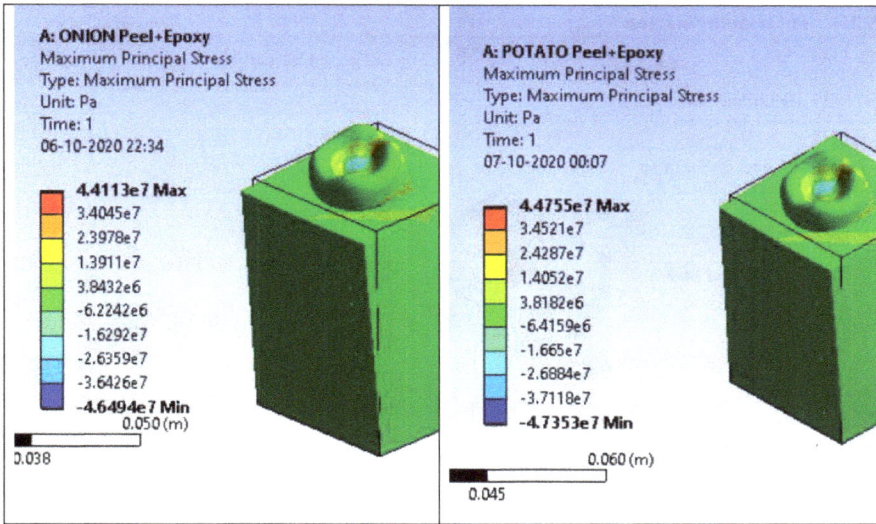

Fig. 8*. Maximum principal stress values*

From the results obtained after virtual analysis of the model using ANSYS 18.2. It is evident that we obtain the maximum and minimum principal stress in the case of Lemon peel/epoxy reinforced composite and onion peel/epoxy reinforced composite obtaining values as $8.64*10^7$ Pa and $4.41*10^7$ Pa. This will act as a way to reconfirm the data that we obtained during our physical testing.

It is evident that potato peel has the maximum deformation occurred in the case of Potato and epoxy and that was a type of force vibration in which vibration of 60Hz along with all the loads was applied to forcefully deform the given body from its nascent position and minimum deformation was observed in the case of Lemon peel and epoxy

Sustainable Natural Fiber Composites

Materials Research Forum LLC

Materials Research Foundations **122** (2022) 77-95

https://doi.org/10.21741/9781644901854-3

6.2   Total deformation

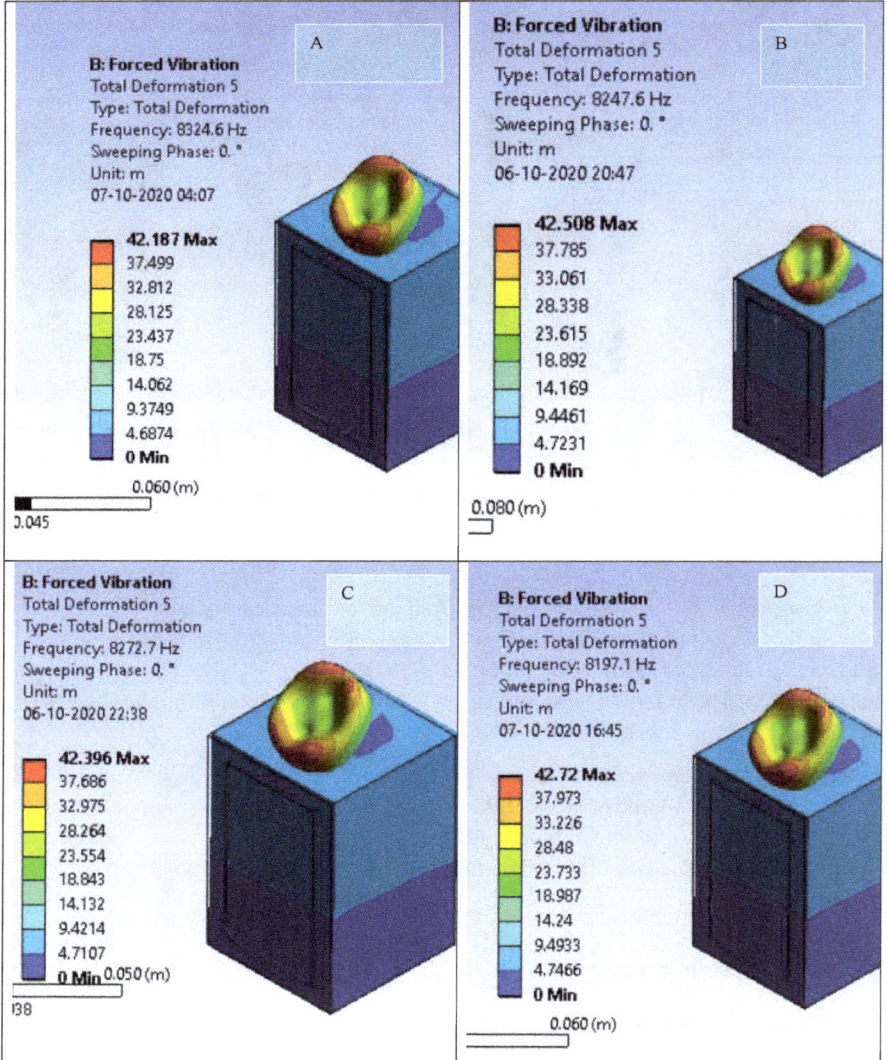

Fig. 9. *Total Deformation A) LEMON PEEL/EPOXY B) CARROT/EPOXY C) ONION/EPOXY D) POTATO/EPOXY*

Sustainable Natural Fiber Composites
Materials Research Foundations **122** (2022) 77-95

Materials Research Forum LLC
https://doi.org/10.21741/9781644901854-3

Acknowledgment

Our team of researchers would like to thank TEQIP for financial assistance. We would like to thank CMTI for analyzing mechanical properties of different specimens. We would also like to convey our gratitude to Prof. B.B. Kotturshettar, Dean(Planning and Development) and Head of School of Mechanical Engineering, Prof. P.G. Tewari (Academic dean and Principal) and Prof. B.L Desai (Registrar) KLE Technological University for his logistics support.

Reference

[1] Cunningham J L. Bond strength of denture teeth to acrylic bases. J Dent 1993; 21: 274–280. https://doi.org/10.1016/0300-5712(93)90106-Z

[2] Darbar U R, Huggett R, Harrison A. Denture fracture a survey. Br Dent J 1994; 176: 342–345. https://doi.org/10.1038/sj.bdj.4808449

[3] Zuckerman G R. A reliable method for securing anterior denture teeth in denture bases. J Prosthet Dent 2003; 89: 603–607. https://doi.org/10.1016/S0022-3913(02)52713-X

[4] Sang-Hui YU, Yoon LE, Seunghan OH, Hye-Won CH, Yutaka OD, Ji-Myung BA. Reinforcing effects of different fibers on denture base resin based on the fiber type, concentration, and combination. Dent Mater J. 2012;3:1039–1046. https://doi.org/10.4012/dmj.2012-020

[5] Farina AP, Cecchin D, Soares RG, et al. Evaluation of Vickers hardness of different types of acrylic denture base resins with and without glass fiber reinforcement. Gerodontol. 2012;29:155–160. https://doi.org/10.1111/j.1741-2358.2010.00435.x

[6] Hamouda IM, Beyari MM. Addition of glass fibers and titanium dioxide nanoparticles to the acrylic resin denture base material: a comparative study with the conventional and high impact types. Oral Health Dent Manag. 2014;13:107–112.

[7] ManamuNagakura, Yasuhiro Tanimoto, Norihiro Nishiyama, Fabrication and physical properties of glass-fiber-reinforced thermoplastics for non-metal-clasp dentures, accepted 11 July 2016 in Wiley Online Library. https://doi.org/10.1002/jbm.b.33761

[8] Dalkiz M, Arslan D, Tuncdemir AR, Bilgin MS, Aykul H. Effect of different palatal vault shapes on the dimensional stability of glass fiber reinforced heat-polymerized acrylic resin denture base material. Eur J Dent. 2012;6:70–78. https://doi.org/10.1055/s-0039-1698933

[9] Fonseca RB, Favarão IN, Kasuya VB, Abrão M, Da Luz N FM, Naves LZ. Influence of glass fiberwt% and silanization on mechanical flexural strength of reinforced acrylics. J Mat Sci Chem Eng. 2014;2:11–15. https://doi.org/10.4236/msce.2014.22003

[10] Mowade TK, Dange SP, Thakre MB, Kamble VD. Effect of fiber reinforcement on impact strength of heat polymerized polymethylmethacrylate denture base resin: in vitro study and SEM analysis. J Adv Prosthodont. 2012;4:30–36. https://doi.org/10.4047/jap.2012.4.1.30

[11] Jassim RK, Radhi AA. Evaluation the biological effect of two types of denture base materials reinforced with silanated glass fiber. J Bagh Coll Dent. 2011;23:26–30.

[12] Yu SH, Ahn DH, Park JS, et al. Comparison of denture base resin reinforced with polyaromatic polyamide fibers of different orientations. Dental Mater J. 2013;32:332–340. https://doi.org/10.4012/dmj.2012-235

[13] Chen SY, Liang WM, Yen PS. Reinforcement of acrylic denture base resin by incorporation of various fibers. J Biomed Mater Res. 2001;58:203–208. https://doi.org/10.1002/1097-4636(2001)58:2<203::AID-JBM1008>3.0.CO;2-G

[14] Jagger DC, Harrison A, Jandt KD. The reinforcement of dentures. J Oral Rehab. 1999;26:185–194. https://doi.org/10.1046/j.1365-2842.1999.00375.x

[15] Ismaeel IJ, Alalwan HA, Mustafa MJ. The effect of the addition of silanated polypropylene fiber to polymethylmethacrylate denture base material on some of its mechanical properties. J Bagh Coll Dent. 2015;27:40–47. https://doi.org/10.12816/0015263

[16] John J, Ann Mani S, Palaniswamy K, Ramanathan A, Razak AA. Flexural properties of poly (Methyl Methacrylate) resin reinforced with oil palm empty fruit bunch fibers: a preliminary finding. J Prosthodont. 2015;24:233–238. https://doi.org/10.1111/jopr.12191

[17] Yu SH, Ahn DH, Park JS, et al. Comparison of denture base resin reinforced with polyaromatic polyamide fibers of different orientations. Dental Mater J. 2013;32:332–340. https://doi.org/10.4012/dmj.2012-235

[18] Xu J, Li Y, Yu T, Cong L. Reinforcement of denture base resin with short vegetable fiber. Dent Mater. 2013;29:1273–1279. https://doi.org/10.1016/j.dental.2013.09.013

[19] Arora P, Singh SP, Arora V. Effect of alumina addition on properties of polymethylmethacrylate – a comprehensive review. Int J Biotech Trends Technol. 2015;9:1–7.

[20] BBC. n.d.. Common limpet. B.B.C. Retrieved August 15, (2008).

[21] Purchon, R. D. The biology of the mollusca. Pergamon Press, New York , (1977). https://doi.org/10.1038/265565a0

[22] Arun Y Patil, N R Banapurmath, B B Kotturshettar, K Lekha, M Roseline, limpet teeth-based polymer nanocomposite: A novel alternative biomaterial for denture base application, book chapter, Fiber-Reinforced Nanocomposites: Fundamentals and Applications. https://doi.org/10.1016/B978-0-12-819904-6.00022-0

[23] Arun Y. Patil, N. R. Banapurmath, Jayachandra S.Y., B.B. Kotturshettar, Ashok S Shettar, G. D. Basavaraj, R. Keshavamurthy, T. M. YunusKhan, Shridhar Mathd, Experimental and simulation studies on waste vegetable peels as bio-composite fillers for light duty applications, Arabian Journal of Engineering Science, Springer-Nature publications, IF:1.518,03 June, 2019. https://doi.org/10.1007/s13369-019-03951-2

[24] Arun Y Patil, Umbrajkar Hrishikesh N Basavaraj G D, Krishnaraja G Kodancha, Gireesha R Chalageri, Influence of Bio-degradable Natural Fiber Embedded in Polymer Matrix, Elsevier, Materials Today Proceedings, Volume 5, 7532–7540, May 2018. https://doi.org/10.1016/j.matpr.2017.11.425

[25] Anirudh Kohli, Annika H, Karthik B, Pavan PK, Lohit P A, Prasad B Sarwad, Arun Y Patil and Basvaraja B Kotturshettar,design and Simulation study of fire-resistant biodegradable shoe,First International Conference on Advances in Physical Sciences and Materials,Journal of Physics: Conference Series 1706 (2020) 012185. https://doi.org/10.1088/1742-6596/1706/1/012185

[26] Anirudh Kohli, Vrishabh Ghalagi, Manoj Divate, Chetan manakatti, Md. Irshad Karigar, Divakar poojar, Tousif pandugol, Arun Y Patil, Santosh Billur and Basavaraja B Kotturshettar, Heat analysis of single point cutting tool coated with different natural bio composite,First International Conference on Advances in Physical Sciences and Materials ,Journal of Physics: Conference Series 1706 (2020) 012186. https://doi.org/10.1088/1742-6596/1706/1/012186

[27] Anirudh Kohli, Ishwar S, Charan M J, C M Adarsha, Arun Y Patil*, Basvaraja B Kotturshettar, Design and Simulation study of pineapple leaf reinforced fiber glass as an alternative material for prosthetic limb ICMSMT 2020 IOP Conf. Series: Materials Science and Engineering 872 (2020) 012118. https://doi.org/10.1088/1757-899X/872/1/012118

Sustainable Natural Fiber Composites
Materials Research Foundations **122** (2022) 96-109

Materials Research Forum LLC
https://doi.org/10.21741/9781644901854-4

Chapter 4

Microstructure Characteristics of Cellulose Fibers in Cement Paste and Concrete - A Review

Syed Habibunnisa[1,a], Ruben Nerella[1,b] and Madduru Sri Rama Chand[2,c]

[1]Department of Civil Engineering, Vignan's Foundation For Science, Technology and Research 522213, Andhra Pradesh, India

[2]Department of Civil Engineering, Sree Chaitanya College of Engineering 505527, Telangana, India

[a]syedhabibbunnisa13@gmail.com, [b]ruebennerella2512@gmail.com, [c]maddurusriram@scce.ac.in

Abstract

This paper explores, present scenario of using natural fibers in concrete, which are having high cellulose content. Cellulose fibers contribute the development of high-quality fiber-reinforced cement composites. In recent days natural fibers attracting many researchers due to its low cost and being largely available in nature with the comparison of artificial fibers. These are accessible in the fibrous formation and extracted from seed pods, herbal leaves, which are become trustful over the regular fibers. Sometimes construction industries also use natural plant cellulose fibers as a secondary raw material. These seem to be a good alternative resource for eco-friendly materials that can be replace the synthetic polymeric fibers. The author aim is to discuss the microstructure of concrete reinforced with cellulose fibers shows a possessive characteristic of good tensile strength and bond properties.

Keywords

Cellulose Fibers, Eco-Friendly, Polymeric Fibers, Mortar, Composites, Fibrous

Contents

Microstructure characteristics of cellulose fibers in cement paste and concrete - A review **96**

1. Introduction...**97**

2. Different types of natural fibers and their chemical composition 97

3.   Mechanical properties of cellulose fibers 98

4.   Cellulose fiber composite materials 99

4.   Manufacturing process of cellulose fiber composites........................104

Conclusion...106

References 106

1.   Introduction

Natural cellulose fibers are also named as nature's fibers. The material which is obtained in fibrous form of origin plant known as cellulose. The main chemical component in the plants is cellulose, and therefore these are also referred as cellulosic fibers. In general, natural cellulosic fibers are included in cotton and hemp, jute and ramie etc. In many research applications, plant fibers are used as alternative sources of steel and/or artificial fibers utilize in composite, like cement paste mortar and/or concrete to the increment of strength properties [1]. The strength and mechanical properties of fibers dependent on a certain percentage of compounds cellulose, lignin, hemicellulose [1]. When compared with the other fibers, cellulose fibers have been used in many applications like textile industries, fiber reinforcement composites, bio composites and polymer composites due to its good bonding properties [2].

2.   Different types of natural fibers and their chemical composition

Main elements in natural plant fibers are cellulose, Hemi-cellulose, lignin, pectin and Ash. Percentage ranges of every fiber components, varies differently based upon the type of fiber. Changes occur in the properties of fibers are mainly depends upon the presence of different components. The compound hemicellulose is accountable for the water absorption, whereas lignin ensures thermal stability [3]. The average chemical compositions of common natural fibers as shown in Table 1. Percentage of cellulose content mainly depend upon the source. Based on the sources fibers are classified as bast-fiber (producing it from bark), core fiber (available in wood) and leaf fiber (produced from leaves) [7]. Depending upon the sources, fibers which are extracted from bark and leaves contains high cellulose percentage, suitable for preparation of reinforcement composites and bio-composites.

Sustainable Natural Fiber Composites   Materials Research Forum LLC
Materials Research Foundations **122** (2022) 96-109   https://doi.org/10.21741/9781644901854-4

Table 1. *Chemical composition of cellulose fibers [7]*

Different fibers		Cellulose percentage	Lignin Percentage	Hemi-cellulose percentage	Pectin percentage	Ash percentage
Bast fiber	Fiber-flax	72.01	3.20	11.1	2.3	0
	Seed-Flax	45	22	25	0	5
	Kenaf	44	17	22.25	0	3.5
	Jute	58.25	19	17.3	0.2	1.25
	Hemp	67	8.35	18.2	0.9	0.8
	Ramie	79.8	1.3	10.85	1.9	0
Core fiber	Kenaf	43	18	21	0	3
	Jute	44.5	22.5	20	0	0.8
Leaf fibers	Abaca	59.5	8	16	0	3
	Sisal	62.5	9	17	10	0.8
	Henequen	77.60	13.11	6	0	0

3. Mechanical properties of cellulose fibers

Cellulose fibers, respond to mechanical stresses. Changes that occur in stress parameter is dependent on type of fiber and chemical composition present in it [4]. Comparison is made between mechanical properties of fibers, with usable fibers like Glass-fiber, Aramid fiber and carbon fibers, as shown in Table 2.

Table 2. *Mechanical properties of cellulose fibers [4]*

Fiber	Density (g/cm^3)	Elongation (%)	Tensile strength (Mpa)	Young's modulus (Gpa)
Cotton	8.75	6.5	442	9.05
Jute	1.38	1.65	596.5	20
Flax	1.45	4.4	992.5	53.8
Hemp	1.48	1.6	725	70
Ramie	1.5	2.9	579	86.0
Sisal	1.415	8	550	23.5
Coir	1.2	22.5	197.5	5.0
Softwood Kraft	1.50	0	1000	40
E-glass	2.50	2.75	2750	70.0
S-glass	2.50	2.8	4570	86.0
Aramid	1.40	3.5	3075	65
carbon	1.40	1.6	4000	235

4.   Cellulose fiber composite materials

Materials which are made with a fiber and matrix with the certain combination are referred as cellulose composite materials. These combination mixes, leads to the creation of a new material which is stronger than, individual fiber. Cellulose fibers can be used in fiber-reinforcement materials, mixing along the polymers in bio- composites also fiber reinforced plastics. Macroscopic features of fibers effects on behavior of produced composite. Table 3 shows, physicaland mechanical properties of cellulose fibers.

Table 3. Physical properties of cellulose fibers [4]

Dimension: length, Diameter	Irregular cross-section of natural plant fibers and fibrilate structure shows beneficial aspects on the anchoring of fibers with a fragile matrix. Length, diameter relation shows a determine factor in transfer efforts to matrix.
Void volume water absorption	Initial moment of immersion shows the absorption phenomenon as high. Due to the presence of large voids in the material. Direct contact results to water shows the negative impact on fiber like shrinkage, binder matrix swelling. High void volume reduce the weight, increase the capacity sound absorption and less thermal conductivity of obtain component.
Tensile capacity	An average similarity like poly-propylene fibers.
Modulus of elasticity	cellulose fibers, classifies as less modulus of elasticity's depend on the usage of components in buildings works as post-cracked stage, due to huge energy absorption.

Microscopic analysis of various cellulose based cement composites reveals a difference in microstructure of fiber composites due to variability content or difference in kinds of fibers [2]. A new micro-computed tomography method is useful for verification of the fabrication process to optimize the percentage of fibers. It is useful for 3D visualization process and distribution of fibers in different fiber cement boards. There is a good bond observed between   cement   matrix   and cellulose fiber in hybrid reinforcement by SEM microphotographs as shown in Figure 1 a) SEM microphotograph of three panels with a scale bar of 500 mm, 200mm and 50mm and b) Panel-B contains Recycled cellulose fibers [2].

There is a good bond between cement matrix and cellulose fibers. Panel A and C contains wood cellulose and PVA fiber. Utilization of waste fibers, as a reinforcing in the cement composites reveals huge growth in the productions of building materials [3]. While a comparison with the cement composites of original wood cellulose fibers along with recycled paper cellulosic fiber composites, there is an optimization in mixture of cellulosic fibers which are produced from recycling of paper. This is because of acceptable behavior between binder and admixtures as shown in Figure 2. The author gives a basic awareness on mortar or plaster and usage of cellulose fibers in composites [3].

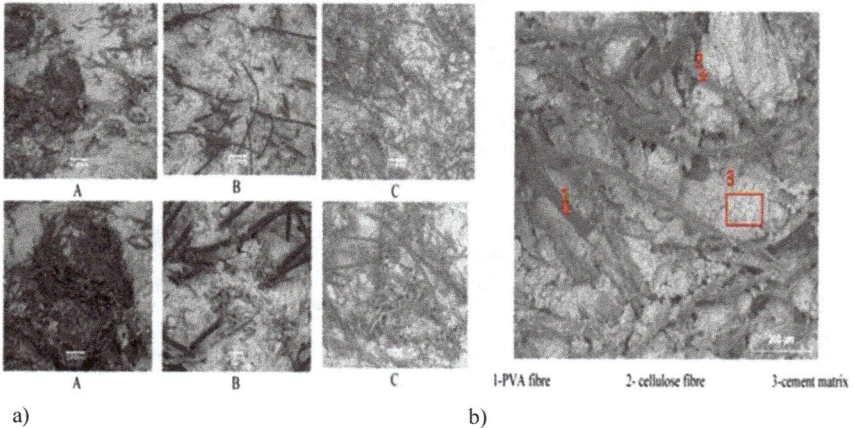

a)   b)

Figure 1. *a) SEM microphotograph of three panels with a scale bar of 500 mm, 200 mm and 50 mm and b) Panel-B contains recycled cellulose from waste paper and PVA fiber [2].*

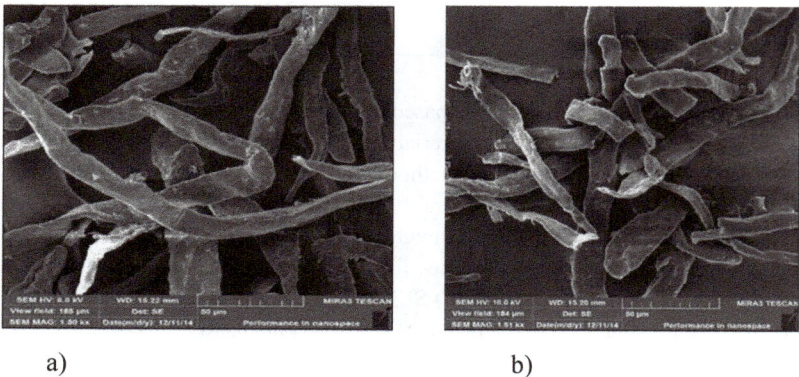

a)   b)

Figure 2. *a) Microstructure view of recycled waste paper and b) microstructure view of woodpulp fibers observed by using scanning electron microscope [3]*

Many investigations were carried out on physical and mechanical properties of fiber-cement composites [4]. The properties like physical and mechanical properties of fiber-cement composites investigated [4]. The effect on using various proportions (0.2,1.0,5.0%) of wood pulp and cellulose fibers of wastage papers reveals the roughly same effect on properties of fiber cement composites, but there is a slight difference in density of composites. Waste paper fibers allow lower water absorption when compared with wood fibers. Compressive strength of waste paper fiber composites shows good results when compared with the composites of wood pulp. Cellulose cement composite shows good mechanical property along with high durability properties has been developed in previous decades [4]. The top view of SEM microphotograph is a hand made with Rps.palustris (Rhodopseudomonas palustris is anoxygenic phototrophic soil bacterium) microbial paper sheet with a scale bar of 2 micro meters. B) Cross-section with a scale of 10 micro meters asshown in Figure 3.

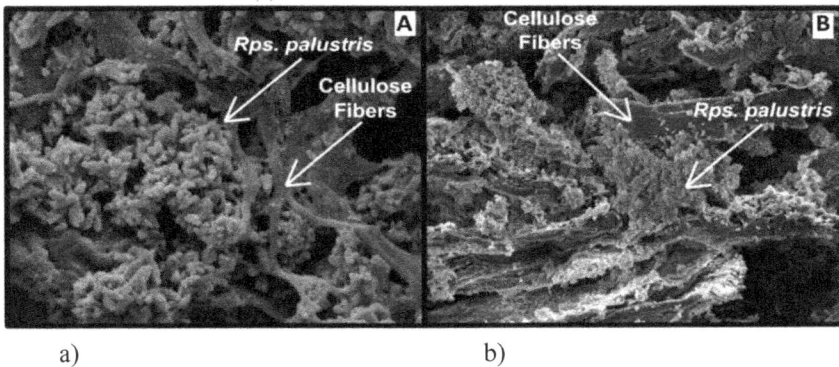

a)   b)

Figure 3. *a) Microstructure of a microbial paper of 2 micro meters and b) cross-section with a scaleof 10 micro meters [5]*

The properties of fiber cement composites are mainly depend upon the nature of cellulose fibers and the treatment process [5]. An attempt of investigation on silica fume and fiber contents, reveals the effects of durability and strength parameters of low relative density oil well cement composites [5]. Result gives a clear idea about cellulose fibers which are responsible for the improvement of strength of the composites. But they are unfavorable for durability parameters like permeability and porosity. Replacing the silica fume 15% into cement specimens, which are reinforced with 8% fibers content provides an improvement in mechanical strength durability of the cement composite. Microstructure

Sustainable Natural Fiber Composites   Materials Research Forum LLC
Materials Research Foundations **122** (2022) 96-109   https://doi.org/10.21741/9781644901854-4

studies were carried on the samples by using SEM. Silica fume capable for the decrement of porosity in the matrix which is filled with pores. It leads to increase in bond in the fiber-cement interface. Because of pozzolans, there is a reaction between silica fume and hydration, compounds, especially C-S-H, is attached in the fiber surface. Lesser porosity in a matrix within the great arrangement of fiber cement interfacial zone provides a further increment of efficiency in fiber reinforcement. It reveals eminent toughness and mechanical strength [5]. The correlation changes between microstructure after aging. It is observed that an improvement in the mechanical performance of cellulose fiber based reinforcement cement composite is observed with aging [6]. The aging conditions promote the carbonation process (natural and accelerated aging of CO_2 in high environments) that reveals the sensing property of matrix overall fibers and petrifaction of fibers. It leads an increment on strength and elastic modulus of reinforcement cement composite. Under normal environment conditions, the results show that there is an accelerated aging and decays the matrix. Which is responsible for reduction of strength due to fiber petrifaction. It is also found that by, XRD and TG analysis, there is a marking increment in Calcium carbonate in the microstructure of concrete. While observing after a natural and accelerating aging process of in a high CO_2 environment. The increment is not necessary follows with a marking changes in C-H content, which is already less to start with it. It may give suggestions on C-S-H. Carbonation part is converted to Calcium carbonate and hydrated as silica gel. In SEM results, it is rare to identify zone along with crystalline C-H morphology and a high ratio of $CaoSio_2$ [6]. Different types of treatment methods can be useful for the improvement of durability in cellulose cement composites. Addition of pozzolans either directly induced into mass of cement or applied to fibers through curing under CO_2 atmosphere. Another one is refining with horrification process of pulps or chemical surface treatment like silanes [7]. Present days natural fibers attracting so many scholars and researchersdue to its low cost and it being largely available in nature [8]. Tensile strength of natural fibers were low compared with synthetic fibers. Strength of fiber depends on the fiber loading. Some of natural fibers are close to the synthetic fibers like hemp, flax, kenaf, abaca, etc. Elastic modulus of natural fibers is mainly depending on the weight ratio of fibers. It increases with fiber weight ratio and increases up to a maximum value after that it will decreases [8]. Some of the natural fibers show higher modulus of elasticity than glass fibers. In comparison with other natural fibers Jute, hemp, flax, pineapple have the highest elastic modulus values. Flexural strength of fiber was depended on the fiber loading. If, fiber load is increased flexural strength is increased up to the optimal level then decreases [9]. Mechanical performance of natural fiber reinforced cementations composite is mainly depended upon interaction between natural fibers and cement based matrices [9]. Various series of pullout test results intend to reproduce meant of tensile failure in fiber or the bond failure throughout interface of fiber matrix. The

Sustainable Natural Fiber Composites   Materials Research Forum LLC
Materials Research Foundations **122** (2022) 96-109   https://doi.org/10.21741/9781644901854-4

outcomes of geometric and mechanical properties control the behavior of sisal, curaua, and jute fibers. However, these are comparable with other kinds of industrial fibers which are already employs as spread reinforcement in fiber reinforced cement composites [9]. Based on simplification of bilinear bond slip law, an inverse identification procedure is applied. Pull out test result based on force slip law shows; there is no possibility for theoccurrence of tensile yielding. The procedure demonstration gives its capability in identification of bond interaction between fibers and matrix [9]. Effects of high initial and subsequent curing temperature on the microstructure of ordinary strength concrete is identified [10]. Electron microscope (SEM) and Environmental electron microscope (ESEM) process are useful for studies on microstructure features of ordinary strength concrete. Specimens are prepared in two different initial temperatures(20°C, 50°C). Comparison of SEM and ESEM pictures of shows many differences in the same concrete. The bulk paste seems to be more micro cracking by observing with SEM which is not observed by ESEM. Due to the drying shrinkage, a gap is visible between the paste and the by SEM. High initial temperatures also promote crystallization. Large crystals in calcium hydroxides seen in aggregate imprints leads an open microstructure characteristic of paste. The Decrement in concrete strength is mainly due to result of weak bond in interfacial transition zone [10]. Formation of zone relates to the moisture movement and chemical reactions in recycled aggregate concrete [11]. The porous interfacial transition zone of microstructure in normal strength concrete may attribute high permeable and absorption capacities of recycled aggregates. The mechanical properties of recycled aggregates concrete shall improve with modification of surface properties and pore structure of recycle aggregates. Strength development in recycled aggregates concrete mainly depends upon microstructure of interfacial transition zone [11]. Long term mechanical properties of cellulose-based fiber-reinforcement of cement mortar incorporated with diatomite as a replaced material to quartz sand were investigated [12]. Microstructure and deformation of high modulus cellulose fiber (fiber B) was distinguished by useful techniques in Raman spectroscopic method and the crystalline nature of fibers are determined by using X-ray diffractions [13]. The cellulose sample of fiber B having abundant uniform orientations across the width. It is a desirable property to maximizing mechanical properties. X-ray diffractions shows fiber B and viscose sample contain differences in microstructure of crystalline distribution and orientation parameters, Which are determined by filament across the widths. Distribution of fibers B sample is greater than viscose fibers. It affects the values of high strength modulus, which are noted. These fibers suggest as high crystal orientation contribute the corresponding increment in it [13]. Concerning the chemical interactions between the fiber matrix, hemp fibers show a trapping nature of calcium on its surface. Pectin fixes the calcium through a formation of complex structureis called as egg boxes. The calcium fixation might responsible for delaying of setting time. The pectin acts as growth inhibitor for calcium silicate hydrate. There is a firm increment in flexural strength due to an optimum fiber content show the composite mechanical behavior. If a composite contains sixteen volume percentage of fibers, flexural strength is maximum and it is high for cement paste around 40%. There is a decrement in elastic modulus of composite/ when compared with the cement paste. Applications on hemp fibers with different treatments improve only flexural strength. With the treatment of alkaline solutions, there is an improvement on flexural strength about 94% when compared with cement paste. Alkali treatments effect strength of fiber along with fiber matrix adhesion in an effective manner. These type of composites shows a consequent interest in building applications [14]. Addition of synthetic and mineral fibers to a base mix of foamed concrete shows improvement in the quality of micro-reinforcement [15]. By using chemical analysis and electron, microscopic methods, relevant experimental result show an effect on cellulose fibers along with structural properties of fiber reinforced foam concretes. The research shows a peculiarity in new appearance formation and densification of binder hydration products in contact zone between fiber cement matrix for the increment in tensile strength of fiber reinforced foam concrete. Physical and mechanical properties in fiber reinforced foam concrete is depending upon addition amount of recycled cellulose fiber base mix. Compression strength increases around 35% when compared with initial samples of high stability foamed concrete mixes. There is a decrement in shrinkage deformation [15]. Microstructural analysis was performed on the fiber cement specimens of virgin cellulose, eucalyptus cellulose, and araucaria cellulose [16]. It has been observed that there is very good interaction between the cement and fibers. These are homogenously distributed in the cement and also exhibit partly lamellar structure [16]. The Physical and mechanical properties along with fracture performance of fly ash basis geopolymers reinforce with cotton fibers (0.3–1.0wt%) shows suitable adding of cotton fiber may improve mechanical properties in geopolymer composites [17]. Particularly, an optimum level of 0.5wt% fiber content shows an increment in flexural strength and fracture toughness. Density parameter of geopolymers composite decreases with fiber content due to void fraction increment. So, there is a chance for fiber agglomerating. SEM results reveals that Composites with less fiber contents show a great fiber matrix interfacial bond, when comparing with high fiber contents [17].

4. Manufacturing process of cellulose fiber composites

Natural/bio-degradable composites perform, similar or better properties compare with natural/synthetic resin composites, mainly depend on the fiber treatments processes like alkali, heat, hot water, saline and saltwater treatment. Natural fiber treatments influence the properties like tensile strength, modulus and flexural strength. It enhances adhesion

Sustainable Natural Fiber Composites   Materials Research Forum LLC
Materials Research Foundations **122** (2022) 96-109   https://doi.org/10.21741/9781644901854-4

between natural fibers and matrix to improve the interface of composites. Strong and better composites achieve only with correct and optimum treatment. Tensile strength variations in natural fiber composite depend ontype of fibers, resin type and manufacture process. Fiber fraction and treatment also effect properties of natural fiber composites. Table 4 shows tensile properties of different natural fiber composites. Composites made with natural fibers and biodegradable resins are important developments that show feasibilities for non-load bearing construction elements but also for structural applications. Thus, natural fiber composites offer precious environment benefits. More research needs in application of natural fiber composites into structural and infrastructure applicability. The issues in cost of fiber and supply of fiber to the mass production of the composites plays main role. Further research is necessary for obtain better strength and modulus properties. It includes optimization of interfacial bond between fiber and resin by means of fiber treatments [18].

Table 4. *Tensile strength properties of cellulose fibers [18]*

S. No	Type of fiber	Resin	Manufacturing process	Tensile Strength (Mpa)	Ref.
1	Flax	Polyester	Vacuum infusion	61.01	
2	Jute	Polyester	-	60.0	18
3	Hemp	Polypropylene	Extrusion & Inj. mold.	50.52	
4	Hemp	Polypropylene	Compression moulding	52.02	
5	Hemp	Polyester	Resin transfer moulding	32.91	
6	Coir	Epoxy	-	17.87	
7	Coir	Polyethylene	Extrusion-compression moulding	26.21	
8	Coir	Polypropylene	Compression moulding	11.02	
9	Sisal	Low density Polyethylene	-	15.58	
10	Sisal	Polyester	-	46.50	
11	Sugar-palm Fiber	Epoxy	Hand lay-up	52.62	
12	Sugar-palm Fiber	Polyester	Hand lay-up	23.48	18

Sustainable Natural Fiber Composites   Materials Research Forum LLC
Materials Research Foundations **122** (2022) 96-109   https://doi.org/10.21741/9781644901854-4

Flax/polyester composite produce with vacuum infusion reveals high Tensile strength, When compared with other fiber composites coir/polypropylene exhibit least tensile strength.

Conclusion

Cellulose component varies from fiber to fiber. Fiber content increment contributes to the development of strength properties of the composites concerning the non-fibrous composites.

The presence of difference in microstructures of cellulose fiber cement composite, mainly depends on variability contents of fiber or different kinds of fibers. There is strong bond between cellulose fiber matrix. It is difficult to analyze a pure, individual fiber. Determination on fiber Properties is mainly depends upon the chemical and structural compositions and also fiber type and its growth circumstances.

The Main influence on mechanical properties of composite is, due to adhesion between matrix and fibers. In contact zone, hydrated binder products in fiber and cement matrix were responsible to boost mechanical strength for fiber reinforced foam concrete.

Further research is necessary for obtaining better strength and modulus properties. It includes optimization in the inter-facial bond between fiber and resin with fiber treatments. The principle involved in fiber composite is utilization of fibers as a reinforcement in matrix of resin.

Further research on replacement of Glass-fibers with cellulose fibers shall be useful for the composite materials. It has to be done because of its advantages.

References

[1]   Majid Ali (March 2012),"Natural fibers as construction materials", Journal of Civil Engineeringand Construction Technology Vol. 3(3), pp. 80-89. https://doi.org/10.5897/JCECT11.100

[2]   K. Schabowicz, et al. (2018), "Microstructural characterization of cellulose fibers in reinforced cement boards", Archives of Civil and Mechanical Engineering Vol. 18(4), pp.1068–1078. https://doi.org/10.1016/j.acme.2018.01.018

[3]   N. Stevulova et al. June (2015), "Cellulose Fibres Used in Building Materials" , Proceedings of Rehva Annual Conference" Advanced HVAC and Natural Gas Technologies, pp.211. https://doi.org/10.7250/rehvaconf.2015.031

[4]   V. Hospodarova et al. (2015), "Possibilities of using cellulose fibers in building materials", IOPConference Series Materials Science and Engineering, Vol 96(1). https://doi.org/10.1088/1757-899X/96/1/012025

[5]   X. W. Cheng et al. (2018) , "A new approach to improve mechanical properties and durability of low-density oil well cement composite reinforced by cellulose fibres in microstructural scale", Construction and Building Materials, Vol.177, pp.499–510. https://doi.org/10.1016/j.conbuildmat.2018.05.134

[6]   A. Bentur, et al. (1989), " The microstructure and ageing of cellulose fiber reinforced cement composites cured in a normal environment", International Journal of Cement Composites and Lightweight Concrete, Vol. 11(2), pp.99–109. https://doi.org/10.1016/0262-5075(89)90120-6

[7]   M. Ardanuy et al. (2015), "Cellulosic fiber reinforced cement-based composites: A review of recent research" , Construction and Building Materials, Vol.79, pp. 115–128. https://doi.org/10.1016/j.conbuildmat.2015.01.035

[8]   K. Srinivas (2017), "A Review on Chemical and Mechanical Properties of Natural Fiber Reinforced Polymer Composites", International Journal of Performability Engineering, Vol. 13(2), pp. 189–200. https://doi.org/10.1016/j.compositesb.2017.12.016

[9]   S. R. Ferreira et al. (2018), "Influence of natural fibers characteristics on the interface mechanics with cement based matrices", Composites Part B: Engineering, Vol.140, pp.183– 196. https://doi.org/10.1016/j.compositesb.2017.12.016

[10]   M. Mouret et al. (1999), "Microstructural features of concrete in relation to initial temperature-SEM and ESEM characterization", Cement and Concrete Research, Vol. 29 (3), pp.369–375. https://doi.org/10.1016/S0008-8846(98)00160-4

[11]   C. S. Poon et al. (2004), " Effect of microstructure of ITZ on compressive strength of concrete prepared with recycled aggregates",Construction and Building Materials, Vol.18(6), pp.461– 468. https://doi.org/10.1016/j.conbuildmat.2004.03.005

[12]   C. Ince et al. (2018), "Long-term mechanical properties of cellulose fiber-reinforced cement mortar with diatomite", Advances in Cement Research, pp.1–36. https://doi.org/10.1680/jadcr.17.00179

[13]   S. J. Eichhorn et al. (2003), "Characterisation of the microstructure and deformation of high modulus cellulose fibres", Polymer, Vol.44(19), pp.5901–5908. https://doi.org/10.1016/S0032-3861(03)00540-8

[14]   D. Sedan (2008), "Mechanical properties of hemp fibre reinforced cement: Influence of the fibre/matrix interaction", Journal of the European Ceramic Society, Vol.28(1), pp.183–192. https://doi.org/10.1016/j.jeurceramsoc.2007.05.019

[15]   A. M. Soydan (2018), "Air-Cured Fiber-Cement Composite Mixtures with Different Types of Cellulose Fibers", Advances in Materials Science and Engineering, Vol. 2018, Article ID 3841514, pp.1-9. https://doi.org/10.1155/2018/3841514

[16]   Valeriy Fedorov (2018), " Influence of cellulose fibers on structure and properties of fiber reinforced foam concrete", MATEC Web of Conferences- 143, vol. 02008. https://doi.org/10.1051/matecconf/201814302008

[17]   T. Alomayri (2013), "Characterisation of cotton fibre-reinforced geopolymer composites", Composites Part B: Engineering, Vol.50, pp.1–6. https://doi.org/10.1051/matecconf/201814302008

[18]   A. Ticoalu (2010), " A review of current development in natural fiber composites for structural and infrastructure applications", Southern Region Engineering Conference, SREC2010-F1-5, pp.1-6.

[19]   Sidney Diamond,(2004) , " The microstructure of cement paste and concrete"- a visual primer Cement & Concrete Composites, Vol.26, pp.919–933. https://doi.org/10.1016/j.cemconcomp.2004.02.028

[20]   Marie-Therese Wisniowsk (2014), " Cellulosic Fibers (Natural)- Cotton Art Resource", cellulosic fibers natural cotton, pp.1-9, (2014).

[21]   I. Oscar (2017), " Microbial Paper: Cellulose Fiber-based Photo-Absorber Producing Hydrogen Gas from Acetate using Dry Stabilized Rhodopseudomonas palustris", Vol.12(2),pp. 4013-4030. https://doi.org/10.15376/biores.12.2.4013-4030

[22]   X.W. Cheng et al.(2018), "A new approach to improve mechanical properties and durability of low-density oil well cement composite reinforced by cellulose fibres in microstructural scale, Construction and Building Materials vol.177, pp. 499–510. https://doi.org/10.1016/j.conbuildmat.2018.05.134

[23]   Zhang Yunsheng, et al. (2010) ,"Composition design and microstructural characterization of calcined kaolin-based geopolymer cement", Applied Clay Science, Vol. 47, pp. 271–275. https://doi.org/10.1016/j.clay.2009.11.002

[24]   Saulo Rocha Ferreira et.al. (2017),"Influence of natural fibers characteristics on the interfacemechanics with cement based matrices", Composites Part B. https://doi.org/10.1016/j.compositesb.2017.12.016

Sustainable Natural Fiber Composites
Materials Research Foundations **122** (2022) 96-109

Materials Research Forum LLC
https://doi.org/10.21741/9781644901854-4

[25]   C.S. Poon, et al. (2004) , "Effect of microstructure of ITZ on compressive strength of concrete prepared with recycled aggregates", Construction and Building Materials, Vol.18, pp. 461- 468. https://doi.org/10.1016/j.conbuildmat.2004.03.005

[26]   P.J. Herrera-Franco et al. (2005) , "A study of the mechanical properties of short natural-fiber reinforced composites", Composites: Part B 36 597–608. https://doi.org/10.1016/j.compositesb.2005.04.001

[27]   Genilson Cunha de Oliveira Filho et.al. (2018), " Effects of hybridization on the mechanical properties of composites reinforced by piassava fibers tissue", Composites Part B. https://doi.org/10.1016/j.compositesb.2018.10.050

[28]   R.S.P. Coutts (1983), "Flax Fibres in as a reinforcement in cement mortars", The International Journal of Cement Composites and Lightweight Concrete, Volume 5, Vol.4, pp.251-262. https://doi.org/10.1016/0262-5075(83)90067-2

Sustainable Natural Fiber Composites

Materials Research Foundations **122** (2022) 110-127

Materials Research Forum LLC

https://doi.org/10.21741/9781644901854-5

Chapter 5

Application of Surface Modification Routes to Coconut Fiber for its Thermoplastic-Based Biocomposite Materials

Ümit Tayfun[1]*, Mehmet Doğan[2]

[1] Department of Basic Sciences, Bartın University, 74110, Bartın, Türkiye

[2] Department of Textile Engineering, Erciyes University, 38039, Kayseri, Türkiye

umit.tayfun@inovasens.com*, mehmetd@erciyes.edu.tr

Abstract

Coconut fibers are high potential reinforcing additive in green polymer composites due to its composition, structure and the ease of availability. Modifications of coconut fiber by chemical and physical methods play key role to achieve desired performance. This review provides a comprehensive description of the influence of varied treatment routes applied to coconut fiber to confer thermal, mechanical and structural behaviors of its eco-composites with reinforcing mechanisms achieved by the help of tunning surface functionality of fibers as well as the chemical interactions with the polymeric matrix. This review covers the related academic research studies dealing with the development of thermoplastic composites containing surface modified coconut fiber. Remarkable future objectives and challenges based on the use of this natural fiber as an effective reinforcing agent in mainly transportation and construction fields are also discussed.

Keywords

Eco-Composites, Coconut Fiber, Surface Treatments, Thermoplastic Composites, Interfacial Interactions

Contents

Application of Surface Modification Routes to Coconut Fiber for its Thermoplastic-Based Biocomposite Materials...110

1.   **Introduction**..111

2.   Surface treatment routes of coconut fiber..**112**

  2.1   Physical treatments 112

  2.2   Chemical treatments 115

  2.3   Biological treatments..115

3.   Thermoplastic composites containing coconut fiber..........................**115**

  3.1   Petroleum-based thermoplastics filled with coconut fiber..............116

  3.2   Bio-based thermoplastics filled with coconut fiber.......................117

4.   Future trends..**118**

References **118**

1.   Introduction

Material researches have been focused on tunning the surface properties of natural fibers for their effective use as environmentally-friendly additives for green composites owing to environmental issues in recent years. This fact is the main motivation for finding suitable surface treatment techniques in academic and technical researches in the polymeric eco-composite field [1-4]. Enrichment of fiber surface using both chemical and physical methods reasonable choice to level up in mechanical resistance of composite material owing to improve compatibility of two phases of composite structure, namely reinforcer and matrix phases. Additionally, these methods are applied due to practical and economical aspects in conventional production steps.

Mechanical responses of polymeric composites are related with some parameters in its structure such as shape, size, dispersion homogeneity and concentration of additive phase. Establishment of chemically and physically stable interface between fiber and polymer matrix is important since the mechanical strength of composite material depends on the load transfer which occurs at polymer-fiber interface during deformation [5-7]. Incompatibility of two phases causes drops down in adhesion and mechanical connection and finally resulted in mechanical deformation stem from the presence of breakage and debonding. Thus, numerous research works were performed on the development of eco-friendly thermoplastic composites involving surface treated lignocellulosic fiber with the driving force of improvement its interfacial interaction with thermoplastic polymers.

Coconut fiber (CF), also known as coir fiber, is one of strongest candidate among natural fibers. CF has varied advantages over other lignocellulosic fibers such as lower moisture absorption capacity and density, resistance to microbial degradation and harder wearing characteristic which is owing to higher lignin composition compared to other natural fibers.

Sustainable Natural Fiber Composites   Materials Research Forum LLC
Materials Research Foundations **122** (2022) 110-127   https://doi.org/10.21741/9781644901854-5

This fiber found wide usage in transportation, textile and construction markets by the help of fabrication of its green composites using numerous thermoplastics. Coconut fiber exhibits sufficient mechanical resistance and biodegradation performance in these application areas thanks to high cellulose and hemicellulose content in its structure [8,9].

Thermoplastic polymers provide varied advantages compared to thermoset polymers such as recyclability due to practical reshaping by heating and the ease of processing using conventional production techniques applied in industrial facilities. The innovations reached by research and development activities and lab-scale research studies based on the natural fiber-reinforced thermoplastic materials can be easily transformed to scale-up production steps since similar devices are used in both laboratory and industrial process levels [10].

In the scope of this review chapter, the findings mainly in the scientific research publications dealing with the effect of varied surface modification routes to basic properties of its thermoplastic composites are summarized. Several surface treatment techniques including physical and chemical methods as well as bio-enzymatic routes with new emerging trends and future perspectives are also discussed. The donation of polymer-compatible segments on coconut surface and its reinforcing effects to composite behaviors are explained in details with interfacial interaction mechanisms.

2. Surface treatment routes of coconut fiber

Surface treatments of coconut fiber are widely applied for the purpose of increasing its interfacial adhesion to polymeric host. CF can be subjected to these treatments using physical or chemical routes. Physical modification techniques introduce physical interlocking on fiber surface and cause no change for the chemical composition of fibers. On the other hand, chemical modification techniques create an intermediate layer and donate reactive moieties on the surface of CF. Such modifications on the surface results in an enhancement its interfacial interaction with polymer phase [11,12].

2.1 Physical treatments

Mercerization or namely alkaline modification is the most favored physical treatment technique to achieve surface roughness on lignocellulosic fibers. Alkali modification is also classified as chemical treatment since some chemical changes occurs on fiber surface depending upon the concentration. Cellulose segment of CF is more reactive content attributed to the presence of orientation of highly packed crystalline parts in its structure. This process results in decrease for diameter of individual fibers in addition to improvement in surface roughness since the removal of the portions such as lignin, hemi-cellulosic, pectin and waxy content on fiber structure [13-16]. Reduction of cellulose and

Materials Research Forum LLC
https://doi.org/10.21741/9781644901854-5

hemi-cellulose constituent of coconut fiber makes it less hydrophilic and more compatible for hydrophilic polymer matrix. The general mechanism of mercerization to lignocellulosic fiber due to the ionization of hydroxyl group to alkoxide group during the reaction of sodium hydroxide with fiber cell is displayed in Eq. 1.

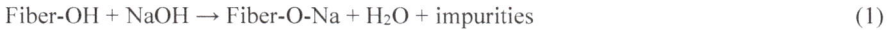

$$\text{Fiber-OH} + \text{NaOH} \rightarrow \text{Fiber-O-Na} + H_2O + \text{impurities} \tag{1}$$

Figure 1 represents the SEM micrographs of neat and alkali-modified coconut fiber. It can be clearly observed from these SEM micrographs that the surfaces of alkali-treated coconut fiber seem to be rougher compared to neat coconut fiber because of the partial removal of soluble content after mercerization.

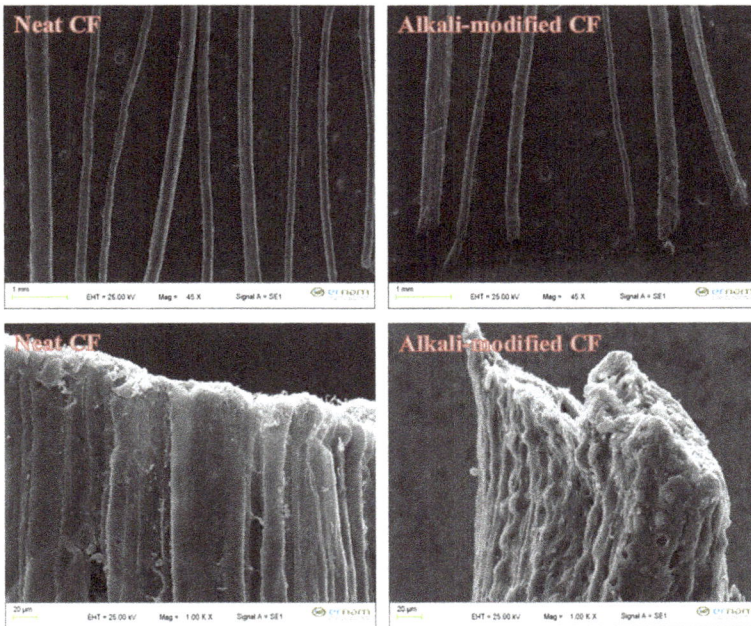

Figure 1.   SEM micrographs of neat and alkali-modified coconut fiber.

Infrared spectroscopy (IR) technique is useful characterization method to demonstrate the difference on surface functionality of natural fibers before and after the modification routes.

Sustainable Natural Fiber Composites   Materials Research Forum LLC
Materials Research Foundations **122** (2022) 110-127   https://doi.org/10.21741/9781644901854-5

According to IR spectra of neat (CF) and alkali-modified (Na-CF) coconut fiber exhibited in Figure 2, intensity of absorption band at 1420 cm^{-1} shows remarkably increasing trend after alkali modification which is stem from the presence of intermediate COO- group on fiber surface. The peak at 1250 cm^{-1} can be seen on the IR spectrum of pristine CF assigns the C–O stretching band of hemi-cellulosic segment. The disappearance of this characteristic band on IR spectrum of Na-CF indicates the removal of hemi-cellulose content after alkali treatment. Reduction of intensities of other absorption peaks related with the oxygen functionalities including C–O and C=O stretching bands seen nearly 900 and 1650 cm^{-1} wavenumbers assign to removal of cellulosic content of CF during mercerization. The weak absorption peak can be observed around 2800 cm^{-1} is the characteristic indication of lignin group due to –CH$_2$ fluctuation vibrations. Similarly, this peak exhibit reduction after alkali-modification stem from the removal of lignin segment [17-21].

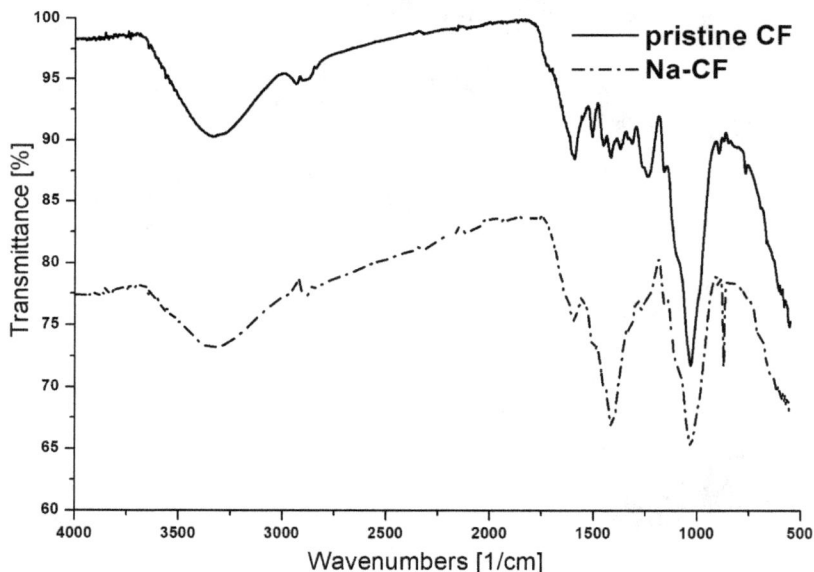

Figure 2.   IR spectra of neat and alkali-modified coconut fiber.

Radiation treatment, plasma and corona discharge modification routes are also applied to coconut fiber as physical treatment methods [22]. As radiation-induced graft

Sustainable Natural Fiber Composites   Materials Research Forum LLC
Materials Research Foundations **122** (2022) 110-127   https://doi.org/10.21741/9781644901854-5

copolymerization approach, synthetic monomers can be grafted to fiber by gamma or UV radiation in order to achieve hydrophilic and polymer-compatible surface [23-25].

2.2   Chemical treatments

Tuning chemical functionality of natural fibers generally performed by reactive agents. Silane modification is the most preferred method since silane coupling agents provide wide diversity with specific functional groups in addition to practical application. The first step of silanization process is hydrolysis of alkoxy groups in which silane modifier is simply mixed in its solvent (water or alcohol) as shown in Eq. 2. As fiber subjected to hydrolyzed silane mixture, silanol groups form. This groups have tendency for the reaction with hydroxyl moieties of cellulosic fibers. Finally, stable chemical segments are formed on fiber surface containing specific hydrocarbon ends (R) in order to interact with polymer chains [26].

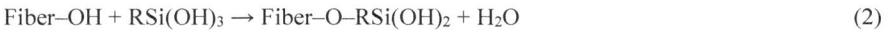

$$RSi(OR)_3 + H_2O \rightarrow RSi(OH)_3 + 3ROH \text{ (Hydrolysis)}$$

$$Fiber–OH + RSi(OH)_3 \rightarrow Fiber–O–RSi(OH)_2 + H_2O \tag{2}$$

Coupling agents other than silane modifiers are also used to donate chemical functionality for coconut fiber surface. Among these compounds, acetic acid [27], sodium sulphite [28], sodium bicarbonate [29], calcium hydroxide [30], furfuryl alcohol [31], ferric nitrate and ammonium chloride [32] are used as chemical modifier agents for coconut fibers.

2.3   Biological treatments

Enzymatic modification is referred to biological treatment method in which gained importance in the last decade. Several enzymes are commonly combined with coconut fiber in order to remove of lignin, pectin, cellulose and hemi-cellulose contents from fiber structure. Additionally, strong interfacial adhesion can be reached between polymer and coconut fiber by the help of varied forms of enzyme such as pectinase, lactase, cellulase and lipase [33-36].

3.   Thermoplastic composites containing coconut fiber

CF are compounded with various thermoplastic matrices in research studies. In most of these works, thermoplastics are filled with short CF as chopped strand form. Conventional production methods are applied to fabricate composite samples including extrusion process followed by hand lay-up, compression or injection molding techniques. Effects of surface treatment methods, fiber length and content of coconut fiber to basically mechanical properties in addition to thermal, structural and water resistance behaviors of thermoplastic-based green composites are reported.

Sustainable Natural Fiber Composites   Materials Research Forum LLC
Materials Research Foundations **122** (2022) 110-127   https://doi.org/10.21741/9781644901854-5

3.1   Petroleum-based thermoplastics filled with coconut fiber

Although the of replacement of petroleum polymers by newly emerged bio-based polymers became trending issue due to the environmental awareness nowadays, the use of recyclable petrol-based thermoplastics containing natural sources seems to be acceptable throughout the intermediate period.

Polypropylene (PP) is one of the most popular thermoplastics in both industrial and academic fields. Research efforts related with CF-filled PP composites are postulated by several material scientists. Haque et al. fabricated PP-based composites involving chemically treated CF using compression molding. They found that chemical treatments caused increase in mechanical and water resistance performance of composites [37,38]. Mir et al. performed chemical treatment to CF using two kinds of salts in acidic media and produced CF/PP composites by compression molding. They reached best results for 20% loading level of chemically treated CF containing sample according to their mechanical and microstructural analyzes [39]. Several researchers suggested the use of CF with other cellulosic fibers for PP-based composites in hybrid form including jute, alfa, oil palm, sisal and bagasse fiber and they compared the effectiveness of CF with respect to other natural fibers by means of mechanical behavior [40-46]. Optimization of alkali treatment parameters for coconut fiber was investigated by Morandim-Giannetti et al using test data of PP/CF composites [47]. They reported that application of mercerization to CF with the concentration of 5.75% of NaOH during 3 hours gave optimum results in the case of mechanical behavior of alkali-treated CF reinforced PP composites. This finding is correlated with Yan et al. in which 5% of NaOH was found to be suitable treatment concentration [48]. Similarly, Hu proposed that modification of CF by high content of NaOH (10%) caused dramatic reductions in tensile properties of CF/PP composites [49]. Bettini and his coworkers found that inclusion lignin with MA-g-PP compatibilizer yield improvement for thermal stability of CF-filled PP composites [50,51]. Ayrilmis et al. also fabricated CF loaded PP with the inclusion of MA-g-PP for the use of automotive interior parts. They investigated that the use of compatibilizer exhibited increase in mechanical and water absorption properties of PP-based composites even at high level of CF loadings [52].

Polyethylene (PE) is another commodity petrol-based thermoplastic widely used in numerous applications. PE-based composites involving CF are produced in a few studies. In one of these works, alkali and silane-treated CF compounded with PE and enhancements in mechanical performance of composites were achieved [53]. Brahmakumar et al. introduced waxy interlayer on CF surface and fabricated PE-based composites. They observed that tuning surface of CF with natural wax resulted in better mechanical resistance of composites with respect to untreated CF [54]. Similar to PP-based composites, CF was incorporated with other lignocellulosic fibers as hybrid form including agave [55], palm [56,57] and banana stem fiber [58] to reach synergistic effect to PE composites.

polyvinylidene fluoride (PVDF) [8], high impact polystyrene (HIPS) [59], ethylene glycol dimethacrylate (EGDMA) [60], ethylene vinyl alcohol (EVA) copolymer [61] and polyvinyl chloride (PVC) [62] are other petrol-based thermoplastics that compounded with coconut fiber according to literature.

3.2   Bio-based thermoplastics filled with coconut fiber

Thermoplastic polymers which derived from natural resources are gain scientific and technical attention due to plastic-waste problem. Biodegradation characteristics of these green thermoplastics is the main driving force especially in production of packaging related products.

Poly (lactic acid) (PLA) has gained significant importance in recent years and it can be available as fiber or extrusion grade according to application areas. In the literature, numbers of research studies conducted dealing with CF-filled PLA composites. Alkaline modification of CF gave significant impact positively on mechanical, thermal and water uptake behavior of PLA-based composites by the help of improved interfacial interactions between fiber and matrix phases [63-67]. Silane-treated CF addition confer mechanical and physical performance of PLA-based composites as well [68,69]. González-López et al. observed increase in psychical and mechanical properties of CF reinforced PLA composites with inclusion of MA-g-PLA modifier [70]. Similarly, Suardana et al. used acrylic acid as modifier for PLA/CF composites and they obtained lower water absorption in addition to mechanical improvements [71].

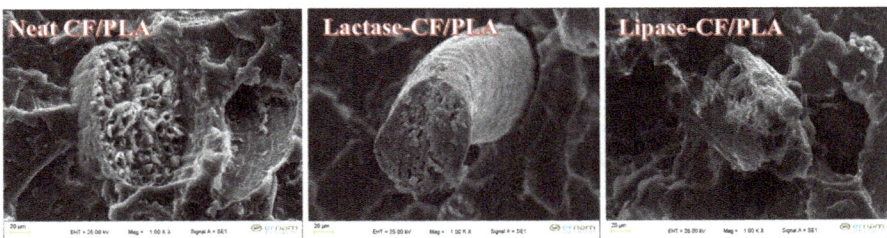

Figure 3.   *SEM micrographs of neat and enzyme-treated coconut fiber containing PLA composites.*

Influence of enzymatic treatments of CF to PLA-based green composites was studied by Coskun and her coworkers [72]. They modified CF surface with lipase, pectinase, lactase, and cellulase enzymes prior to fabrication of composites. They found that lipase and lactase treatment routes resulted in remarkable enhancement in mechanical behavior of PLA. SEM micrographs of PLA composites filled with neat CF, lactase and lipase modified CF are visualized in Figure 3. It can be observed from SEM images that enzymatic treatment caused increase in interfacial adhesion between PLA and CF phases in addition to physical roughness of fiber surface.

Polybutylene succinate (PBS) is another biodegradable thermoplastic. Nam et al. applied mercerization process to CF using 5% NaOH solution in order to obtain compatible surface with PBS matrix [73]. Mechanical parameters of PBS-based composites were found to be relatively higher for alkali-treated CF compared to untreated one according to their findings.

4.   Future trends

Based on the recent demands on environmentally-friendly composites, newly emerged modification routes of natural fibers are established for development of high-performance materials. Number of research efforts in this area rising up with various technical solutions. Recent progress in the engineered use lignocellulosic fiber show that more innovative approach for sustainable products by the help of digital and systems. For example, a novel system for using water-based solutions of bio-based polymers and cellulosic compounds to produce biocomposite structures owing to combination of lightweight and mechanically strong characteristics are achieved according to study performed by Mogas-Soldevila and Oxman [74]. Recently, new fabrication processes are paid attention such as additive manufacturing in order to optimize rheology, flow behavior and physical specifications of natural fiber reinforced thermoplastics to develop sustainable 3D parts. The prospects of process methods are generated with more accuracy thanks to digital simulations. Thermoplastic composites involving coconut fiber have potential for fabrication biomedical devices, sustainable structures and cost-effective composite parts. The variation of new biodegradable thermoplastics opens ways to obtain greener composite materials by effective use of coconut fiber.

References

[1]   M.A. Fuqua, S. Hou, C.A. Ulven, Natural fiber reinforced composites, Polym. Rev. 52 (2012): 259. https://doi.org/10.1080/15583724.2012.705409

[2]   A.K. Mohanty, M. Misra, L.T. Drzal, Natural Fibers, Biopolymers and Biocomposites, Taylor&Francis, Florida, 2005. https://doi.org/10.1201/9780203508206.ch1

[3]   O. Faruk, A.K. Bledzki, H.-P. Fink, M. Sain, Progress report on natural fiber reinforced composites, Macromol. Mater. Eng. 299 (2014) 9-26. https://doi.org/10.1002/mame.201300008

[4]   S. Ebnesajjad, Handbook of Biopolymers and Biodegradable Plastics: Properties, Processing and Applications, William Andrew Publishing, New York, 2012.

[5]   M. Galbe, G. Zacchi, Pretreatment: The key to efficient utilization of lignocellulosic materials, Biomass Bioenergy., 46 (2012) 70-78. https://doi.org/10.1016/j.biombioe.2012.03.026

[6]   R. Latif, S. Wakeel, N.Z. Khan, A.N. Siddiquee, S.L. Verma, Z.A. Khan, Surface treatments of plant fibers and their effects on mechanical properties of fiber-reinforced composites: A review, J. Reinf. Plast. Compos. 38 (2019) 15-30. https://doi.org/10.1177/0731684418802022

[7]   O. Faruk, A.K. Bledzki, H.-P. Fink, M. Sain, Biocomposites reinforced with natural fibers: 2000–2010, Prog. Polym. Sci. 37(2012) 1552-1596. https://doi.org/10.1016/j.progpolymsci.2012.04.003

[8]   L.Q.N. Tran, C. Fuentes, C. Dupont-Gillain, A. Van Vuure, I. Verpoest, Understanding the interfacial compatibility and adhesion of natural coir fibre thermoplastic composites, Compos. Sci. Technol. 80 (2013) 23-30. https://doi.org/10.1016/j.compscitech.2013.03.004

[9]   A.A. Owodunni, R. Hashim, O.F.A. Taiwo, M.H. Hussin, M.H.M. Kassim, Y. Bustami, O. Sulaiman, M.H.M. Amini, S. Hiziroglu, Flame-retardant properties of particleboard made from coconut fibre using modified potato starch as a binder, J. Phys. Sci. 31 (2020) 129-143. https://doi.org/10.21315/jps2020.31.3.10

[10]   J.D. Muzzy, Thermoplastic-properties, in: A. Kelly., C. Zweben (Eds.), Comprehensive Composite Materials, Elsevier Science, Amsterdam, 2000.

[11]   A.K. Bledzki, J. Gassan, Composites reinforced with cellulose based fibres, Prog. Polym. Sci. 24 (1999) 221-274. https://doi.org/10.1016/S0079-6700(98)00018-5

[12]   M. Baiardo, G. Frisoni, M. Scandola, A. Licciardelo, Surface chemical modification of natural cellulose fibers, J. Appl. Polym. Sci. 83 (2002) 38. https://doi.org/10.1002/app.2229

[13]   M.M. Kabir, H. Wang, K.T. Lau, F. Cardona, Chemical treatments on plant-based natural fibre reinforced polymer composites: An overview, Compos. Part B Eng. 43 (2012) 2883-2892. https://doi.org/10.1016/j.compositesb.2012.04.053

[14]   X. Li, L.G. Tabil, S. Panigrahi, Chemical treatments of natural fiber for use in natural fiber-reinforced composites: A review, J. Polym. Environ. 15 (2007) 25. https://doi.org/10.1007/s10924-006-0042-3

[15]   R. Vinayagamoorthy, Influence of fibre pretreatments on characteristics of green fabric materials, Polym. Polym. Compos. (2020) e-published: July 27 DOI: 10.1177/0967391120943461

[16]   A.K. Mohanty, M. Misra, L.T. Drzal, Surface modifications of natural fibers and performance of the resulting biocomposites: An overview, Compos. Interface. 8 (2001) 313. https://doi.org/10.1163/156855401753255422

[17]   C.G. Mothe, I.C. Miranda, Characterization of sugarcane and coconut fibers by thermal analysis and FTIR, J. Therm. Anal. Calorim. 97 (2009) 661-665. https://doi.org/10.1007/s10973-009-0346-3

[18]   T.F. Salem, S. Tirkes, A.O. Akar, U. Tayfun, Enhancement of mechanical, thermal and water uptake performance of TPU/jute fiber green composites via chemical treatments on fiber surface, e-Polymers 20 (2020)133-143. https://doi.org/10.1515/epoly-2020-0015

[19]   A.I.S. Brígida, V.M.A. Calado, L.R.B. Gonçalves, M.A.Z. Coelho, Effect of chemical treatments on properties of green coconut fiber, Carbohyd. Polym. 79 (2010) 832-838. https://doi.org/10.1016/j.carbpol.2009.10.005

[20]   R.K. Samal, B.B. Panda, S.K. Rout, M. Mohanty, Effect of chemical modification on FTIR spectra. I. Physical and chemical behavior of coir, J. Appl. Polym. Sci. 58 (1995) 745-752. https://doi.org/10.1002/app.1995.070580407

[21]   M.K.D. Rambo, A.R. Alves, W.T. Garcia, M.M.C. Ferreira, Multivariate analysis of coconut residues by near infrared spectroscopy, Talanta 138 (2015) 263-272. https://doi.org/10.1016/j.talanta.2015.03.014

[22]   A.G. Adeniyi, D.V. Onifade, J.O. Ighalo, A.S. Adeoye, A review of coir fiber reinforced polymer composites, Compos. Part B Eng.176 (2019) 107305. https://doi.org/10.1016/j.compositesb.2019.107305

[23]   M.A. Khan, S. Rahaman, A. Al-Jubayer, J.M.M. Islam, Modification of jute fibers by radiation-induced graft copolymerization and their applications, V.K. Thakur (Ed.), Cellulose-Based Graft Copolymers: Structure and Chemistry, CRC Press, Florida, 2015.

[24]   O. Owolabi, T. Czvikovszky, Composite materials of radiation-treated coconut fiber and thermoplastics, J. Appl. Polym. Sci. 35 (1988):573-582. https://doi.org/10.1002/app.1988.070350302

[25]   Y. El Moussi, B. Otazaghine, A.S. Caro-Bretelle, R. Sonnier, A. Taguet, N. Le Moigne, Controlling interfacial interactions in LDPE/flax fibre biocomposites by a combined chemical and radiation-induced grafting approach, Cellulose 27 (2020) 6333-6351. https://doi.org/10.1007/s10570-020-03221-7

[26]   M.D. Banea, J.S. Neto, D.K. Cavalcanti, Recent trends in surface modification of natural fibres for their Use in green composites. T. Sabu, B. Preetha (Eds.), Green Composites, Springer, Singapore, 2021. https://doi.org/10.1007/978-981-15-9643-8_12

[27]   H. Essabir, M. Bensalah, D. Rodrigue, R. Bouhfid, A. Qaiss, Structural, mechanical and thermal properties of bio-based hybrid composites from waste coir residues: Fibers and shell particles, Mech. Mater. 93 (2016) 134-144. https://doi.org/10.1016/j.mechmat.2015.10.018

[28]   D. Madyira, A. Kaymakci, Mechanical characterization of coir epoxy composites and effect of processing methods on mechanical properties, COMA international conference on competitive manufacturing, (2016) 187-192.

[29]   M. Hasan, M.E. Hoque, S.S. Mir, N. Saba, S. Sapuan, Manufacturing of coir fibre-reinforced polymer composites by hot compression technique, M. Jawaid, M.S. Salit, M.E. Hoque, N.B. Yusoff (Eds.) Manufacturing of Natural Fibre Reinforced Polymer Composites, Springer, Switzerland, 2015. https://doi.org/10.1007/978-3-319-07944-8_15

[30]   R. Siakeng, M. Jawaid, H. Ariffin, M.S. Salit, Effects of surface treatments on tensile, thermal and fibre-matrix bond strength of coir and pineapple leaf fibres with poly lactic acid, J. Bionics. Eng. 15 (2018) 1035-1046. https://doi.org/10.1007/s42235-018-0091-z

[31]   S.K. Saw, G. Sarkhel, A. Choudhury, Surface modification of coir fibre involving oxidation of lignins followed by reaction with furfuryl alcohol: Characterization and stability, Appl. Surf. Sci. 257 (2011) 3763–3769. https://doi.org/10.1016/j.apsusc.2010.11.136

[32]   A. Khan, M.A. Ahmad, S. Joshi, S.A. Al Said, Abrasive wear behavior of chemically treated coir fibre filled epoxy polymer composites, Am. J. Mech. Eng. Autom. 1 (2014) 1-5.

[33] T. Gurunathan, S. Mohanty, S.K. Nayak, A review of the recent developments in biocomposites based on natural fibers and their application perspectives, Compos. Part A Appl. Sci. 77 (2015) 1-25. https://doi.org/10.1016/j.compositesa.2015.06.007

[34] P. Widsten, A. Kandelbauer, Adhesion improvement of lignocellulosic products by enzymatic pre-treatment, Biotechnol. Adv. 26 (2008) 379-386. https://doi.org/10.1016/j.biotechadv.2008.04.003

[35] S.K. Ramamoorthy, M. Skrifvars, A. Persson, A review of natural fibers used in biocomposites: Plant, animal and regenerated cellulose fibers, Polym. Rev. 55 (2015) 107-162. https://doi.org/10.1080/15583724.2014.971124

[36] D. Verma, P.C. Gope, A. Shandilya, A. Gupta, M.K. Maheshwari, Coir fibre reinforcement and application in polymer composites, J. Mater. Environ. Sci, 4 (2013) 263-276.

[37] M.M. Haque, M.S. Islam, M.N. Islam, Preparation and characterization of polypropylene composites reinforced with chemically treated coir, J. Polym. Res. 19 (2012) 9847. https://doi.org/10.1007/s10965-012-9847-z

[38] M.N. Islam, M.M. Haque, M.M. Huque, Mechanical and morphological properties of chemically treated coir-filled polypropylene composites, Ind. Eng. Chem. Res. 48 (2009) 10491-10497. https://doi.org/10.1021/ie900824c

[39] S.S. Mir, N. Nafsin, M. Hasan, N. Hasan, A. Hassan, Improvement of physicomechanical properties of coir-polypropylene biocomposites by fiber chemical treatment, Mater. Des. 52 (2013) 251-257. https://doi.org/10.1016/j.matdes.2013.05.062

[40] E. Zainudin, L.H. Yan, W. Haniffah, M. Jawaid, O.Y. Alothman, Effect of coir fiber loading on mechanical and morphological properties of oil palm fibers reinforced polypropylene composites. Polym. Compos. 35 (2014) 1418-1425. https://doi.org/10.1002/pc.22794

[41] F. Arrakhiz, M. Malha, R. Bouhfid, K. Benmoussa, A. Qaiss, Tensile, flexural and torsional properties of chemically treated alfa, coir and bagasse reinforced polypropylene, Compos. Part B Eng. 47 (2013) 35-41. https://doi.org/10.1016/j.compositesb.2012.10.046

[42] H.U. Zaman, M.A. Khan, R.A. Khan, Comparative experimental measurements of jute fiber/polypropylene and coir fiber/polypropylene composites as ionizing radiation, Polym. Compos. 33 (2012) 1077-1084. https://doi.org/10.1002/pc.22184

[43]   S. Siddika, F. Mansura, M. Hasan, Physico-mechanical properties of jute-coir fiber reinforced hybrid polypropylene composites, Eng. Technol. 73 (2013) 1145-1149.

[44]   P. Sudhakara, D. Jagadeesh, Y. Wang, C.V. Prasad, A.K. Devi, G. Balakrishnan, B. Kim, J. Song, Fabrication of Borassus fruit lignocellulose fiber/PP composites and comparison with jute, sisal and coir fibers, Carbohydr. Polym. 98 (2013) 1002-1010. https://doi.org/10.1016/j.carbpol.2013.06.080

[45]   Haydaruzzaman, A. Khan, M. Hossain, M.A. Khan, R.A. Khan, Mechanical properties of the coir fiber-reinforced polypropylene composites: Effect of the incorporation of jute fiber, J. Compos. Mater. 44 (2010) 401-416. https://doi.org/10.1177/0021998309344647

[46]   A. Arya, J.E. Tomlal, G. Gejo, J. Kuruvilla, Commingled composites of polypropylene/coir-sisal yarn: effect of chemical treatments on thermal and tensile properties, e-Polymers 15 (2015) 169-177. https://doi.org/10.1515/epoly-2014-0186

[47]   A. de Araújo Morandim-Giannetti, C.G. Pasquoto, T.M. Sombra, B.C. Bonse, S.H.P. Bettini, Polypropylene/chemically treated coir composites: optimizing coir delignification conditions using central composite design, Cellulose 25 (2018) 1159-1170. https://doi.org/10.1007/s10570-017-1617-y

[48]   L. Yan, N. Chouw, L. Huang, B. Kasal, Effect of alkali treatment on microstructure and mechanical properties of coir fibres, coir fibre reinforced-polymer composites and reinforced-cementitious composites, Constr. Build. Mater. 112 (2016) 168-182. https://doi.org/10.1016/j.conbuildmat.2016.02.182

[49]   H. Gu, Tensile behaviours of the coir fibre and related composites after NaOH treatment, Mater. Des. 30 (2009) 3931-3934. https://doi.org/10.1016/j.matdes.2009.01.035

[50]   S.H. Bettini, A.C. Biteli, B.C. Bonse, A.D.A. Morandim-Giannetti, Polypropylene composites reinforced with untreated and chemically treated coir: Effect of the presence of compatibilizer, Polym. Eng. Sci. 55 (2015) 2050-2057. https://doi.org/10.1002/pen.24047

[51]   A.A. Morandim-Giannetti, J.A.M. Agnelli, B.Z. Lanças, R. Magnabosco, S.A. Casarin, S.H. Bettini, Lignin as additive in polypropylene/coir composites: thermal, mechanical and morphological properties, Carbohydr. Polym. 87 (2012) 2563–2568. https://doi.org/10.1016/j.carbpol.2011.11.041

[52]   N. Ayrilmis, S. Jarusombuti, V. Fueangvivat, P. Bauchongkol, R.H. White, Coir fiber reinforced polypropylene composite panel for automotive interior applications, Fiber. Polym. 12 (2011) 919-926. https://doi.org/10.1007/s12221-011-0919-1

[53]   F. Arrakhiz, M. Achaby, A. Kakou, S. Vaudreuil, K. Benmoussa, R. Bouhfid, O. Fassi-Fehri, A. Qaiss, Mechanical properties of high density polyethylene reinforced with chemically modified coir fibers: impact of chemical treatments, Mater. Des. 37 (2012) 379-383. https://doi.org/10.1016/j.matdes.2012.01.020

[54]   M. Brahmakumar, C. Pavithran, R. Pillai, Coconut fibre reinforced polyethylene composites: Effect of natural waxy surface layer of the fibre on fibre/matrix interfacial bonding and strength of composites, Compos. Sci. Technol. 65 (2005) 563-569. https://doi.org/10.1016/j.compscitech.2004.09.020

[55]   A.A. Perez-Fonseca, M. Arellano, D. Rodrigue, R. Gonzalez-Núñez, J.R. Robledo-Ortíz, Effect of coupling agent content and water absorption on the mechanical properties of coir-agave fibers reinforced polyethylene hybrid composites, Polym. Compos. 37 (2016) 3015-3024. https://doi.org/10.1002/pc.23498

[56]   R. Chollakup, W. Smitthipong, W. Kongtud, R. Tantatherdtam, Polyethylene green composites reinforced with cellulose fibers (coir and palm fibers): Effect of fiber surface treatment and fiber content, J. Adhes. Sci. Technol. 27 (2013) 1290-1300. https://doi.org/10.1080/01694243.2012.694275

[57]   H. Essabir, R. Boujmal, M.O. Bensalah, D. Rodrigue, R. Bouhfid, A. Qaiss, Mechanical and thermal properties of hybrid composites: oil-palm fiber/clay reinforced high density polyethylene, Mech. Mater. 98 (2016) 36-43. https://doi.org/10.1016/j.mechmat.2016.04.008

[58]   K.G. Arifuzzaman, M. Alam Shams, M.R. Kabir, M. Gafur, M. Terano, M. Alam, Influence of chemical treatment on the properties of banana stem fiber and banana stem fiber/coir hybrid fiber reinforced maleic anhydride grafted polypropylene/low-density polyethylene composites, J. Appl. Polym. Sci. 128 (2013) 1020-1029. https://doi.org/10.1002/app.38197

[59]   K.C.C. Carvalho, D.R. Mulinari, H.J.C. Voorwald, M.O.H. Cioffi, Chemical modification effect on the mechanical properties of hips/coconut fiber composites, BioResources 5 (2010) 1143-1155.

[60]   J.K. Roy, N. Akter, H.U. Zaman, K.M. Ashraf, S. Sultana, Shahruzzaman, N. Khan, M.A. Rahman, T. Islam, M.A. Khan, R.A. Khan, Preparation and properties of coir fiber-reinforced ethylene glycol dimethacrylate-based composite, J. Thermoplast. Compos. Mater. 27 (2014) 35-51. https://doi.org/10.1177/0892705712439568

[61] M.F. Rosa, B.S. Chiou, E.S. Medeiros, D.F. Wood, T.G. Williams, L.H. Mattoso, W.J. Orts, S.H. Imam, Effect of fiber treatments on tensile and thermal properties of starch/ethylene vinyl alcohol copolymers/coir biocomposites, Bioresource Technol. 100 (2009) 5196-5202. https://doi.org/10.1016/j.biortech.2009.03.085

[62] J.L. Leblanc, C.R. Furtado, M.C. Leite, L.L. Visconte, A.M. de Souza, Effect of the fiber content and plasticizer type on the rheological and mechanical properties of poly (vinyl chloride)/green coconut fiber composites, J. Appl. Polym. Sci. 106 (2007) 3653-3665. https://doi.org/10.1002/app.26567

[63] Y. Dong, A. Ghataura, H. Takagi H, H.J. Haroosh, A.N. Nakagaito, K.T. Lau, Poly lactic acid (PLA) biocomposites reinforced with coir fibers: Evaluation of mechanical performance and multifunctional properties, Compos. Part A Appl. Sci. 63 (2014) 76-84. https://doi.org/10.1016/j.compositesa.2014.04.003

[64] Z. Sun, L. Zhang, D. Liang, W. Xiao, J. Lin, Mechanical and thermal properties of PLA biocomposites reinforced by coir fibers, Int. J. Polym. Sci. 2017 (2017) 1-8. https://doi.org/10.1155/2017/2178329

[65] J. Duan, H. Wu, W. Fu, M. Hao, Mechanical properties of hybrid sisal/coir fibers reinforced polylactide biocomposites, Polym. Compos. 39 (2018) E188-E199. https://doi.org/10.1002/pc.24489

[66] T.H. Nam, S. Ogihara, S. Kobayashi, Interfacial, mechanical and thermal properties of coir fiber reinforced poly (lactic acid) biodegradable composites, Adv. Compos. Mater. 21 (2012) 103-122. https://doi.org/10.1163/156855112X629540

[67] R.B. Yusoff, H. Takagi, A.N. Nakagaito, Tensile and flexural properties of polylactic acid-based hybrid green composites reinforced by kenaf, bamboo and coir fibers, Ind. Crops Prod. 94 (2016) 562-573. https://doi.org/10.1016/j.indcrop.2016.09.017

[68] L. Zhang, Z. Sun, D. Liang, J. Lin, W. Xiao, Preparation and performance evaluation of PLA/coir fibre biocomposites, BioResources 22 (2017) 7349-7362.

[69] R. Siakeng, M. Jawaid, H. Ariffin, M.S. Salit, Effects of surface treatments on tensile, thermal and fibre-matrix bond strength of coir and pineapple leaf fibres with poly lactic acid, J. Bionic. Eng. 15 (2018) 1035-1046. https://doi.org/10.1007/s42235-018-0091-z

[70] M.E. González-López, A.A. Pérez-Fonseca, R. Manríquez-González, M. Arellano, D. Rodrigue, J.R. Robledo-Ortíz, Effect of surface treatment on the physical and

mechanical properties of injection molded poly (lactic acid)-coir fiber biocomposites, Polym. Compos. 40 (2019) 2132-2141. https://doi.org/10.1002/pc.24997

[71] N.P.G. Suardana, I.P. Lokantara, J.K. Lim, Influence of water absorption on mechanical properties of coconut coir fiber/polylactic acid biocomposites, Mater. Phys. Mech. 12 (2011) 113-125.

[72] K. Coskun, A. Mutlu, M. Dogan, E. Bozaci, Effect of various enzymatic treatments on the mechanical properties of coir fiber/poly (lactic acid) biocomposites, J. Thermoplast. Compos.Mater. 34 (2021) 1066-1079. https://doi.org/10.1177/0892705719864618

[73] L. Mogas-Soldevila, N. Oxman, Water-based engineering & fabrication: Large-scale additive manufacturing of biomaterials, MRS Online Proceedings Library (2015) 1800. https://doi.org/10.1557/opl.2015.659

[74] N. Gama, A. Ferreira, A. Barros-Timmons, 3D printed cork/polyurethane composite foams, Mater. Design 179 (2019) 107905. https://doi.org/10.1016/j.matdes.2019.107905

[75] Y. Zhong, U. Kureemun, L.Q. Tran, H.P. Lee, Natural plant fiber composites-constituent properties and challenges in numerical modeling and simulations, Int. J. Appl. Mech. 9 (2017) 1750045. https://doi.org/10.1142/S1758825117500454

[76] S. Jayavani, H. Deka, T.O. Varghese, S.K. Nayak, S.K., Recent development and future trends in coir fiber-reinforced green polymer composites: Review and evaluation, Polym. Compos. 11 (2016) 3296-3309. https://doi.org/10.1002/pc.23529

[77] P. Sahu, M.K. Gupta, Water absorption behavior of cellulosic fibres polymer composites: A review on its effects and remedies, J. Ind. Text. (2020) e-publised: Nov. 26. https://doi.org/10.1177/1528083720974424

[78] N.S.N. Arman, R.S. Chen, S. Ahmad, Review of state-of-the-art studies on the water absorption capacity of agricultural fiber-reinforced polymer composites for sustainable construction, Construct. Build. Mater. 302 (2021) 124174. https://doi.org/10.1016/j.conbuildmat.2021.124174

[79] A.R. Torun, A.S. Dike, E.C. Yildiz, I. Saglam, N. Choupani, Fracture characterization and modeling of Gyroid filled 3D printed PLA structures, Mater. Test. 63 (2021) 397-401. https://doi.org/10.1515/mt-2020-0068

[80] M. Jawaid, H.A. Khalil, A. Hassan, E. Abdallah, Bi-layer hybrid biocomposites: Chemical resistant and physical properties, BioResources 7 (2012) 2344-2355. https://doi.org/10.15376/biores.7.2.2344-2355

[81] S.K. Mary, M.S. Thomas, R.R. Koshy, P.K. Pillai, L.A. Pothan, S. Thomas, Adhesion in biocomposites: A critical review, Rev. Adhes. Adhes. 8 (2020) 527-553.

[82] V.K. Balla, K.H. Kate, J. Satyavolu, P. Singh, J.G.D. Tadimeti, Additive manufacturing of natural fiber reinforced polymer composites: Processing and prospects, Compos. Part B Eng. 174 (2019) 106956. https://doi.org/10.1016/j.compositesb.2019.106956

[83] A.S. Mangat, S. Singh, M. Gupta, R. Sharma, Experimental investigations on natural fiber embedded additive manufacturing-based biodegradable structures for biomedical applications, Rapid Prototyp. J. 24 (2018) 1221-1234. https://doi.org/10.1108/RPJ-08-2017-0162

[84] K.F. Hasan, P.G. Horváth, M. Bak, T. Alpár, A state-of-the-art review on coir fiber-reinforced biocomposites. RSC Adv. 11 (2021) 10548-10571. https://doi.org/10.1039/D1RA00231G

[85] M. Rafiee, R. Abidnejad, A. Ranta, K. Ojha, A. Karakoc, J. Paltakari, Exploring the possibilities of FDM filaments comprising natural fiber-reinforced biocomposites for additive manufacturing, Environ. Pollut. 8 (2021) 9. https://doi.org/10.3934/matersci.2021032

[86] U. Bongarde, B. Khot, A review on coir fiber reinforced polymer composite, Int. J. Eng. Technol. 6 (2019) 793-795.

[87] M. Wróbel-Kwiatkowska, M. Kropiwnicki, W. Rymowicz, Green biodegradable composites based on natural fibers, V.K. Thakur, M.K. Thakur, M.R. Kessler (Eds.) Handbook of Composites from Renewable Materials, John Wiley & Sons, New Jersey, 2017. https://doi.org/10.1002/9781119441632.ch93

[88] M.E. Lamm, L. Wang, V. Kishore, H. Tekinalp, V. Kunc, J. Wang, D.J. Gardner, S. Ozcan, Material extrusion additive manufacturing of wood and lignocellulosic filled composites, Polymers 12 (2020) 2115. https://doi.org/10.3390/polym12092115

Sustainable Natural Fiber Composites
Materials Research Foundations **122** (2022) 128-153

Materials Research Forum LLC
https://doi.org/10.21741/9781644901854-6

Chapter 6

Durability Against Fatigue and Moisture of Natural Fibre Composite

Bidita Salahuddin[1*], Zhao Sha[2], Shazed Aziz[3], Shaikh N. Faisal[4], Mohammad S. Islam[2*]

[1]ARC Centre of Excellence for Electromaterials Science and Intelligent Polymer Research Institute, University of Wollongong, Innovation Campus, Squires Way, North Wollongong, NSW 2522, Australia

[2]School of Mechanical and Manufacturing Engineering, the University of New South Wales, Sydney, NSW 2052, Australia

[3]School of Chemical Engineering, The University of Queensland, QLD 4072, Australia

[4]School of Electrical and Data Engineering, University of Technology Sydney, Sydney, NSW 2007, Australia

* bbs622@uowmail.edu.au, m.s.islam@unsw.edu.au

Abstract

The development of high-performing engineering materials fabricated using constituents of natural origins is gradually increasing around the globe. In the last few decades, there have been notable research achievements in green materials science through the development of natural fibre reinforced composites (NFCs). The advantages of these materials over synthetic fibre composites support extensive range of potential applications, with added benefits of low environmental impact and inexpensive throughput. Significant effort has gone into improving their performance to extend the capabilities and applications of this materials. Although there is a range of performance limitations which can be seen from NFCs, durability against fatigue and moisture of these materials are amongst the major concerns. This chapter aims to provide an overview of the effects of fatigue and moisture on NFCs by discussing factors, testing technology and protection techniques, and their effect on the durability and performance of the composites.

Keywords

Natural Fibre Polymer Composites (NFCs), Durability, Fatigue, Moisture, Hygroscopicity, Degradation

Contents

Durability Against Fatigue and Moisture of Natural Fibre Composite 128

1.   Introduction...129

2.   Fatigue and natural fibre composites 131

      2.1   Fatigue testing methodology 131

      2.2   Factors affecting the fatigue properties 132

      2.3   Damage development and property degradation............................134

      2.4   Fatigue improving techniques 136

3.   Moisture absorption phenomena 137

      3.1   Impact of moisture absorption on natural fibre composites...........137

      3.2   Moisture related degradation mechanisms 138

      3.3   Moisture protection technologies 140

Conclusions and recommendations...142

References 142

1.   Introduction

Over the past decades, interest in researching natural fibre reinforced composites (NFCs) has been gradually increased due to their inherent benefits over using artificial fibres, including low density, satisfactory mechanical properties at low cost, and low environmental impact [1-5]. NFCs are often used in targeted applications to attain recyclability, biodegradability, high stiffness-to-weight ratio, thermal insulation, and CO_2 neutrality over their conventional counterparts, for example, carbon and glass fibres [6-8]. NFCs are emerging materials that can also be a prospective alternative to the metal or ceramic based materials in applications involving automotive, aerospace, marine, and electronic industries [9,10]. Although NFCs exhibit many desired properties that are expected to form structural composite materials, they can also show inconsistent property distribution across the range of fabricated composites[11]. The development of more advanced processing strategies of NFCs could be a potential solution to overcome their weaknesses.

In general, NFCs hold high tensile and bending strength, excellent ductility and resistance to cracking, and thus enhanced impact strength and toughness [12,13]. Their durability of withstanding load is a requirement and that corresponds to its resistance to deterioration stemming from external and internal reasons. There are some factors that affect mechanical performance of NFCs, for instance, fibre selection that includes fibre type, extraction method, harvest time, treatment, aspect ratio, fibre volume fraction, choice of matrix, dispersion factor, fibre-matrix interfacial stress, fibre orientation, porosity, and composite manufacturing process [14]. Adequate knowledge on the persistent behaviour of NFCs particularly subjected by cyclic (fatigue) loading often limits their application in engineering design. Failure in these composites categorized by amassing of numerous damage modes, for example, fibre failure, matrix failure, and debonding between the polymer matrix and the reinforcing fibres are different from monolithic materials where failures take place with the initiation and propagation of a leading fracture scenario [15-17]. Difference in properties and surface characteristics lead NFCs to further complicate this behaviour. In fact, an independent, more specifically, synergistic manner is often obvious in their damage mechanisms.

Hygroscopicity of NFCs is an undesirable property owing to their chemical constituents. The moisture absorption of NFCs has numerous undesirable effects on their properties and that causes negative impact on their long-term performance [18-20]. To illustrate this, the reduction in their mechanical properties happens because of their increasing moisture content. It leads to biodegradation as well as change in dimensions by providing necessary conditions. In addition, hygroscopicity causes swelling that deteriorates performance and reliability. The inherent large dissimilarities in fibre characteristics and properties contributing to large scatter in NFCs [21-23]. The decrease in mechanical properties is often caused due to the poor wettability (hydrophilic fibres and hydrophobic matrices) of natural fibres that leads to weak fibre-matrix interfacial bonding. Increasing fibre loading causes increasing moisture absorption [24,25]. The correlation between the moisture absorption and fibre loadings of pineapple leaves fibre reinforced low density polyethylene (LDPE) composites has been studied and found that there is a linear relationship between the moisture absorption and fibre loadings. To enhance their moisture resistance, compatibilizers, coupling materials, or several chemical modifications were considered although a major concern still exists for their applications in open air [26].

This chapter entails the durability against the fatigues and moisture absorption phenomena of NFCs. Firstly, it discusses the fatigue and its different testing methodology. Then it highlights the factors affecting the fatigue properties as well as damage development and property degradation. Besides, this chapter also explains different fatigue tolerance technologies. While discussing moisture absorption of NFCs, emphasis is also given on the degree of moisture absorption into NFCs and moisture related degradation mechanisms. Lastly, it features the comparison between real life and accelerated ageing environments and a description on moisture protection technologies.

Sustainable Natural Fiber Composites   Materials Research Forum LLC
Materials Research Foundations **122** (2022) 128-153   https://doi.org/10.21741/9781644901854-6

2.   Fatigue and natural fibre composites

At large, fatigue within NFCs is a multifaceted property usually measured at the commencement of cracks, depending on the integrated ductility of both matrix and fibre. In fact, the damage remains diffuse and propagates by fibre–matrix delamination or crack bridging and is directed by the fibre–matrix interaction [27]. The use of NFCs is increasing continuously, however, some of their specific features still lack good in-depth understanding. To exemplify this, viscoelastic, viscoplastic or time-dependent behaviours are considered because of the presence of creep and fatigue [28]. Due to the lack of methodical and detailed information, creep and fatigue properties of NFCs are less explored compared to glass or carbon fibre reinforced plastics.

The following sections will discuss the fatigue testing methodology, factors that influence the fatigue properties, damage tolerance and property degradation phenomena, and fatigue improving techniques in brief.

2.1   Fatigue testing methodology

Up until now, many investigations have been conducted on the static mechanical properties of NFCs [29-31], however, merely very limited studies associated to fatigue. An investigation on the fatigue behaviour of hemp fibre reinforced HDPE composites was conducted using fatigue-life (S–N) curves at distinctive fibre volume fractions. For this purpose, the normalization of the developed S–N curves into one normalized S–N curve was studied from a proposed newly modified stress level. The fatigue-life response of these materials was simulated from the developed generalized fatigue behaviour model. The outcomes from this model revealed the capability of predicting the fatigue behaviour of the NFCs at distinctive fibre fractions and fatigue stress ratios [32]. In another study, the fatigue life evaluation of multi-layered, jute fibre fabric reinforced epoxy matrix composites was performed via tension fatigue tests with a constant fatigue stress ratio ($R = 0.1$), and results acquired from the tests were used to plot S-N curve. It was attributed that a broad study and analysis of fatigue behaviour of composite was required before its use or application in any form [33]. Two biodegradable polymers: thermoplastic starch and polylactide blend filled with 10 wt% short flax fibres were employed to analyse and compare fatigue properties of composites using accelerated fatigue tests with increasing amplitude. The fatigue stress of the materials was evaluated by applying dissipated energy, strain, and the temperature on the sample surface as characteristic values and that were measured using Lehr's method. The temperature rise was not visible while adding flax fibres caused improvement in fatigue strengths of polylactide-based composite. However, thermoplastic starch filled with flax fibres presented lower fatigue strengths than that of neat polymer, shortened fatigue life and greater susceptibility to cyclic creep [34]. It has

Sustainable Natural Fiber Composites   Materials Research Forum LLC
Materials Research Foundations **122** (2022) 128-153   https://doi.org/10.21741/9781644901854-6

been assessed that the potential use of plant fibre composites (PFCs) in fatigue critical components (like rotor blades) as component fatigue life prediction is not possible due to the remarkable lack of fatigue data. However, three aspects of this study were to examine the fatigue loads on a 3.5-meter study blade, construct S-N diagrams and constant-life diagrams of PFCs, and predict the design life of a hemp/polyester blade and found that PFC blade can be able to satisfy the 20-year design life criteria, inclusive of a 1.50 safety factor in the prediction of fatigue life [35]. The examination of cyclic behaviour of sisal fibre reinforced thermoset composites under both tension–tension fatigue and fully reversed loading showed that alkali treated fibre composites exhibited better fatigue performance owing to enhanced fibre–matrix adhesion [36]. A study on the fatigue behaviour of unreinforced sisal natural fibres illustrated that conventional empirical fatigue-life models functioned well to correlate fatigue response [37]. The flexural fatigue behaviour of wood flour reinforced high density polyethylene was evaluated based on a statistical model using a Weibull distribution to assess the fatigue behaviour of these materials [38].

2.2   Factors affecting the fatigue properties

There are several factors, for example, polymer type (brittle or soft matrix)/structure, type of filler and its content, molecular weight (MW) and crosslinking that affect the fatigue properties of NFCs [28, 39, 40]. The interfacial strength parameters such as brittle matrix (epoxy) can influence on fatigue behaviour of composites. The crack propagation is controlled by these parameters. It is important to examine the textile architecture, individual effects of fibre type, interphase properties, fibre properties and their content on the fatigue behaviour of the materials. Debonding and fictional sliding happen promptly upon crack extension due to the poor interfacial bonding between the phases. This leads to the bridge the crack while the long fibres remain intact [41, 42]. To solve this problem, a stronger interface is important to restrain sliding and contributes to fibre fracture instead of bridging the crack. A decrease in fibre modulus often leads to fatigue in unidirectional composites [43-45]. During alkali treatment, shrinkage of fibres plays a considerable impact on the fibre structure and thus on the mechanical properties of the fibres, thereby on the composites. As can be seen from Fig. 1, remarkable difference in the tensile modulus of alkali (NaOH) treated flax fibre yarn (altered from 1.5 to 0.2 times) in comparison with their untreated counterpart once the shrinkage differed from 0 to 26%. This treatment modifies the Hermans orientation factor (0.96 for zero-shrunk fibres and 0.653 for 26% shrunk fibres). An increase in the shrinkage of the fibres caused a higher specific damping capacity and reduction of required stress to initiate damage in a composite system at a given loading while decreased shrinkage increase the impact stiffness [46].

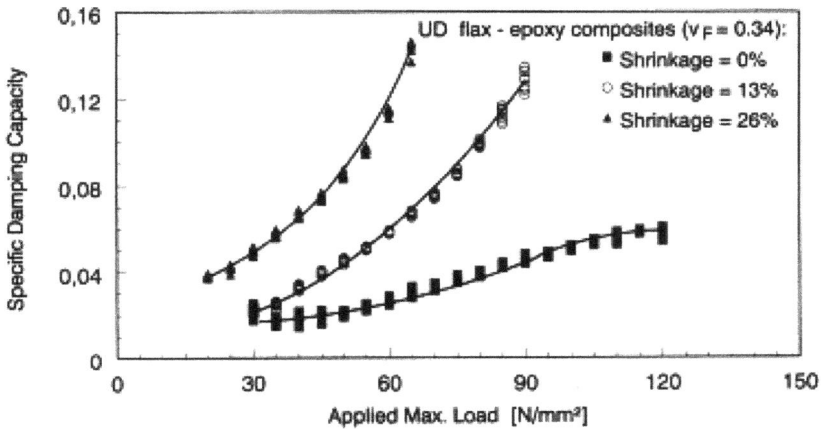

Figure 1. Specific damping capacity as a function of applied maximum load for UD epoxy composites which contain flax fibres of various fibre tensile strength and modulus at 10^4 load cycles/load level [46].

Jute fibre was modified with alkali, maleic anhydride grafted polypropylene (MAH-PP) and silane and obtained that surface modification caused understandable effects on the fatigue behaviour of jute fibre reinforced polypropylene composites (Fig. 2). The fatigue behaviour connected with specific damping properties, hysteresis, dynamic modulus, and accumulated dissipated energy of jute–PP composites was examined to study the effect of MAH-PP content on this behaviour by employing intelligent dynamic mechanical testing. The damage resistance was augmented to upper maximal stresses and the progress (rate) of damage was decreased with improved fibre–matrix adhesion. This fibre–matrix adhesion contributes to the reduction of the progress of the damage with increasing fibre content. The growth in dynamic strength was visible because untreated jute–PP composites scarcely exhibit any modification in amassed dissipated energy, however, the MAH-PP coupling agent enhanced the stress transfer [47].

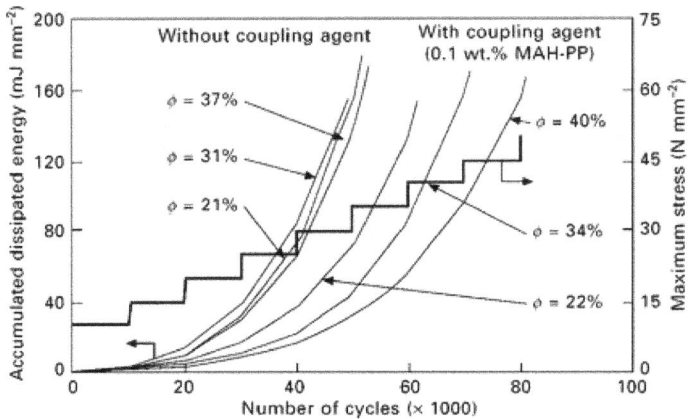

Figure 2. Effect of fibre content on the specific damping capacity of MAH-PP modified jute-PP composites, test frequency 10 Hz, R 0.1, 10^4 cycles/stress level, φ fibre content (vol%) [47].

2.3 Damage development and property degradation

NFCs are prone to damage from the initiation and propagation of damage. Initiation of damages is typically triggered by impact followed by propagation and growth because of cyclic thermomechanical service loads, or it results from manufacturing defects [48-50]. Machining (eg, cutting) can be another source of initiation of damages in these materials [16, 51]. The influence of the fibre and the interface parameters on the tension–tension fatigue behaviour of NFCs was investigated in one study. The combination of fibres (jute and flax in yarns and woven form) and matrix (epoxy resin, polyester resin, and polypropylene) were used in this study. While increasing applied load, a difference in the damping of the materials was linked to the distinct effect of these parameters. Higher critical loads for damage initiation and larger failure loads were demonstrated in this study while NFCs engaging greater fibre strength and modulus, stronger interfacial adhesion between the phases or higher amounts of fibre in the matrix. As depicted in Fig. 3A, the initiation of the damage initiation is at 45 N mm^{-2} for the jute–epoxy composite revealed by the noticeable increase in the specific damping capacity. Damage propagation led to the rise in the specific damping capacity at increasing mean stress and eventually fractured at 120 N mm^{-2}. Flax–epoxy composites illustrated similar type of behaviour. With higher cellulose content, it showed much higher specific damping capacity values than the jute–epoxy composite. The stress was transferred by cellulose microfibrils from one layer to the other layer and that reduces the function of the cementing matrix (hemicellulose and lignin) as a load carrying component (similarly to the laminate composite fibres). Delamination crack growth resistance was mainly determined by the cellulose microfibrils. In short, the viscoelastic deformation of the fibre itself caused the loss of bigger part of the energy in the internal friction. Also, micromechanical degradation due to the structural breakdown of the microfibrils caused the damage of the fatigued fibres [52].

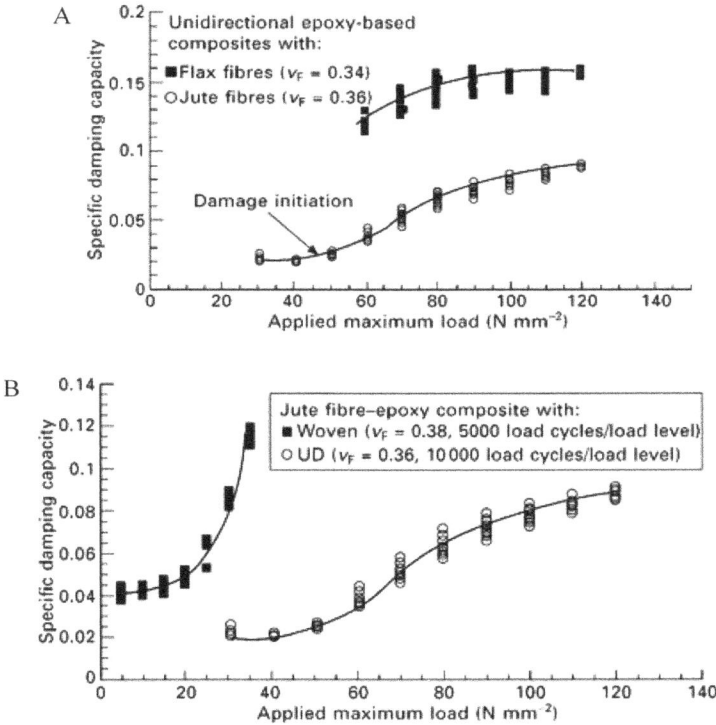

Figure 3. Specific damping capacity as a function of applied maximum load for; (A) unidirectional flax and jute epoxy composites at 10^4 load cycles/load level, (B) woven and unidirectional (UD) reinforced jute-epoxy composites [52].

Mainly, the damping mechanism results from the viscoelastic nature of matrix and fibre, thermoelastic damping, the Coulomb friction damping because of slip between the unbonded, debonded, and bonded region of fibre-matrix interface and the energy dissipation happening at delamination or cracks [53]. Fig. 3B illustrated that the speedy damage development throughout the initial loading cycles is predominantly attributed to the matrix cracking along the off-axis plies and perpendicular to the loading direction alongside woven fibres in the brittle epoxy matrix. Over the time, these microcracks coalesce and enlarge with interfaces and delaminate leading to the failure of composites having the fracture of the fibres. However, the dominant mechanism of composite failure is interfacial debonding in unidirectional composites. The interfacial interaction between the matrix and the fibre is dependent on crack propagation in composites. Debonding and frictional sliding takes place at a comparatively weaker interface. A robust interface between the phases subdues the interface sliding and subsidises to fibre fracture along with cumulative critical stress for damage propagation.

The interaction between the fibres and the matrix was enhanced using coupling agents in NFCs. The adding of MAH-PP copolymer as a coupling agent in jute–PP composites confirmed reduced damage propagation with improved critical loadings [52].

2.4   Fatigue improving techniques

The applications of NFCs have been expanding in various fields including aerospace, transportation, military, electronic, and health care. To meet the requirements, it is important to have satisfactory mechanical performance, fire retardancy, and chemical resistance in these materials. Numerous approaches have been implemented that includes refining the fracture toughness of the ply interfaces by using epoxy elastomer blends; and decreasing the variance of elastic properties as well as stress concentrations at the interfaces between the laminated plies to tackle the fatigue damage of these materials. These materials also have shortage of other essential functional properties, for instance, high electrical and thermal conductivity for electrostatic dissipation and lightning strike protection.

Presently, it is understood that the finest way to attain multifunctional properties in a polymer is to blend it with nanoscale fillers. Because polymer nanocomposites hold three main characteristics that involves the decrease of nanoscopic confinement of matrix polymer chains, dissimilarity in properties of nanoscale inorganic constituents and nanoparticle composition and formation of a considerable polymer or particle interfacial area. In advanced aircraft, these materials can be used as adhesives to join aircraft components such as composite to composite, metal to metal and metal to composite components. However, poor adhesion between fibre and polymer matrix can happen because thermoplastics are hydrophobic substances which are not compatible with hydrophilic wood fibres. Chemical coupling agent (polymers) in small quantities can be a solution for the treatment of the surface to increase bonding between the treated surface and other surfaces. A study showed that dynamic strength of ~40% increased to be found with 0.1 wt.% of coupling agent in jute–PP composites as illustrated in Fig. 4 [47].

Materials Research Foundations **122** (2022) 128-153   https://doi.org/10.21741/9781644901854-6

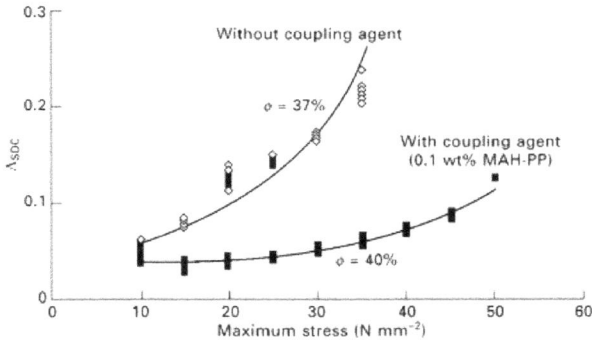

Figure 4. Influence of surface treatment on the specific damping capacity (SDC) of jute-PP composites, test frequency 10 Hz, R 0.1, 10^4 cycles/stress level, ϕ fibre content (vol%) [47].

3.   Moisture absorption phenomena

NFCs is constituted of cellulose, hemicellulose, lignin, and pectin [7, 54, 55]. Of them, cellulose is considered as the key framework element of the fibre structure. It exerts stiffness, strength, and structural stability to the fibres. Nevertheless, the highest amount of moisture was absorbed by cellulose and hemicellulose, the major constituents of natural fibres. The effective shielding was provided by lignin for hemicellulose/cellulose against severe environmental circumstances, for example, humidity and temperature [24, 56-59]. The following sections will explain the effect of moisture absorption on natural fibre composites, degradation mechanisms and protection technologies in brief.

3.1   Impact of moisture absorption on natural fibre composites

Two main models are frequently used to describe the absorption behaviour including (i) pseudo-Fickian behaviour, where the weight gains of water cannot be able to reach equilibrium after absorption; and (ii) linear Fickian behaviour, where the weight gains of water can gradually be able to attain equilibrium after quick initial absorption [57, 60, 61]. In one study, the impact of moisture on rice hulls (40 mesh size), HDPE, and maleic anhydride polyethylene (MAPE) compounded composites has been studied. For this reason, moisture absorption as a function of $\sqrt{t}$ has been investigated and noted that there was no level-off for the equilibrium moisture absorption for composites having 40%, 50%, 55%, and 60% fibre loadings (Fig. 5) [62].

Sustainable Natural Fiber Composites   Materials Research Forum LLC
Materials Research Foundations **122** (2022) 128-153   https://doi.org/10.21741/9781644901854-6

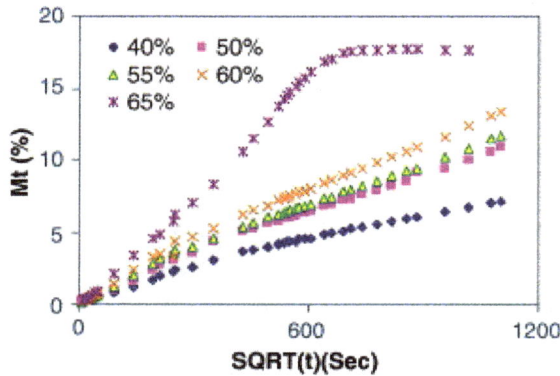

Figure 5. Moisture absorption behaviour of composites with different fibre loadings [62].

Biodegradation takes place due to the microbial attack caused by the higher moisture absorption. Moisture can pass inside the NFCs via various mechanisms, as an example, micro gaps within the polymer chains can be used for the diffusion of water [63]. There are two classes to classify the water absorbed in the NFCs that includes bound water (bounded to the polar groups of the polymers) and free water (pass freely across the cracks and voids) [64]. Once NFCs are exposed to moisture, the water molecules diffuse into the composite and connect onto hydrophilic groups of natural fibres, creating an intermolecular bonding of hydrogen with the fibres that alleviates the interfacial adhesion at the fibre/matrix interface, therefore, degradation of composites happen [65, 66].

3.2   Moisture related degradation mechanisms

The degradation of NFCs take place through the stress created by the swelling of cellulosic fibres at the interfacial regions that causes micro-cracking mechanism in the matrix adjacent to the swollen fibre. This accentuates transport and capillarity via micro-cracks. The increase in bound water happens while the free water decreases due to the absorption of excessive water. As a consequence, water-soluble substances initiate to leach from fibre resulting eventual debonding between the matrix and fibre [67-74]. The diffusion mechanism of water molecules into the NFCs is illustrated in Fig. 6.

Sustainable Natural Fiber Composites   Materials Research Forum LLC
Materials Research Foundations **122** (2022) 128-153   https://doi.org/10.21741/9781644901854-6

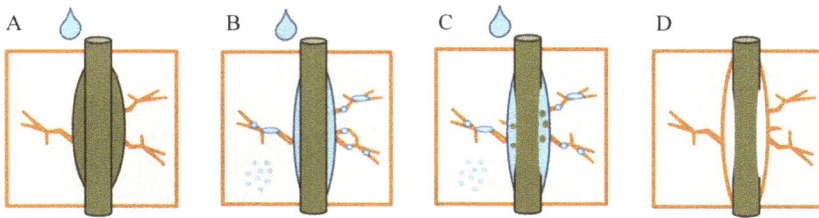

Figure 6. The interfacial adhesion caused by degradation due to moisture absorption;
(A) development of micro-cracks due to expansion of swollen fibre, (B) water molecules
diffused in the bulk matrix flow along the fibre–matrix interface, (C) water-soluble
components leach from fibre, (D) ultimate matrix–fibre debonding takes place [68].

The increase in all three dimensions of the laminated composite is caused due to the moisture absorption. However, the thickness of these materials has higher increase compared to width and length of these materials. Two directions such as swelling of the fibre itself; and (ii) modifications in the fibre's density across the weight of absorbed moisture are used for water uptake of NFCs. The polar and hydroxyl groups of NFCs cause the water uptake [68]. Delamination of the laminated composite, debonding between matrix and fibres, and physical damage to the polymeric matrix augmented the mechanical and thermal degradation of NFCs [75-83]. The mechanical interlocking between the matrix and fibre is also influenced by minor swelling of NFCs. The mobility of the side groups and molecular chains increase due to the water uptake result in the reversible plasticization of the polymer matrix. This leads to the increase in fracture toughness while reducing the fatigue durability, strength, and stiffness of NFCs [84-90]. A study based on sisal fibre-reinforced polypropylene (PP) composite showed that the tensile strength and modulus are decreased while increasing immersion time in the water at 90°C. However, the initial increasing immersion time enhance the impact strength. The intermolecular interaction can significantly be changed by the diffusion of water molecules into types of polymer films that enhance the thermal conductivity and hence decrease thermal insulation properties of NFCs [91].

Another study revealed that the outcomes of the impact strength of hybrid phenol formaldehyde (PF) fabricated with Areca Fine Fibres (AFFs) and Calotropis Gigantea Fibre (CGF) composite at dry and wet conditions (Fig. 7A). The impact strength amplified as hybrid fibre content increased up to 35 wt.% and then declined for the dry composite specimens whereas the impact strength of composite reduced after 35 wt.% of hybrid fibres at wet condition. It is obvious that the impact strength of hybrid composites at dry condition is higher than the impact strength of hybrid composites at wet condition. The diffusion of water molecules into the interface region led to the development of a greater number of microcracks as a consequence of fibres' swelling, as shown in Fig. 7B and 7C [92].

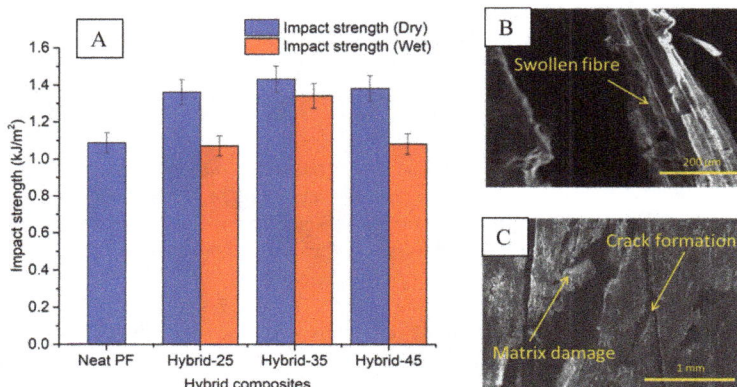

Figure 7. (A) Effects of water absorption on the impact strength of CGF/AFF/PF hybrid composites based on the weight percentages of fibre, (B, C) SEM images of fractured surfaces of the 35 wt.% hybrid composites after the impact test (wet condition) [92].

3.3   Moisture protection technologies

The hydrophilic nature of NFCs introduces noteworthy changes to the surface morphology. The moisture absorption of NFCs can be changed due to the chemical and physical treatments of natural fibres, fibre distribution, fibre size and shape, fibre loading, linear mass density of natural yarns, permeability of the fibre, hydrophilicity of each composite's constitute, void content, duration of exposure to humid environment, fabrication method, and temperature [62, 93-100]. To illustrate this, the hemp fibre-reinforced polylactic acid (PLA) having non-woven form has greater amount of moisture absorption rather than aligned fibre composite structure. Suitable chemical treatments, additives, nano coatings, and bio-based coatings can be used to enhance the surface hydrophobicity of NFCs. Commonly, coatings are only applicable in posing short to medium-term protection against moisture. In fact, the decrease in the diffusion of water molecules can be 50% by choosing suitable application and material characteristics of the coating. A larger amount of moisture can be absorbed with higher content of voids in NFCs compared to the ones without voids. To overcome this problem, it is important to maintain NFCs under vacuum and low humidity to decrease the void content throughout the manufacturing process. Fig. 8 depicts the percentage of water absorption outcomes for coated and uncoated flax/phenolic

Sustainable Natural Fiber Composites   Materials Research Forum LLC
Materials Research Foundations **122** (2022) 128-153   https://doi.org/10.21741/9781644901854-6

laminates. In this study, it has been examined that the amount of water absorption was increased at increasing immersion time for woven flax fibre reinforced phenolic laminates having three different bio-based coatings namely: water resistant market product (FIRESHELL® (F1E)), polyurethane (PU) and poly (furfuryl alcohol) (PFA). Nonetheless, the susceptibility to water uptake caused a substantial increase in water absorption for F1E coated and uncoated flax/phenolic laminates with increasing time as shown in (Fig. 8) [101].

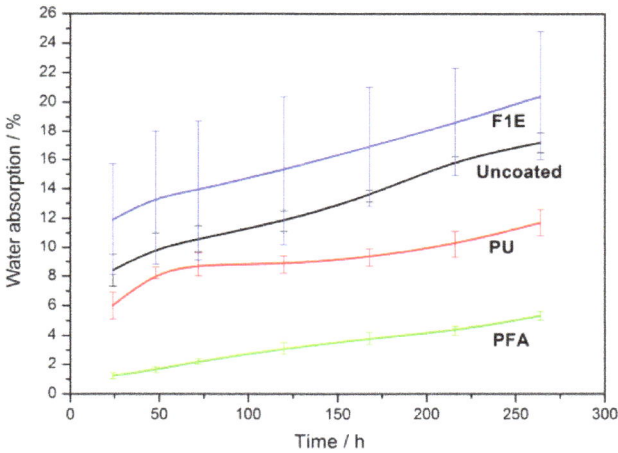

Figure 8. Water absorption behaviour of flax/phenolic laminates coated with different bio-based coatings [101].

Excessive moisture content can be removed through the drying of NFCs. The water uptake is interrelated with the nature of the fibre. Ramie and cotton fibres have a higher ability to absorb moisture relative to flax and hemp fibres [66]. An efficient technique to lessen the moisture absorption of NFCs is to hybridize the natural fibres with other synthetic or natural fibres. In addition, the equilibrium moisture absorption and the rate of water diffusion is increased at increasing temperatures because of the activation of the diffusion process. In some cases, the water molecules attach to the fibre surface throughout the amorphous waxy materials, leading to the water sorption acceleration. The decreasing temperatures can cause increasing diffusion coefficient in many cases. The material durability under different weathering conditions i.e., temperature, rain, air and life cycle effect throughout the composite design stage is sometimes considered [55, 102-105].

Conclusions and recommendations

Natural fibre composites (NFCs) have well-acceptance over synthetic fibre composites in many applications because of their high performance and extensive range of application flexibilities.

These materials are environmentally beneficial and commercially feasible as well as sustainable. Nevertheless, there are some concerns regarding these materials for durability and long-term performance. The fatigue properties of these materials in general depend on the various factors such as polymer type, molecular weight, crosslinking, type of filler and its content, and the interfacial strength. It is important to understand the fatigue behaviour of NFCs since it activates failure mechanism. Cumulative evaluation of property degradation mechanisms throughout the process and developing the capability to predict the life of NFCs under distinct conditions can improve these materials with more dependable and desired characteristics.

The high moisture absorption behaviour of NFCs is responsible for the rate of degradation of the mechanical properties that is relatively high in comparison with that of synthetic fibre composites. Several researchers have been working on developing moisture protection technologies to provide materials with improved mechanical properties. A more extensive fundamental study of the fatigue and moisture absorption of NFCs for durability is necessary with the enhanced potential of investigating their applications in different field such as construction and automotive.

References

[1] V.K. Balla, K.H. Kate, J. Satyavolu, P. Singh, J.G.D. Tadimeti, Additive manufacturing of natural fiber reinforced polymer composites: Processing and prospects, Composites Part B: Engineering 174 (2019) 106956. https://doi.org/10.1016/j.compositesb.2019.106956

[2] M. Li, Y. Pu, V.M. Thomas, C.G. Yoo, S. Ozcan, Y. Deng, K. Nelson, A.J. Ragauskas, Recent advancements of plant-based natural fiber–reinforced composites and their applications, Composites Part B: Engineering 200 (2020) 108254. https://doi.org/10.1016/j.compositesb.2020.108254

[3] V. Paul, K. Kanny, G.G. Redhi, Mechanical, thermal and morphological properties of a bio-based composite derived from banana plant source, Composites Part A: Applied Science and Manufacturing 68 (2015) 90-100. https://doi.org/10.1016/j.compositesa.2014.08.032

Materials Research Foundations **122** (2022) 128-153   https://doi.org/10.21741/9781644901854-6

[4] Y.G. Thyavihalli Girijappa, S. Mavinkere Rangappa, J. Parameswaranpillai, S. Siengchin, Natural Fibers as Sustainable and Renewable Resource for Development of Eco-Friendly Composites: A Comprehensive Review, Frontiers in Materials 6(226) (2019). https://doi.org/10.3389/fmats.2019.00226

[5] R.T.S. Freire, J.C. dos Santos, T.H. Panzera, L.J. da Silva, Chapter Five - Recent research and developments in hybrid natural fiber composites, in: A. Khan, S.M. Rangappa, S. Siengchin, M. Jawaid, A.M. Asiri (Eds.), Hybrid Natural Fiber Composites, Woodhead Publishing 2021, pp. 91-112. https://doi.org/10.1016/B978-0-12-819900-8.00008-8

[6] A. Azman, M.R.M. Asyraf, K. Abdan, M. Petru, C. Ruzaidi, S. Sapuan, w.s. wan nik, M. Ishak, I. R.A, S. Mat Jusoh, Natural Fiber Reinforced Composite Material for Product Design: A Short Review, Polymers 13 (2021) 1-24. https://doi.org/10.3390/polym13121917

[7] M. Zwawi, A Review on Natural Fiber Bio-Composites, Surface Modifications and Applications, Molecules 26(2) (2021) 404. https://doi.org/10.3390/molecules26020404

[8] K. Amulya, R. Katakojwala, S. Ramakrishna, S. Venkata Mohan, Low carbon biodegradable polymer matrices for sustainable future, Composites Part C: Open Access 4 (2021) 100111. https://doi.org/10.1016/j.jcomc.2021.100111

[9] S.S.R. Raj, J.E.R. Dhas, C.P. Jesuthanam, Challenges on machining characteristics of natural fiber-reinforced composites – A review, Journal of Reinforced Plastics and Composites 40(1-2) (2020) 41-69. https://doi.org/10.1177/0731684420940773

[10] P. Peças, H. Carvalho, H. Salman, M. Leite, Natural Fibre Composites and Their Applications: A Review, Journal of Composites Science 2(4) (2018) 66. https://doi.org/10.3390/jcs2040066

[11] S. Mat Jusoh, M.M.H.M. Hamdan, H.Y. Sastra, S. Sapuan, Study of Interfacial Adhesion of Tensile Specimens of Arenga Pinnata Fiber Reinforced Composites, Multidiscipline Modeling in Materials and Structures 3 (2007) 213-224. https://doi.org/10.1163/157361107780744360

[12] H.-y. Cheung, M.-p. Ho, K.-t. Lau, F. Cardona, D. Hui, Natural fibre-reinforced composites for bioengineering and environmental engineering applications, Composites Part B: Engineering 40(7) (2009) 655-663. https://doi.org/10.1016/j.compositesb.2009.04.014

[13] M. Muneer Ahmed, H.N. Dhakal, Z.Y. Zhang, A. Barouni, R. Zahari, Enhancement of impact toughness and damage behaviour of natural fibre reinforced composites and

their hybrids through novel improvement techniques: A critical review, Composite Structures 259 (2021) 113496. https://doi.org/10.1016/j.compstruct.2020.113496

[14] K.L. Pickering, M.G.A. Efendy, T.M. Le, A review of recent developments in natural fibre composites and their mechanical performance, Composites Part A: Applied Science and Manufacturing 83 (2016) 98-112. https://doi.org/10.1016/j.compositesa.2015.08.038

[15] P.J. Herrera-Franco, A. Valadez-González, Mechanical properties of continuous natural fibre-reinforced polymer composites, Composites Part A: Applied Science and Manufacturing 35(3) (2004) 339-345. https://doi.org/10.1016/j.compositesa.2003.09.012

[16] M.-P. Ho, H. Wang, J.-H. Lee, C.-k. Ho, K.-t. Lau, J. Leng, D. Hui, Critical factors on manufacturing processes of natural fibre composites, Composites Part B: Engineering 43(8) (2012) 3549-3562. https://doi.org/10.1016/j.compositesb.2011.10.001

[17] O.I. Okoli, G. Smith, Failure modes of fibre reinforced composites: The effects of strain rate and fibre content, Journal of Materials Science 33 (1998) 5415-5422.

[18] M. Alhijazi, B. Safaei, Q. Zeeshan, M. Asmael, A. Eyvazian, Z. Qin, Recent Developments in Luffa Natural Fiber Composites: Review, Sustainability 12 (2020) 7683. https://doi.org/10.3390/su12187683

[19] L. Mohammed, M.N.M. Ansari, G. Pua, M. Jawaid, M.S. Islam, A Review on Natural Fiber Reinforced Polymer Composite and Its Applications, International Journal of Polymer Science 2015 (2015) 243947. https://doi.org/10.1155/2015/243947

[20] T. Mokhothu, M. John, Review on hygroscopic aging of cellulose fibres and their biocomposites, Carbohydrate Polymers 131 (2015) 337-354. https://doi.org/10.1016/j.carbpol.2015.06.027

[21] A. Célino, S. Fréour, F. Jacquemin, P. Casari, The hygroscopic behavior of plant fibers: a review, Front Chem 1 (2014) 43-43. https://doi.org/10.3389/fchem.2013.00043

[22] H. Dhakal, Z.Y. Zhang, M. Richardson, Effect of water absorption on the mechanical properties of hemp fibre reinforced unsaturated polyester composites, Composites Science and Technology 67 (2007) 1674-1683. https://doi.org/10.1016/j.compscitech.2006.06.019

[23] A. Moudood, A. Rahman, A. Ochsner, M. Islam, G. Francucci, Flax fiber and its composites: An overview of water and moisture absorption impact on their

Materials Research Foundations **122** (2022) 128-153   https://doi.org/10.21741/9781644901854-6

performance, Journal of Reinforced Plastics and Composites 38 (2018). https://doi.org/10.1177/0731684418818893

[24] M.M. Kabir, H. Wang, K.T. Lau, F. Cardona, Chemical treatments on plant-based natural fibre reinforced polymer composites: An overview, Composites Part B: Engineering 43(7) (2012) 2883-2892. https://doi.org/10.1016/j.compositesb.2012.04.053

[25] C.H. Lee, A. Khalina, S.H. Lee, Importance of Interfacial Adhesion Condition on Characterization of Plant-Fiber-Reinforced Polymer Composites: A Review, Polymers 13(3) (2021) 438. https://doi.org/10.3390/polym13030438

[26] O. Daramola, A. Adediran, Mechanical Properties and Water Absorption Behaviour of Treated Pineapple Leaf Fibre Reinforced Polyester Matrix Composites, Leonardo Journal of Sciences (2017) 15-30.

[27] J. Zangenberg, P. Brøndsted, J. Gillespie, Fatigue damage propagation in unidirectional glass fibre reinforced composites made of a non-crimp fabric, Journal of Composite Materials 48 (2013) 2711-2727. https://doi.org/10.1177/0021998313502062

[28] M. Misra, S. Ahankari, A. Mohanty, Creep and fatigue of natural fibre composites, 2011, pp. 289-332. https://doi.org/10.1533/9780857092281.2.289

[29] S. Chikkol Venkateshappa, A. Arifulla, N. Goutham, T. Santhosh, H. Jaeethendra, R. Ravikumar, S. Anil, D. Kumar, J. Ashish, Static bending and impact behaviour of areca fibers composites, Materials & Design 32 (2011) 2469-2475. https://doi.org/10.1016/j.matdes.2010.11.020

[30] J. Fajrin, Compressive Properties of Tropical Natural Fibers Reinforced Epoxy Polymer Composites, Jurnal Ilmu dan Teknologi Kayu Tropis 14 (2016).

[31] T. Raja, A. Palanivel, M. Karthik, M. Sundaraj, Evaluation of mechanical properties of natural fibre reinforced composites - A review, International Journal of Mechanical Engineering and Technology 8 (2017) 915-924.

[32] A. Fotouh, J. Wolodko, M. Lipsett, Fatigue of natural fiber thermoplastic composites, Composites Part B: Engineering 62 (2014) 175–182. https://doi.org/10.1016/j.compositesb.2014.02.023

[33] P. N H, C. K N, Pavan, O. Anand, Fatigue Behaviour and Life Assessment of Jute-epoxy Composites under Tension-Tension Loading, IOP Conference Series: Materials Science and Engineering 225 (2017) 012017. https://doi.org/10.1088/1757-899X/225/1/012017

[34] A. Liber-Kneć, P. Kuźniar, S. Kuciel, Accelerated Fatigue Testing of Biodegradable Composites with Flax Fibers, Journal of Polymers and the Environment 23(3) (2015) 400-406. https://doi.org/10.1007/s10924-015-0719-6

[35] D. Shah, Fatigue characterisation of plant fibre composites for rotor blade applications, JEC Composites Magazine, No. 73: Special JEC Asia 49 (2012) 51-54.

[36] A. Towo, M. Ansell, Fatigue of sisal fibre reinforced composites: Constant-life diagrams and hysteresis loop capture, Composites Science and Technology 68 (2008) 915-924. https://doi.org/10.1016/j.compscitech.2007.08.021

[37] J. Huang, G. Tian, P. Huang, Z. Chen, Flexural Performance of Sisal Fiber Reinforced Foamed Concrete under Static and Fatigue Loading, Materials (Basel) 13(14) (2020) 3098. https://doi.org/10.3390/ma13143098

[38] H.-S. Yang, P. Qiao, M. Wolcott, Flexural Fatigue and Reliability Analysis of Wood Flour/High-density Polyethylene Composites, Journal of Reinforced Plastics and Composites 29 (2010) 1295-1310. https://doi.org/10.1177/0731684409102753

[39] T. Gurunathan, S. Mohanty, S.K. Nayak, A review of the recent developments in biocomposites based on natural fibres and their application perspectives, Composites Part A: Applied Science and Manufacturing 77 (2015) 1-25. https://doi.org/10.1016/j.compositesa.2015.06.007

[40] K.M.F. Hasan, P.G. Horváth, T. Alpár, Potential Natural Fiber Polymeric Nanobiocomposites: A Review, Polymers 12(5) (2020) 1072. https://doi.org/10.3390/polym12051072

[41] Z. Mahboob, H. Bougherara, Fatigue of flax-epoxy and other plant fibre composites: Critical review and analysis, Composites Part A: Applied Science and Manufacturing 109 (2018) 440-462. https://doi.org/10.1016/j.compositesa.2018.03.034

[42] F. Bensadoun, V. Ignaas, A.W. Vuure, Interlaminar fracture toughness of flax-epoxy composites, Journal of Reinforced Plastics and Composites 36 (2016). https://doi.org/10.1177/0731684416672925

[43] 7 - Defects and damage and their role in the failure of polymer composites, in: E.S. Greenhalgh (Ed.), Failure Analysis and Fractography of Polymer Composites, Woodhead Publishing, pp. 356-440, (2009). https://doi.org/10.1533/9781845696818.356

[44] D.U. Shah, Damage in biocomposites: Stiffness evolution of aligned plant fibre composites during monotonic and cyclic fatigue loading, Composites Part A: Applied

Science and Manufacturing 83 (2016) 160-168.
https://doi.org/10.1016/j.compositesa.2015.09.008

[45] H.A. Aisyah, M.T. Paridah, S.M. Sapuan, R.A. Ilyas, A. Khalina, N.M. Nurazzi, S.H. Lee, C.H. Lee, A Comprehensive Review on Advanced Sustainable Woven Natural Fibre Polymer Composites, Polymers 13(3) (2021) 471. https://doi.org/10.3390/polym13030471

[46] J. Gassan, A study of fibre and interface parameters affecting the fatigue behaviour of natural fibre composites, Composites Part A: Applied Science and Manufacturing 33(3) (2002) 369-374. https://doi.org/10.1016/S1359-835X(01)00116-6

[47] K.-P. Mieck, A. Nechwatal, C. Knobelsdorf, Faser-Matrix-Haftung in Kunststoffverbunden aus thermoplastischer Matrix und Flachs, 2. Die anwendung von funktionalisiertem polypropylen, Die Angewandte Makromolekulare Chemie 225(1) (1995) 37-49. https://doi.org/10.1002/apmc.1995.052250104

[48] J. Beaugrand, S. Guessasma, J.-E. Maigret, Damage mechanisms in defected natural fibers, Scientific Reports 7(1) (2017) 14041. https://doi.org/10.1038/s41598-017-14514-6

[49] V. Mahesh, S. Joladarashi, S.M. Kulkarni, A comprehensive review on material selection for polymer matrix composites subjected to impact load, Defence Technology 17(1) (2021) 257-277. https://doi.org/10.1016/j.dt.2020.04.002

[50] M. Mehdikhani, L. Gorbatikh, I. Verpoest, S. Lomov, Voids in fiber-reinforced polymer composites: A review on their formation, characteristics, and effects on mechanical performance, Journal of Composite Materials 53 (2018) 002199831877215. https://doi.org/10.1177/0021998318772152

[51] A. Lotfi, H. Li, D.V. Dao, G. Prusty, Natural fiber–reinforced composites: A review on material, manufacturing, and machinability, Journal of Thermoplastic Composite Materials 34 (2019) 089270571984454. https://doi.org/10.1177/0892705719844546

[52] J. Gassan, Naturfaserverstarkte Kunststoffe-Korrelation zwischen Struktur und Eigenschaften der Fasern und deren Composites, PhD thesis, Institut fur Werkstofftechnik, University of Kassel (1997).

[53] J. Gassan, A. Bledzki, Possibilities to Improve the Properties of Natural Fiber Reinforced Plastics by Fiber Modification – Jute Polypropylene Composites –, Applied Composite Materials 7 (2000) 373-385. https://doi.org/10.1023/A:1026542208108

[54] N. Chand and M. Fahim (Editors), *Tribology of Natural Fiber Polymer Composites*, Woodhead Publishing. p. 1-58, (2008). https://doi.org/10.1533/9781845695057.1

[55] S. Amiandamhen, M. Meincken, L. Tyhoda, Natural Fibre Modification and Its Influence on Fibre-matrix Interfacial Properties in Biocomposite Materials, 2020. https://doi.org/10.1007/s12221-020-9362-5

[56] K.M.F. Hasan, P.G. Horváth, Z. Kóczán, T. Alpár, Thermo-mechanical properties of pretreated coir fiber and fibrous chips reinforced multilayered composites, Scientific Reports 11(1) (2021) 3618. https://doi.org/10.1038/s41598-021-83140-0

[57] M. Sorieul, A. Dickson, S.J. Hill, H. Pearson, Plant Fibre: Molecular Structure and Biomechanical Properties, of a Complex Living Material, Influencing Its Deconstruction towards a Biobased Composite, Materials (Basel) 9(8) (2016) 618. https://doi.org/10.3390/ma9080618

[58] V.K. Thakur, M.K. Thakur, Processing and characterization of natural cellulose fibers/thermoset polymer composites, Carbohydrate Polymers 109 (2014) 102-117. https://doi.org/10.1016/j.carbpol.2014.03.039

[59] I. Van de Weyenberg, T. Chi Truong, B. Vangrimde, I. Verpoest, Improving the properties of UD flax fibre reinforced composites by applying an alkaline fibre treatment, Composites Part A: Applied Science and Manufacturing 37(9) (2006) 1368-1376. https://doi.org/10.1016/j.compositesa.2005.08.016

[60] C. Hansen, The significance of the surface condition in solutions to the diffusion equation: explaining "anomalous" sigmoidal, Case II, and Super Case II absorption behavior, European Polymer Journal 46 (2010) 651-662. https://doi.org/10.1016/j.eurpolymj.2009.12.008

[61] A. Azizan, M. Johar, S.S. Karam Singh, S. Abdullah, S.S.R. Koloor, M. Petrů, K.J. Wong, M.N. Tamin, An Extended Thickness-Dependent Moisture Absorption Model for Unidirectional Carbon/Epoxy Composites, Polymers 13(3) (2021) 440. https://doi.org/10.3390/polym13030440

[62] W. Wang, M. Sain, P. Cooper, Study of moisture absorption in natural fiber plastic composites, Composites Science and Technology - Compos Sci Technol 66 (2006) 379-386. https://doi.org/10.1016/j.compscitech.2005.07.027

[63] E. Manaila, G. Craciun, D. Ighigeanu, Water Absorption Kinetics in Natural Rubber Composites Reinforced with Natural Fibers Processed by Electron Beam Irradiation, Polymers 12(11) (2020) 2437. https://doi.org/10.3390/polym12112437

[64] N.A. Zulkarnain, Degradability of bamboo fibre reinforced polyester composites, 2014.

[65] Y. Zhou, M. Fan, L. Chen, Interface and bonding mechanisms of plant fibre composites: An overview, Composites Part B: Engineering 101 (2016) 31-45. https://doi.org/10.1016/j.compositesb.2016.06.055

[66] A. Al-Maharma, N. Al-Huniti, Critical Review of the Parameters Affecting the Effectiveness of Moisture Absorption Treatments Used for Natural Composites, 3 (2019) 40. https://doi.org/10.3390/jcs3010027

[67] H.M. Akil, L.W. Cheng, Z.A. Mohd Ishak, A. Abu Bakar, M.A. Abd Rahman, Water absorption study on pultruded jute fibre reinforced unsaturated polyester composites, Composites Science and Technology 69(11) (2009) 1942-1948. https://doi.org/10.1016/j.compscitech.2009.04.014

[68] Z.N. Azwa, B.F. Yousif, A.C. Manalo, W. Karunasena, A review on the degradability of polymeric composites based on natural fibres, Materials & Design 47 (2013) 424-442. https://doi.org/10.1016/j.matdes.2012.11.025

[69] N. Banik, V. Dey, G.R.K. Sastry, An Overview of Lignin & Hemicellulose Effect Upon Biodegradable Bamboo Fiber Composites Due to Moisture, Materials Today: Proceedings 4(2, Part A) (2017) 3222-3232. https://doi.org/10.1016/j.matpr.2017.02.208

[70] O.A.Z. Errajhi, J.R.F. Osborne, M. Richardson, H. Dhakal, Water absorption characteristics of aluminised E-glass fibre reinforced unsaturated polyester composites, Composite Structures 71 (2005) 333-336. https://doi.org/10.1016/j.compstruct.2005.09.008

[71] J. Gassan, A.K. Bledzki, Effect of cyclic moisture absorption desorption on the mechanical properties of silanized jute-epoxy composites, Polymer Composites 20 (1999) 604-611. https://doi.org/10.1002/pc.10383

[72] N. Jauhari, R. Misra, H. Thakur, Natural Fibre Reinforced Composite Laminates – A Review, 2015. https://doi.org/10.1016/j.matpr.2015.07.304

[73] Y. Pan, Z. Zhong, A nonlinear constitutive model of unidirectional natural fiber reinforced composites considering moisture absorption, Journal of the Mechanics and Physics of Solids 69 (2014). https://doi.org/10.1016/j.jmps.2014.04.007

[74] D.N. Saheb, J.P. Jog, Natural fiber polymer composites: A review, Advances in Polymer Technology 18(4) (1999) 351-363. https://doi.org/10.1002/(SICI)1098-2329(199924)18:4<351::AID-ADV6>3.0.CO;2-X

[75] H.S. Abdo, R. Abuzade, O. Adekomaya, M.G. Akhil, K.I. Alzebdeh, A.G. Arsha, V. Babaahmadi, A. Balea, A. Bergeret, V. Bermudez, F. Berzin, C.W. Bielawski, K. Bilisik, Á. Blanco, L.G. Blok, R. Brüll, A. Caggiano, A.M.B. da Silva, B.D.S. Deeraj, M. Delgado-Aguilar, D. Devapal, R. Devasia, F.X. Espinach, R. Fangueiro, S. Fathima, D.P. Ferreira, E. Fuente, G. George, W. Gindl-Altmutter, T. Gries, R. Haas, I. Hamerton, P. Huber, I. Improta, J.S. Jayan, K. Joseph, C. Jubsilp, F. Julian, S.V. Kanhere, N.S. Karaduman, Y. Karaduman, K.A. Khalil, H. Lammer, K. Lebelo, M.L. Longana, Z. Lu, S. Magagula, A.R. Mahendran, T. Majozi, V. Manoj, G. Marmol, A.B. Martins, M.J. Mochane, M.L. Mohammed, M. Mohapi, M.C. Monte, P. Mora, T.S. Motsoeneng, P. Mutjé, M. Naeimirad, M.M.A. Nassar, C. Negro, R.E. Neisiany, L. Nele, A.A. Ogale, K. Oksman, H. Oliver-Ortega, B.C. Pai, A. Painuly, F. Pursche, T. Quadflieg, T.P.D. Rajan, R.S. Rajeev, S. Rimdusit, R.M.C. Santana, F. Sarasini, S. Appukuttan, F. Sbardella, J.S. Sefadi, C. Sergi, K.J. Sreejith, M. Sreejith, O. Stolyarov, R.J. Tapper, Q. Tarrés, J. Tirillò, B. Vergnes, R. Wilson, G. Wuzella, Contributors, in: K. Joseph, K. Oksman, G. George, R. Wilson, S. Appukuttan (Eds.), Fiber Reinforced Composites, Woodhead Publishing, pp. xiii-xix, (2021).

[76] H.S. Abdo, M.L. Mohammed, K.A. Khalil, 21 - Fiber-reinforced metal-matrix composites, in: K. Joseph, K. Oksman, G. George, R. Wilson, S. Appukuttan (Eds.), Fiber Reinforced Composites, Woodhead Publishing, pp. 649-667, (2021). https://doi.org/10.1016/B978-0-12-821090-1.00010-7

[77] O. Adekomaya, T. Majozi, 23 - Industrial and biomedical applications of fiber reinforced composites, in: K. Joseph, K. Oksman, G. George, R. Wilson, S. Appukuttan (Eds.), Fiber Reinforced Composites, Woodhead Publishing pp. 753-783, (2021). https://doi.org/10.1016/B978-0-12-821090-1.00004-1

[78] M.G. Akhil, A.G. Arsha, V. Manoj, T.P.D. Rajan, B.C. Pai, P. Huber, T. Gries, 17 - Metal fiber reinforced composites, in: K. Joseph, K. Oksman, G. George, R. Wilson, S. Appukuttan (Eds.), Fiber Reinforced Composites, Woodhead Publishing, pp. 479-513, (2021). https://doi.org/10.1016/B978-0-12-821090-1.00024-7

[79] K.I. Alzebdeh, M.M.A. Nassar, 9 - Polymer blend natural fiber based composites, in: K. Joseph, K. Oksman, G. George, R. Wilson, S. Appukuttan (Eds.), Fiber Reinforced Composites, Woodhead Publishing, pp. 215-239, (2021). https://doi.org/10.1016/B978-0-12-821090-1.00011-9

[80] A. Balea, E. Fuente, M.C. Monte, Á. Blanco, C. Negro, 20 - Fiber reinforced cement based composites, in: K. Joseph, K. Oksman, G. George, R. Wilson, S. Appukuttan (Eds.), Fiber Reinforced Composites, Woodhead Publishing, pp. 597-648, (2021). https://doi.org/10.1016/B978-0-12-821090-1.00019-3

[81] A. Bergeret, 3 - Surface treatments in fiber-reinforced composites, in: K. Joseph, K. Oksman, G. George, R. Wilson, S. Appukuttan (Eds.), Fiber Reinforced Composites, Woodhead Publishing, pp. 47-81, (2021). https://doi.org/10.1016/B978-0-12-821090-1.00020-X

[82] F. Berzin, B. Vergnes, 5 - Thermoplastic natural fiber based composites, in: K. Joseph, K. Oksman, G. George, R. Wilson, S. Appukuttan (Eds.), Fiber Reinforced Composites, Woodhead Publishing, pp. 113-139, (2021). https://doi.org/10.1016/B978-0-12-821090-1.00015-6

[83] K. Bilisik, 18 - Aramid fiber reinforced composites, in: K. Joseph, K. Oksman, G. George, R. Wilson, S. Appukuttan (Eds.), Fiber Reinforced Composites, Woodhead Publishing, pp. 515-559, (2021). https://doi.org/10.1016/B978-0-12-821090-1.00003-X

[84] Front Matter, in: K. Joseph, K. Oksman, G. George, R. Wilson, S. Appukuttan (Eds.), Fiber Reinforced Composites, Woodhead Publishing, pp. i-iii, (2021).

[85] A.M.B. da Silva, A.B. Martins, R.M.C. Santana, 10 - Biodegradability studies of lignocellulosic fiber reinforced composites, in: K. Joseph, K. Oksman, G. George, R. Wilson, S. Appukuttan (Eds.), Fiber Reinforced Composites, Woodhead Publishing, pp. 241-271, (2021). https://doi.org/10.1016/B978-0-12-821090-1.00006-5

[86] R. Devasia, A. Painuly, D. Devapal, K.J. Sreejith, 22 - Continuous fiber reinforced ceramic matrix composites, in: K. Joseph, K. Oksman, G. George, R. Wilson, S. Appukuttan (Eds.), Fiber Reinforced Composites, Woodhead Publishing, pp. 669-751, (2021). https://doi.org/10.1016/B978-0-12-821090-1.00022-3

[87] S. Fathima, B.D.S. Deeraj, S. Appukuttan, K. Joseph, 12 - Carbon fiber and glass fiber reinforced elastomeric composites, in: K. Joseph, K. Oksman, G. George, R. Wilson, S. Appukuttan (Eds.), Fiber Reinforced Composites, Woodhead Publishing, pp. 307-340, (2021). https://doi.org/10.1016/B978-0-12-821090-1.00005-3

[88] J.S. Jayan, S. Appukuttan, R. Wilson, K. Joseph, G. George, K. Oksman, 1 - An introduction to fiber reinforced composite materials, in: K. Joseph, K. Oksman, G. George, R. Wilson, S. Appukuttan (Eds.), Fiber Reinforced Composites, Woodhead Publishing2021, pp. 1-24. https://doi.org/10.1016/B978-0-12-821090-1.00025-9

[89] H. Oliver-Ortega, F. Julian, F.X. Espinach, Q. Tarrés, M. Delgado-Aguilar, P. Mutjé, 6 - Biobased polyamide reinforced with natural fiber composites, in: K. Joseph, K. Oksman, G. George, R. Wilson, S. Appukuttan (Eds.), Fiber Reinforced Composites, Woodhead Publishing, pp. 141-165, (2021). https://doi.org/10.1016/B978-0-12-821090-1.00008-9

[90] M. Sreejith, R.S. Rajeev, 25 - Fiber reinforced composites for aerospace and sports applications, in: K. Joseph, K. Oksman, G. George, R. Wilson, S. Appukuttan (Eds.), Fiber Reinforced Composites, Woodhead Publishing, pp. 821-859, (2021). https://doi.org/10.1016/B978-0-12-821090-1.00023-5

[91] A. Orue, A. Eceiza, C. Peña-Rodriguez, A. Arbelaiz, Water Uptake Behavior and Young Modulus Prediction of Composites Based on Treated Sisal Fibers and Poly(Lactic Acid), Materials (Basel) 9(5) (2016) 400. https://doi.org/10.3390/ma9050400

[92] S. Sanjeevi, V. Shanmugam, S. Kumar, V. Ganesan, G. Sas, D.J. Johnson, M. Shanmugam, A. Ayyanar, K. Naresh, R.E. Neisiany, O. Das, Effects of water absorption on the mechanical properties of hybrid natural fibre/phenol formaldehyde composites, Scientific Reports 11(1) (2021) 13385. https://doi.org/10.1038/s41598-021-92457-9

[93] P. Benard, E. Kroener, P. Vontobel, A. Kaestner, A. Carminati, Water percolation through the root-soil interface, Advances in Water Resources 95 (2016) 190-198. https://doi.org/10.1016/j.advwatres.2015.09.014

[94] H.N. Dhakal, Z.Y. Zhang, M.O.W. Richardson, Effect of water absorption on the mechanical properties of hemp fibre reinforced unsaturated polyester composites, Composites Science and Technology 67(7) (2007) 1674-1683. https://doi.org/10.1016/j.compscitech.2006.06.019

[95] A. Espert, F. Vilaplana, S. Karlsson, Comparison of water absorption in natural cellulosic fibres from wood and one-year crops in polypropylene composites and its influence on their mechanical properties, Composites Part A: Applied Science and Manufacturing 35(11) (2004) 1267-1276. https://doi.org/10.1016/j.compositesa.2004.04.004

[96] C.M. Hansen, A. Björkman, The Ultrastructure of Wood from a Solubility Parameter Point of View, 52(4) (1998) 335-344. https://doi.org/10.1515/hfsg.1998.52.4.335

[97] S. Kalia, A. Dufresne, B.M. Cherian, B.S. Kaith, L. Avérous, J. Njuguna, E. Nassiopoulos, Cellulose-Based Bio- and Nanocomposites: A Review, International Journal of Polymer Science 2011 (2011) 837875. https://doi.org/10.1155/2011/837875

[98] N. Sgriccia, M.C. Hawley, M. Misra, Characterization of natural fiber surfaces and natural fiber composites, Composites Part A: Applied Science and Manufacturing 39(10) (2008) 1632-1637. https://doi.org/10.1016/j.compositesa.2008.07.007

[99] N. Soykeabkaew, N. Arimoto, T. Nishino, T. Peijs, All-cellulose composites by surface selective dissolution of aligned ligno-cellulosic fibres, Composites Science and Technology 68(10) (2008) 2201-2207. https://doi.org/10.1016/j.compscitech.2008.03.023

[100] Y. Wang, Q. Wei, S. Wang, W. Chai, Y. Zhang, Structural and water diffusion of poly(acryl amide)/poly(vinyl alcohol) blend films: Experiment and molecular dynamics simulations, Journal of Molecular Graphics and Modelling 71 (2017) 40-49. https://doi.org/10.1016/j.jmgm.2016.11.001

[101] T. Mokhothu, M. John, Bio-based coatings for reducing water sorption in natural fibre reinforced composites, Scientific Reports 7 (2017). https://doi.org/10.1038/s41598-017-13859-2

[102] Index, in: K. Joseph, K. Oksman, G. George, R. Wilson, S. Appukuttan (Eds.), Fiber Reinforced Composites, Woodhead Publishing, pp. 861-883, (2021). https://doi.org/10.1016/B978-0-12-821090-1.09999-3

[103] V. Chaudhary, P. Bajpai, S. Maheshwari, Effect of moisture absorption on the mechanical performance of natural fiber reinforced woven hybrid bio-composites, Journal of Natural Fibers 17 (2018) 1-17. https://doi.org/10.1080/15440478.2018.1469451

[104] J. Sudeepan, A Review of Chemical Treatments on Natural Fibers-Based Hybrid Composites for Engineering Applications, in: K. Kumar, J.P. Davim (Eds.), Composites and Advanced Materials for Industrial Applications, IGI Global, Hershey, PA, USA, 2018, pp. 16-37. https://doi.org/10.4018/978-1-5225-5216-1.ch002

[105] B.P. Chang, A.K. Mohanty, M. Misra, Studies on durability of sustainable biobased composites: a review, RSC Advances 10(31) (2020) 17955-17999. https://doi.org/10.1039/C9RA09554C

Sustainable Natural Fiber Composites
Materials Research Foundations **122** (2022) 154-198

Materials Research Forum LLC
https://doi.org/10.21741/9781644901854-7

Chapter 7

Resol–Vegetable Fibers Composites

Wei Ni[1,3,4,a], Lingying Shi[2,b]

[1]State Key Laboratory of Vanadium and Titanium Resources Comprehensive Utilization, PANGANG Group Research Institute, ANSTEEL Research Institute of Vanadium & Titanium (Iron & Steel), Chengdu 610031, China

[2]College of Polymer Science and Engineering, State Key Laboratory of Polymer Materials Engineering, Sichuan University, Chengdu 610065, China

[3]Vanadium and Titanium Resource Comprehensive Utilization Key Laboratory of Sichuan Province, Panzhihua University, Panzhihua 617000, China

[4]Material Corrosion and Protection Key Laboratory of Sichuan Province, Sichuan University of Science and Engineering, Zigong 643000, China

[a] niwei@iccas.ac.cn, max.ni@hotmail.com, [b] shilingying@scu.edu.cn

Abstract

In this chapter, we provide a detailed review on the reinforcement effect of a broad range of vegetable fibers (VFs) in thermosetting phenolic resin (resol type) based composites. The different varieties of VFs, surface modification techniques, processing conditions and their effects on the overall performances (i.e., the strengths and weaknesses), are assessed to advance future studies and applications of natural fiber-reinforced plastic composites.

Keywords

Vegetable Fibers, Natural Fibers, Resol, Phenolic Resins, Thermosets, Composites, Reinforcement

Contents

Resol–Vegetable Fibers Composites..**154**

1.   **Introduction**..**155**

2.   **Properties, mechanisms, and characterizations** **157**

      2.1   Fibers content, distribution, and configuration 158

      2.2   Interfacial adhesion and chemical modifications 161

2.3   Mechanical properties 164

2.3.1 Tensile, impact, and flexible strength 165

2.3.2 Hardness 167

2.3.3 Friction and wear properties 167

2.4   Other properties 168

2.5   Characterizations 171

2.6   Theoretical modeling 171

2.7   Processing techniques 172

3.   Categories and applications 173

3.1   Cellulosic/lignocellulosic fibers 173

3.2   Combined fibers 174

3.2.1 Natrual/natural fibers 175

3.2.2 Natural/synthetic fibers 176

3.3   Foams 176

3.4   Applications 177

Conclusions and perspectives 178

Acknowledgements 179

References 180

1.   Introduction

Natural composites, or called green composites, have wide applications in many fields including structural and non-structural applications such as infrastructure, automotive body panels, owing to the merits of natural fibers such as biodegradability, low cost, low weight, environmental friendliness, good specific mechanical properties, easier handling, and lower production energy. They are regarded as an economical and renewable alternative, compared to most synthetic/manmade polymeric and inorganic fibers, e.g., steel, glass, carbon or mineral fibers [1-10]. Robust vegetable fibers (VFs, well known as natural fibers or biofibers), with higher mechanical advantages and beyond than polymeric matrices, are expected to develop sustainable structural composites with high performance for scientific and industrial communities [6,11-15,1]. Furthermore, compared to conventional composites, vegetable fiber-reinforced polymer composites have been intensively studied in recent years owing to their significant mechanical performance (e.g., high moduli, higher

Materials Research Foundations **122** (2022) 154-198

https://doi.org/10.21741/9781644901854-7

specific strength), low densities, economic/eco-friendly benefits than synthetic fibers, wide applications, and growing global market size (over USD 4.46 billion in 2016) [16-24].

Resol, a curable phenolic resin or phenol-formaldehyde resin (PF), is the first commercial synthetic resin and a typical thermoset polymer (i.e., a highly crosslinked dense polymer via base-catalyzed step-growth polymerization of phenol with formaldehyde; molar ratio F:P > 1, usually around 1.5), beyond epoxy and unsaturated polyester as well as thermoplastic novolak (or novolac) [25-30]. Vegetable fibers, including various types such as bast, leaf, seed or fruit, straw, grass, and wood, have been adopted to reinforce the polymeric matrix (Fig. 1) [1,15]; and resol, as a matrix material, is somewhat different to thermoplastics and has its own unique merits (e.g., excellent mechanical strength, dimensional/thermal/chemical stability, heat insulation, flame resistance, high rigidity, durability, and exceptional adhesive properties) [11,31,26,32,33,27,29,34]. These vegetable fibers include wood [35], bamboo [36-38], cotton [11], flax, kenaf [39], hemp, sisal [11,3,40], bagasse [11,41], and their microfibrillated celluloses (MFC, or cellulose nanofibers), *etc.* [42,43,21]

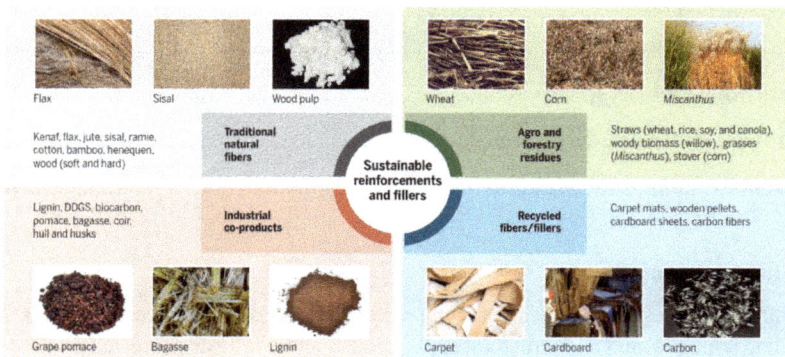

Fig. 1. *Fibers and fillers from renewable and sustainable resources. Reproduced with permission from American Association for the Advancement of Science 2018 (Ref. [1]).*

Many effects have been made to overcome several typical drawbacks of natural composites including the higher moisture adsorption, lower mechanical properties and durability, inferior fire resistance, quality variation, incompatible with established manufacturing practices and long-term durability [2,44]. And by particular surface treatment of natural fibers and the improvement of fiber–matrix interfacial interactions, most concerns have been addressed, the extended application of natural fiber composites including VFs–resol

Sustainable Natural Fiber Composites   Materials Research Forum LLC
Materials Research Foundations **122** (2022) 154-198   https://doi.org/10.21741/9781644901854-7

have promoted the concept of sustainability [2,4,44-46]. Moreover, the technical processing for these vegetable fiber-reinforced polymer composites is also improved.

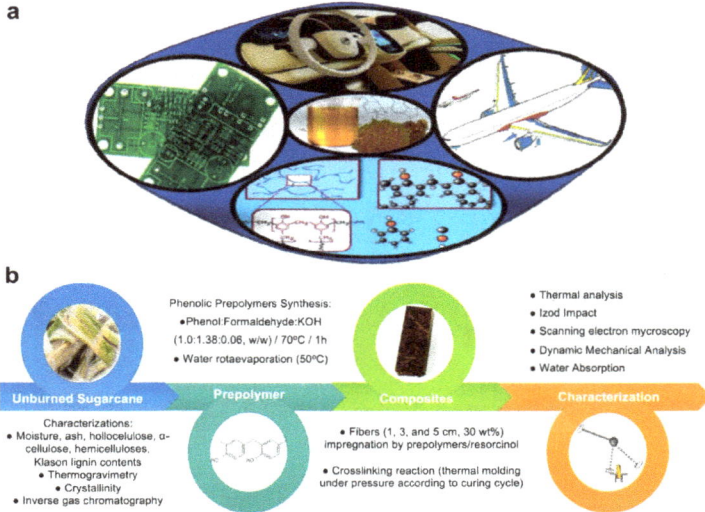

Fig. 2. *(a) Structures and applications of phenolic resin and its fiber-reinforced composites. (b) Information regarding the synthesis, preparation, and characterization of vegetable fiber (unburned sugarcane bagasse) reinforced phenolic thermoset composites. a, Reproduced with permission from Bentham Science Publishers 2018 (Ref. [29]); b, Reproduced with permission from Springer Science+Business Media, LLC, part of Springer Nature 2020 (Ref. [41]).*

2.   Properties, mechanisms, and characterizations

Resol is referred to be a "one-step" resin as it cures without a crosslinker unlike the "two-step" resin novolak; the high crosslinking (3D network) gives it many advantages including hardness, good thermal stability, chemical imperviousness, and biodegradation [25,47,29]. Unfortunately, resol is usually brittle/fragile thermosetting plastic (i.e., low impact strength, related to the high crosslinking density) [48,26]. However, its specific mechanical performances and beyond could now be significantly reinforced with various fibers, of which the natural vegetable fibers (generally lignocellulosic and comprised of microfibrils with oriented cellulose in a hemicellulose and lignin matrix) are of tremendous and increasing interest [49,11,29]. These typical VFs including palm fibers (oil, sugar, or date palm fiber) [49-55], cotton [11,56], sisal [11,57,40,58-62], jute [63], ramie [64], sugarcane

Materials Research Forum LLC

https://doi.org/10.21741/9781644901854-7

bagasse [11,48,41,20], pineapple leaf fiber [33,65,39], grass fiber [66], and wood fibers [67,68,35]. The low-cost and fire-resistant engineering materials made of cheap resol and VFs could be obtained for structural materials and other functional applications (Fig. 2) [11,29].

The final performances of the composites are obviously controlled by the properties and quantities of the related component materials and the characteristics of the interfacial region between the matrix (resol) and the reinforcement (VFs) (Fig. 3) [11,69,70]. These related aspects are detailed as follows.

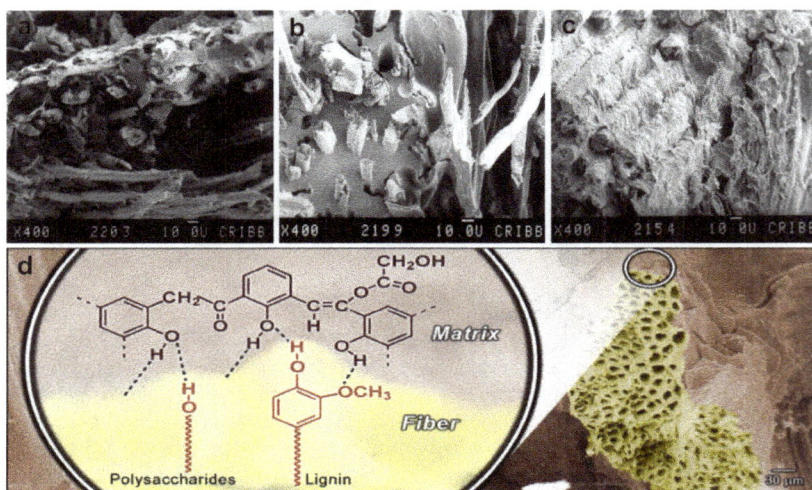

Fig. 3. *(a–c) Cross-sectional SEM images of resol/raw cotton, resol/clean cotton, and resol/bagasse composites with ~50% fiber volume fraction (SEM ×400), revealing well-adhered broken fibers or pulled-out fibers/bundles from the fracture surface. (d) Schematic illustration of the interactions between the polar groups of lignocellulosic fiber (sisal) and glyoxal–phenolic resin matrix using an SEM image. a–c, Reproduced with permission from John Wiley & Sons, Inc. 2000 (Ref. [11]); d, Reproduced with permission from Elsevier Ltd. 2009 (Ref. [191]).*

2.1   Fibers content, distribution, and configuration

The lignocellulosic VFs have their own inherent mechanical properties, of which the angle between the axis of VF and the microfibrils (subunits) has a notable effect, i.e., the smaller the angel, the higher the elastic modulus and resilience (or the smaller elongation at break),

Materials Research Forum LLC
https://doi.org/10.21741/9781644901854-7

thus for a generally enhanced performance [11]. And the VFs with higher crystallinity (e.g., cotton fiber vs. lyocell fiber) usually show elevated mechanical performance [71].

The fiber length and fiber loading have an effect on the mechanical properties of the resol composites, e.g., the tensile, flexural, and impact properties *etc.* [49,67,72,71,36,73,62,55] Specifically, very short fibers will have few contact points with the resol matrix, which may lead to a pull-out mechanism (i.e., debonding from the matrix), thus the reinforcement is insignificant or even deteriorated. On the contrary, very long fibers may be more difficult for a homogeneous distribution (i.e., resulting in agglomeration or entanglement), thus medium length and/or better arrangement are required for desired mechanical performances [48,41,74,75,69,63,62]. Usually, the fiber volume fraction (V_f) has a volcano trend on the mechanical properties, i.e., at a medium or optimized V_f value, the mechanical properties such as flexural strength and modulus are maxima (e.g., some typical lignocellulosic fiber-reinforced resol composites) [11,76,67,35,72,50,63,73]. For example, the tensile, compressive, and flexural strengths of fiber-reinforced resol composites with 10 wt.% and 30 wt.% fiber loading (i.e., lignocellulosic *Saccharum cilliare*) are around 2 times and 3 times that of resol, while the composites with 40 wt.% fiber loading show lowered mechanical strengths than that with 30 wt.% fiber loading. Because the much higher fiber loading may increase the agglomeration (or fiber–fiber contact and difficulty of impregnation, or higher content of voids), and therefore hinder the stress transfer from matrix to reinforcing fiber, resulting in deterioration of mechanical performances [76,77] [62]. However, there are also some exceptions, e.g., some VFs such as sisal/resol and lyocell/resol composites and some ordered/woven fiber composites exhibit a continuous rise in these mechanical indices as the V_f increases up to a quite high value (e.g. > 70%) [11,71,62,38].

The stacking sequence of fiber layers, i.e., the pattern of arrangement, in the (hybrid) composites also have effect on the mechanical behaviors [74]. Optimized fibers with unidirectional "stitched" fabrics (i.e., the orientation of the fibers beyond the interfacial interaction and the nature of the matrix) are of great advantage for enhanced mechanical properties [6]. The mechanical performances along the fiber orientation are usually linearly enhanced (to a certain maximum with the increase of V_f), and the optimized properties are found when placing high strength fibers as skin layers [74].

Furthermore, the diameter of VFs also greatly affects the mechanical performances. By splitting the fibers into finer filaments (i.e., reduction in fiber diameter), macro VFs may evolve into microfibers or nanofibers, e.g., via steam explosion and acid hydrolysis [78,79,60]. The isolated cellulose nanofibers show the highest surface acidity (or surface area), best thermal stability and optimized mechanical properties including tensile, flexural and impact strengths (i.e., increase by 142%, 280%, and 133%, respectively, compared to that of resol matrix). That is to say, a competitive composite with desired strength could be attained by incorporation of a smaller volume fraction of nanofibrils (Fig. 4 and Table 1) [78]. These microfibrillated cellulose (MFC) or cellulose nanofibers (with extremely rigid crystalline regions holding a Young's modulus as high as 138 GPa [80]) based resol (nano)composites have showed greatly enhanced mechanical, e.g., toughness, bending/flexural/impact strength, than the macrofibers, untreated pulp fiber and even synthetic aramid/glass fiber composites, owing to many advantages including enhanced fiber ductility and better interfacial adhesion. And they may demonstrate a wide variety of other applications [81-83,79,84,60].

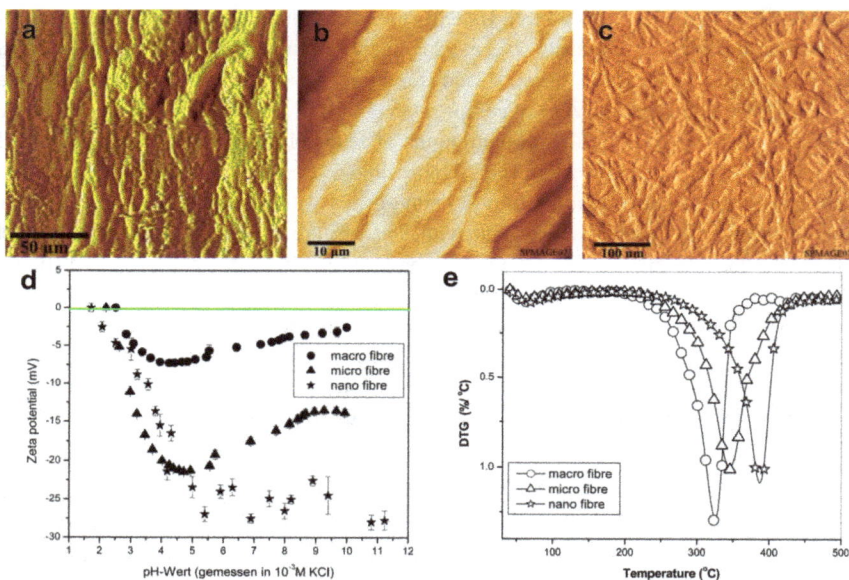

Fig. 4. *(a–c) AFM images of macro banana fiber, micro banana fiber, and nano banana fiber, respectively. (d) pH dependence of zeta potential of macro, micro and nanocellulose fibers. (e) Derivative thermogravimetry (DTG) curves of macro, micro and nanocellulose fibers. Reproduced with permission from Wiley Periodicals, Inc. 2013 (Ref. [78]).*

Table 1. *The Effect of fiber diameter (macro-, micro- and nanocellulose) on the mechanical properties of banana fiber (BF) reinforced phenol-formaldehyde (PF) composites with different fiber loading (wt.%).*

		Tensile Strength (MPa)	Flexural Strength (MPa)	Impact Strength (kJ m^{-2})
	PF	7±2	10±2	12±1.8
	R-BF	540±9	/	/
Macrocellulose reinforced PF	10 % fiber	13±1.6	25±1.9	13±2
	20 % fiber	16±2.7	34±2.3	17±1.9
Microcellulose reinforced PF	10 % fiber	14±3.2	20±1.7	14±1.8
	20 % fiber	18±3	25±2.3	20±1.9
Nanocellulose reinforced PF	10 % fiber	18±1.9	36±2.2	25±1.8
	20 % fiber	17±1.5	38±1.8	28±2.2

Note: R-BF = raw banana fiber (macrocellulose, diameter 80±2 μm), PF = phenol-formaldehyde resol type resin.

2.2   Interfacial adhesion and chemical modifications

The character of interfacial region, besides the inherit properties of each component materials, greatly affects the final properties of the composites [11]. Some resins inherently have good affinity with the reactive groups of VFs, thus a superficial treatment to the fibers is less necessary to improve the adhesion [11,46]. Resol has a very strong hydrogen bonding and even chemical bonding with lignocellulosic VFs (e.g., serving as excellent adhesives via forming chemical bonds with the phenol-like lignin component), thus it is more attractive than the use of cellulose fiber composites due to the much better interfacial adhesion [11,27,26]. It is noteworthy that the interfacial sheer strength (ISS) values are higher for VFs (e.g., banana fiber) in resol than that for glass fiber, indicating a strong adhesion between the lignocellulosic fiber and resols [69].

Most VFs are derived from lignocellulosic raw materials, and these surface modifications mainly include alkali treatment (or called mercerization) [48,81,85], silane treatment [86-88], chlorine dioxide treatment (i.e., delignification of fibers) [61], acetylation, grafting reaction and graft copolymerization [89-94], and lignin modification [95,96], as well as physical treatments (e.g., ionized air, plasma, ozone, thermal treatment, steam explosion, adsorption of compatibilizer/modifier) [20,3,97,98], *etc.* [99-102,24,63,46] Mercerization (moderate) is a well-known strategy to alter the surface morphology and chemical constituents of VFs, the interfacial compatibility, and the mechanical performance of VFs–resol composites [103,81,104,63,52,53,38]. The hydrophobic organosilane treatment (i.e., silanization), especially using silanes with higher number of polar groups to form potential strong covalent bonding with hydroxyl groups of VFs, will directly affect the physicochemical properties (Fig. 5). It potentially improves the performances of VFs–resol composites related to (decreased) moisture absorbance, water uptake, solvent absorption, thermal conductivity, and (increased) chemical resistance beyond the enhanced mechanical properties including interfacial compatibility/adhesion [86,103,105-107,72,108-111,87,88]. The combination of two or more treatments such as mercerization/silanization, or enzyme/alkali treatment (Fig. 6) could have a synergistic effect for enhanced mechanical performances [109-113]. Generally, the modified fibers with closer/similar dispersive components and acid–base characteristics to that of resol will have better fiber–matrix interaction at the composite interface and the improved mechanical properties thereof [95,114]. The drawbacks of VFs like moisture absorption, low ageing/weathering resistance, and hydrophilicity that weakens fiber/hydrophobic-matrix interface may be simultaneously overcome [115,63,52].

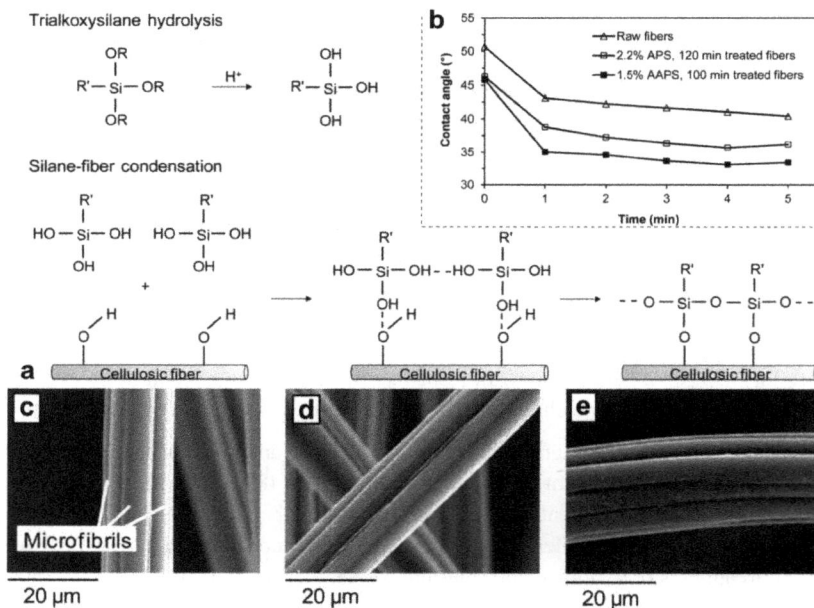

Fig. 5. *(a) Chemical treatment of cellulosic fibers with silanes. (b) Contact angles for the raw cellulosic fibers, fibers treated with 2.2% APS for 120 min, and fibers treated with 1.5% AAPS for 100 min. SEM images of the raw and the treated viscose cellulosic fibers: (c) raw fiber; (d) 2.5% APS silane, 105 min; (e) 1.7% AAPS silane, 105 min. APS= (3-aminopropyl) trimethoxysilane, AAPS = 3-(2-aminoethylamino) propyltrimethoxysilane. The treatment of the fibers with silanes reduces the contact angle between cellulosic fibers and phenolic resin due to the improvement in fiber wettability. Reproduced with permission from Wiley Periodicals, Inc. 2015 (Ref. [106]).*

Sustainable Natural Fiber Composites   Materials Research Forum LLC
Materials Research Foundations **122** (2022) 154-198   https://doi.org/10.21741/9781644901854-7

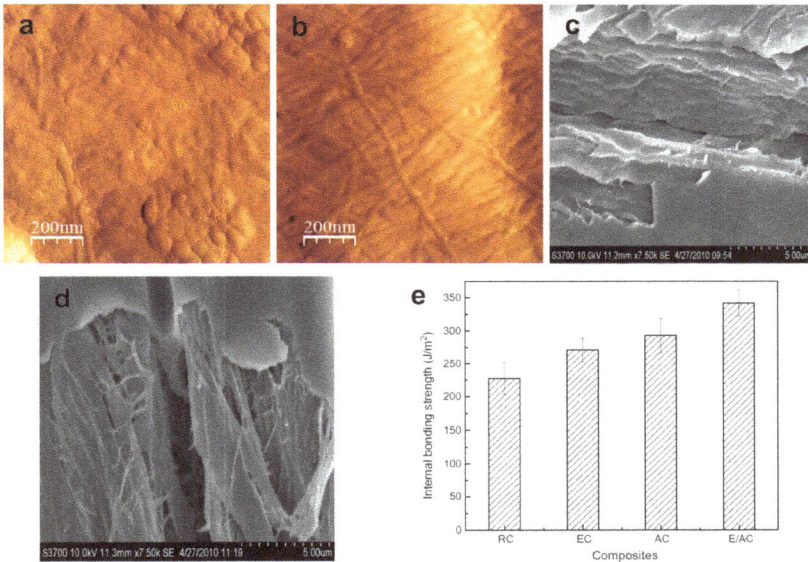

Fig. 6. *(a and b) AFM images of sisal fibers (SFs) surfaces: the raw pulp fiber (RF) and the enzyme (laccase)/alkali-modified fiber (E/AF). (c and d) SEM images of the tensile failure of fibers: RF and E/AF. (e) Effect of treatments on the internal bonding strength of composites. The E/AF shows a stronger interfacial strength between resin and fiber benefited from the removal of surface lignin. Reproduced with permission from Elsevier Ltd. 2010 (Ref. [112]).*

Furthermore, some derived resols are also synthesized to extend the family of VFs-based thermoset composites, such as lignophenolic resins/VFs (i.e., lignin, hydroxymethylated lignin, or lignosulfonate–formaldehyde resins reinforced by bagasse, sisal, wood, flax/kenaf, or flax/wood fiber) (Fig. 7) [20,116-121], phenol–furfural/sisal [122], glyoxal–phenolic resin/sisal [123], tannin–phenolic resin/sisal or coir fiber [77,124], cashew nut shell liquid (CNSL)–formaldehyde resin/sisal [125], vegetable oil–phenolic resin/lignocellulosic fibers (Fig. 8) [126], and urea–formaldehyde/rice straw [127]. These abundant renewable monomers/precursors obtained from agricultural or forestry waste residues are more attractive for decent composites avoiding the use of formaldehyde.

Sustainable Natural Fiber Composites
Materials Research Foundations **122** (2022) 154-198

Materials Research Forum LLC
https://doi.org/10.21741/9781644901854-7

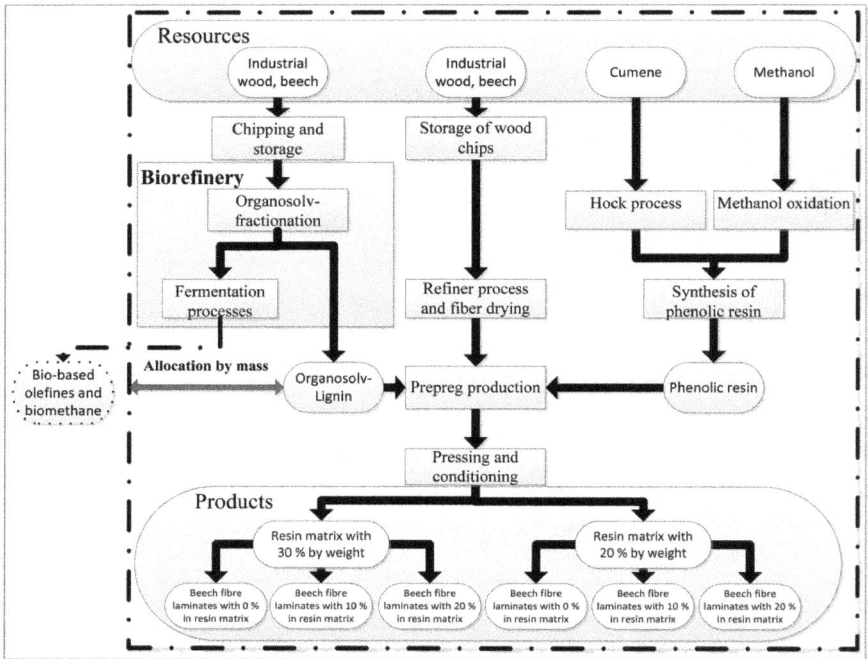

Fig. 7. *System boundaries and alternative product compositions (key unit processes) of the life cycle assessment for assessing the technical and environmental performance of beech wood-based fiber laminates with lignin-based phenolic resin systems, for example. Reproduced with permission from Elsevier B.V. 2018 (Ref. [120]).*

2.3 Mechanical properties

Since the VFs have notable higher specific mechanical strengths than that of resol and most polymers besides their other significant merits abovementioned, these VFs are thus utilized as ideal filler and reinforcement in resol and other polymeric matrices [11]. Not only the VFs, but also the surface conditions of VFs, i.e., the interfacial adhesion between the fiber and resol matrix, greatly impact the mechanical performance of the composites [11]. A good interfacial adhesion can endow the composite with higher mechanical values of strength and stiffness, *etc.* [11,128]

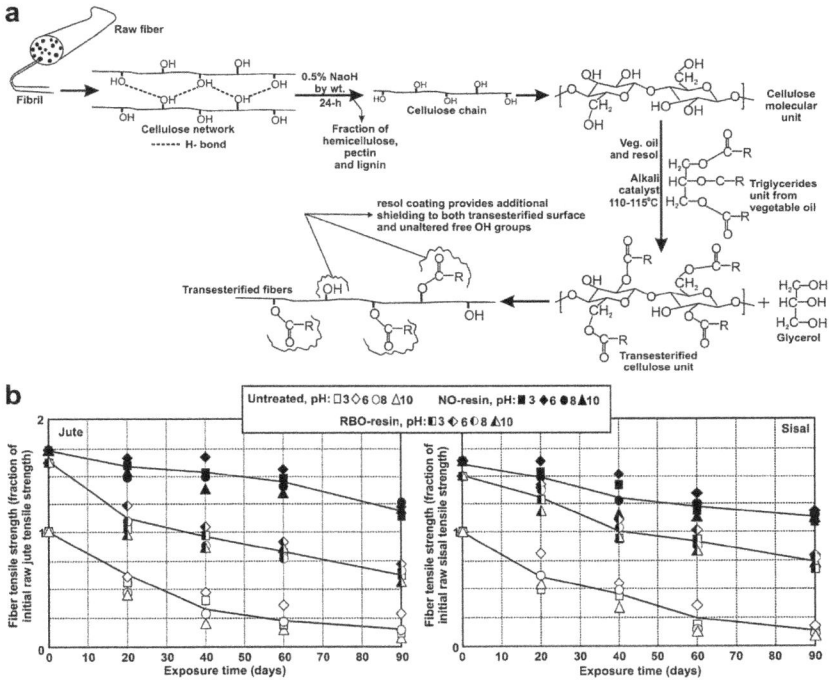

Fig. 8. *(a) Schematic illustration of probable chemical alteration (transesterification) of fibers during chemical process. (b) Durability/Degradation of tensile strength of lignocellulosic fibers (LCFs; jute and sisal, untreated and treated, alkali and vegetable oil–phenolic resin treatment) upon exposure to different pH environment (pH 3–10) during 90 days. NO–resin = neem oil–phenolic resin, RBO–resin = rice bran oil–phenolic resin. Reproduced with permission from Elsevier Ltd. 2011 (Ref. [126]).*

2.3.1 Tensile, impact, and flexible strength

The Young's modulus (E) and tensile stress (σ) of VFs are usually significantly higher than those of resol (e.g., $E_{hemp\ fiber} = 35$ GPa, $E_{flax\ fiber} = 58$ GPa, $E_{stinging\ nettle\ fiber} = 87$ GPa, $E_{banana\ fiber} = 29$ GPa, $E_{wood} = 11$ GPa; $\sigma_{cotton\ fiber} \approx 539$ MPa, $\sigma_{sisal\ fiber} \approx 548$ MPa, $\sigma_{banana\ fiber} \approx 540$ MPa, $\sigma_{bamboo} \approx 350–500$ MPa, $\sigma_{pine} \approx 40$ MPa; $E_{resol} \approx 0.2$ or $2–4$ GPa, and $\sigma_{resol} \approx 7$ or $20–40$ MPa. *Note*: some data may vary according to different references) [11,129-133,78,108,128,134,10,70,135,102,136]. The Young's modulus (or elastic modulus, essentially the stiffness to resist bending and stretching), as well as the strength of resol, is reinforced by incorporation of VFs. An efficient criterion ($E^{1/3}/\rho$, ρ is density of material) even shows that the typical resol–VFs composites are much better than steel and polyethylene and can compete with materials such as aluminum and magnesium, exhibiting great advantages in application where weight (or low density) is an important factor [137,11,128].

Ideally, tensile modulus (Young's modulus) or compressive modulus of elasticity is equivalent to the flexural or bending modulus of elasticity; in reality, these values may be somewhat different, especially for polymers [138,132,69]. For resol–VFs composites, the flexural moduli and strengths depend mainly on the degree of disorder (or uneven distribution) of the VFs under the same conditions. The resol–VFs (e.g., cotton fiber, bagasse) will have a maximum value at a medium fiber volume fraction, because the higher disorder (at higher loads) results in incomplete wetting of fibers and increase the void content and fiber–fiber contact in the composites [11]. However, for the highly ordered sisal fibers, the resol–sisal fiber composites show a positive correlation with the fiber fraction, due to the better wetting features of the non-woven mats made of highly entangled long sisal fibers [11].

Generally, the introduction and the increase of VFs in the resol–VFs composite will enhance the impact strength of the composite (positive increase as a function of the volume fraction of VFs), due to the merits of VFs; the fibers in the thermosetting polymer suffer a break (rather than a pull-out mechanism) during the impact, revealing a good interfacial adhesion [48]. To be more specific, the reinforcement of the adhesion at the fiber–matrix interface will increase the impact strength due to the enhanced storage modulus and loss modulus, provided by the higher moduli of VFs (> resol) as well as the inhibition of the relaxation process in the composite (usually leading to wider and slightly lower glass transition temperature peak, T_g, for the composites owing to the suppressed crosslinking by VFs and the structurally heterogeneous feature) [48,41,128,139]. Thus, a synergic implication (i.e., balance) between these properties would be an ideal resolution. Overall, these mechanical performances, including tensile, compressive, and flexural strength, *etc.*, are usually simultaneously enhanced via the incorporation of VFs [68].

Furthermore, the modification of the fiber–matrix interaction at the interface would enhance the overall mechanical properties of the composites [33,95,94]. For example, the saline treatment of VFs is an efficiency way to enhance the flexural strength/modulus, impact strength (energy adsorption), and/or tensile strength/modulus [33]. The mercerization (or called alkali treatment) of VFs will significant increase the impact strength of resol–VFs composites (e.g., increased by 72% compared to original resol) [48]. However, it should be mentioned that in some cases the structure would be partially degraded by the chemical treatment of VFs (e.g., via oxidation, or overtreatment), thus decreasing the tensile strength and/or impact strength of the composites reinforced with them [140,89,63]. So optimization of fiber surface modification should be particularly considered to intensify the fiber–matrix interaction without deterioration of mechanical properties [140]. In addition, some other components such as inorganic/organic fillings may also be used to tune the mechanical, electrical, thermal, or water resistance properties, e.g., silicone-containing monomers, polymers [141,142].

2.3.2 Hardness

The incorporation of VFs into brittle and hard thermosetting resol matrix, however, usually decreases the hardness of the composites despite that the normal mechanical properties, e.g., the impact strength, are enhanced [48,134]. The decreased hardness with the increasing fiber percentage could be a consequence of the hygroscopicity of VFs, i.e., the absorbed water molecules on the surface of VFs act as a kind of plasticizer; and the larger the fiber fraction in the composite, the higher humidity adsorption and the lower hardness [48]. And for the hybrid composites (e.g., glass-fiber/OPEFB-fiber/resol composites, OPEFB = oil palm empty fruit bunch), the hardness usually follows the same trend as the reducing density, i.e., the increasing ratio of VFs decreases the hardness [134]. It should be mentioned that there are also a few exceptions [143].

2.3.3 Friction and wear properties

VFs-reinforced friction composites are attracting increasing interest due to the specific mechanical properties (including tribological behavior), light weight, low cost, eco-friendliness and renewable nature of their ingredients [107,144]. The incorporation of VFs will usually enhance the wear resistance (or reduce the specific abrasion), although lower than that of commercially available brake pads [143]. Furthermore, tribological behavior of the composite is temperature-dependent, that is, the introduction of VFs (e.g., sisal) into resol usually does not dramatically alter the friction coefficient and wear rate until *ca.* 200 °C; above that temperature (especially > 250 °C), with the increase of fiber content, the friction coefficient is significantly decreased (due to the formation of carbon powder and some film on the friction surface), and the wear rate remarkably increased (due to the thermal decomposition and weakened interfacial adhesion) [145,146].

The friction and wear properties of resol–VFs composites can be further improved by surface treatment (to enhance interfacial adhesion via alkali/borax treatment, e.g.) or by incorporation of specific functional components (e.g., nanosilica, $BaSO_4$, and polysulfone) [147,145,148,149,146,51,150]. By adding polysulfone the tribological performance of the composites (e.g., sisal fiber/polysulfone/phenolic resin) is improved along with the heat resistance, showing a smoother worn surface with 5% polysulfone in the composites [147].

2.4   Other properties

Density. The density of VFs is usually slightly higher than that of resol, however, it does not definitely mean that the higher VFs fraction, the higher density for the resol–VFs composites. Because the measured density of the composites are usually lower than the theoretical values owing to the presence of voids, and the higher VFs fraction may simultaneously render more voids due to the incomplete wetting [11].

Shrinkage. The incorporation of VFs into resol matrix induces a slight increase in the shrinkage of the molded piece (e.g., resol/40–70% bagasse fiber composites show a shrinkage of 0.54% compared to that of 0.40% for resol), probably due to the interfacial interaction during the cure process [48]. And fiber treatment such as thermal treatment can reduce the standard deviation and the coefficient of variation due to lowered moisture content, favoring the production of structural resol composites with high precision [3].

Water absorption (or moisture resistance). The resol–VFs composites usually render higher water absorption and thickness swelling compared to the original resol (i.e., the higher the fraction of hydrophilic lignocellulosic VFs, the higher the water absorption), despite that the larger fiber proportion presents better mechanical performances such as impact strength [48,151,65]. The hydrophilic nature of VFs could be reduced and better adhesion could form between fiber and phenolic matrix/binder via chemical/surface modification of the fiber (e.g., alkaline treatment and/or silanization process, esterification, graft copolymerization); the water/solvent/chemical resistance may simultaneously be improved [107,103,105,90,95,152,91,153,97,126,66]. The composites with longer VFs, however, show less water absorption, which will be advantageous for applications when exposed to moisture [41]. The addition of a small quantity of inorganic nanoclay could further reduce the water absorption and improve the mechanical properties such as flexural strength and modulus [154,155].

Environmental ageing. The composite materials may be subject to various conditions such as thermal ageing, water ageing, soil burial, and outdoor weathering, resulting in an adverse influence of properties [156,152]. The higher the relative humidity, the higher the water absorption and the more susceptible the mechanical properties, due to the extra moisture absorbed by cracks and interphases through the swelling of fibers [157,64]. And the degradation of properties would usually level off over exposure time (room temperature, 50–98% relative humidity during 6 months), indicating a low degree of permanent degradation occurred due to exposure [157]. Generally, the chemical modification of fibers reduces the degradability of the resol–VFs composites, and the reduced diameter of VFs (from macrofiber, through microfibril, to nanofibril) improves the environmental ageing of the reinforced polymer composites, i.e., becomes more resistant to all ageing processes

[75,152]. Developing optimal hydrophobic nature at the fiber–matrix interface and a resin-rich layer on the external surface of composite can be an alternative solution for wet environmental applications [158]. However, the environmental ageing is not always negative for the VFs-reinforced composites, for example, the water ageing of banana fiber/resol composites shows an enhanced tensile strength up to 25%, due to the enhanced stress transfer at the fiber–matrix interface via the swelling of VFs; while for glass fiber-reinforced composites, it could be an opposite effect [152]. In addition, the degradation in mechanical performances is mainly dependent on the exposure time and temperature related to various environmental conditions [74,156]. Some UV stabilizers may also be added to enhance the outdoor performance but result in compromised strength, so a balance between strength and durability requirements should be considered [159].

Thermal decomposition and fire resistance. Resol type phenolic thermosets are well resistant to high temperatures and can generate high ratio of char during pyrolysis, while the VFs (usually lignocellulosic) show somewhat lower thermal decomposition temperatures (*ca.* 300 °C), resulting in reduced thermal stability of the composites. However, the thermal stability of VFs could be enhanced after covering and protection by the resol matrix [26,132,160,60,161,3]. Therefore, the thermal stability of the resol–VFs composites depends on not only the fiber type (composition, e.g., content of lignin, cellulose, and hemicellulose) but also the interfacial interaction between VFs and resol [26]. The modified/treated VFs–resol composites with better fiber–matrix adhesion usually show slightly higher thermal stability and therefore an improved thermal resistance than that of untreated VFs–resol composites [100,132,60,85,88]. Resol has a high critical oxygen index (45–70%), thus it is difficult to ignite and keep burning, which made it one of the lowest smoke-producing plastics with much less toxic smoke than conventional flame-retardant counterparts [48]. Some organic phosphorus compounds such as [9,10-dihydro-9-oxa-10-phosphaphenanthrene-10-oxide (DOPO)–silane coupling agents] or inorganic functional nanofillers such as layered double hydroxides (LDHs) could further be introduced into the composites, via fiber or matrix modification, to improve the flame retardancy (Fig. 9) [88,162,87,97].

Electric properties. The dielectric constant (ε') of the resol–VFs composite usually decrease with the elevated fiber loading (because $\varepsilon'_{\text{lignocellulosic fiber}} < \varepsilon'_{\text{resol}}$) and frequency (due to the suppressed orientation of molecule at high frequencies). And the modification of VFs, with lowered amount of hydroxyl group/water and decreased hydrophilicity, usually results in further decrease of dielectric constant, thus rendering the composite with an even lower dielectric constant [115,163]. The volume resistivity of VFs, however, increases after modification (due to reduced moisture content and increased fiber–matrix adhesion), compared to the decrease for the composites of untreated VFs with increased

Materials Research Forum LLC

https://doi.org/10.21741/9781644901854-7

fiber loading and frequency. Therefore, the good mechanical and electrically insulating properties can be simultaneously achieved for wider applications [115,97]. And the dielectric loss factor (ε'') is usually decreased as well when adding chemically treated VFs in resol matrix with good interfacial adhesion (i.e., regional rigidification via hindrance in chain mobility), however, depending on the specific process and VFs (e.g., alkaline treated VFs), the results may be different or even contradictive [115,164].

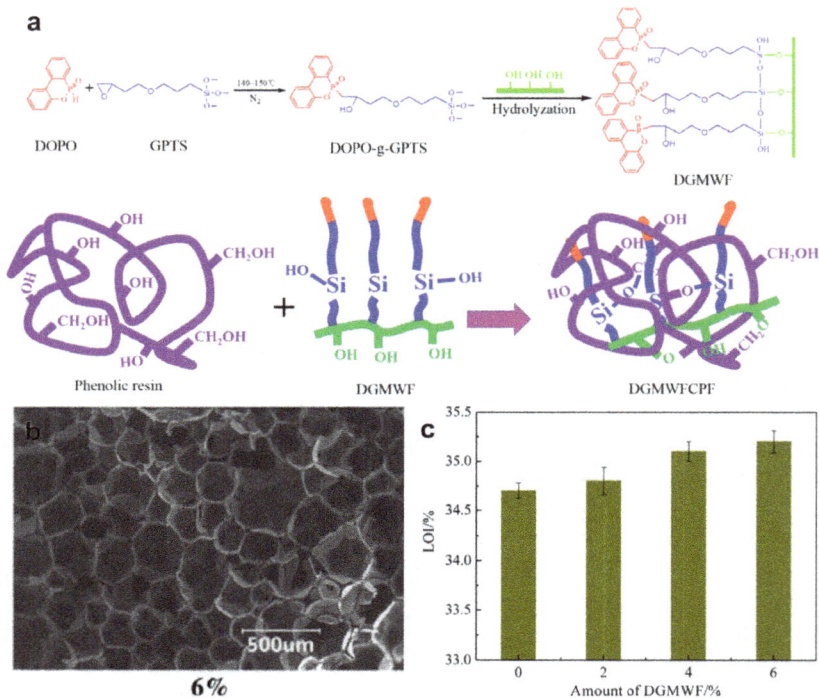

Fig. 9. *Scheme of DOPO-g-GPTS modified wood fiber composite phenolic foams. (b) SEM images of composite phenolic foam with 6% DGMWF added. (c) Limited oxygen indexes (LOIs) of phenolic foams, showing further improved the flame retardancy (LOIs > 27%) by adding DGMWF. DOPO-g-GPTS = 9,10-dihydro-9-oxa-10-phosphaphenanthrene-10-oxide (DOPO) grafted γ-glycidoxy propyl trimethoxy silane (GPTS). DGMWF = DOPO-g-GPTS modified wood fiber. Reproduced with permission from Wiley Periodicals, Inc. 2018 (Ref. [88]).*

Sustainable Natural Fiber Composites   Materials Research Forum LLC
Materials Research Foundations **122** (2022) 154-198   https://doi.org/10.21741/9781644901854-7

2.5   Characterizations

Fracture surfaces, fiber–matrix adhesion, fiber pull-out, and fiber surface topography are typical features to analyze and estimate the performance of the composite and the structure–performance relation [49]. These structural and functional (e.g., thermal, mechanical) characterizations of the composites can be achieved by scanning electron microscopy (SEM) [11], atomic force microscopy (AFM) [112], Fourier transform infrared spectroscopy (FTIR) [106], energy dispersive X-ray spectroscopy (EDS) [106], X-ray photoelectron spectroscopy (XPS) [61], inverse gas chromatography (IGC) [41,95,93], thermogravimetry/differential thermal analysis (TGA/DTA), differential mechanical thermoanalysis (DMTA) [48], contact angle measurement [106], and various mechanical tests such as impact, hardness test *etc*. [10,24]

It is noteworthy that the IGC can provide information regarding the polarity of the groups of the matrix prepolymers or fiber constituents (i.e., dispersive and acid–base properties of fiber surface), which can benefit the intermolecular interaction at the interface of fiber–matrix region [41,95,93]. Some advanced/joint methods are also exploited for better investigation, e.g., ultra-depth-of-field microscopy, and SEM in conjunction with EDS (SEM-EDS) to access the distribution and penetration of (labeled) resin in the VFs, and solid-state cross-polarization/magic angle spinning (CP/MAS) ^{13}C NMR spectra, FTIR images, atomic force microscope infrared spectroscopy (AFM-IR), contact resonance atomic force microscope (CR-AFM) measurement, and nanoindentation to analyze the molecular-scale and nanoscale interactions between resin and VFs wall (Fig. 10) [165,166].

2.6   Theoretical modeling

Theoretical predictions, as compared with the experimental results, should be conducted for better material design. Several theories have been proposed for modeling the mechanical properties such as tensile strength of these rigid matrices embedded with randomly oriented fibers, Parallel/Series model and Hirsch model are useful. And the latter one is more suitable for the VFs–resol systems of which the interaction between the components (evident from interfacial shear strength via pull-out tests) is taken into consideration, thus showing a better agreement with the experimental results [69,134]. It should be noted that due to the presence of voids in the composite, the actual density of the composite is usually lower than the theoretical value, thus a fair criterion relating strength and other applicable properties with density would be better for comparison [11,62]. Comprehensive assessment of the technical and environmental performance (or environmental impact) of the reinforced composites could also be conducted by a Life Cycle Assessment (LCA) for optimal development of this added-value chain under the frame of emerging bioeconomy [120].

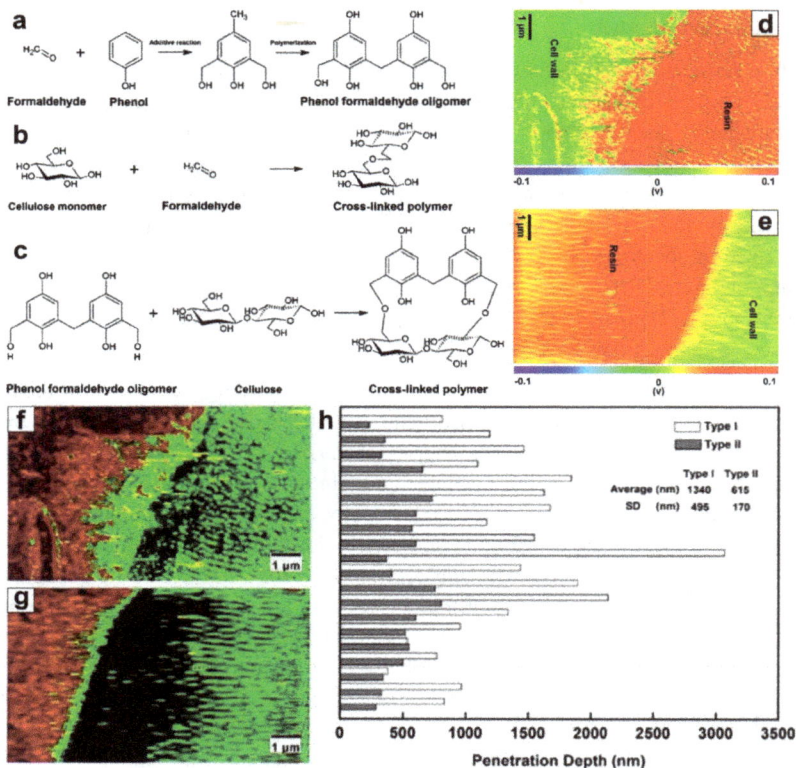

Fig. 10. *Schematic chemical reactions of wood cell-wall polymers and phenol-formaldehyde resin (PF): (a) synthesis of PF; (b) reaction of cellulose and residual formaldehyde; (c) reaction of cellulose and resin oligomer. (d and e) Corresponding IR amplitude images: C=C bond (1603 cm−1) on type I and type II locations, respectively. (f and g) Segmentation images of type I and type II interphase region; (h) measurement of penetration depth of resin into cell wall. IR = infrared spectroscopy, SD = standard deviation. Reproduced with permission from The Royal Society of Chemistry 2016 (Ref. [166]).*

2.7   Processing techniques

The manufacturing and machinability (or called machining process, i.e., secondary operations to weld together or glue) as well as machining process monitoring affect the performances of natural fiber reinforced composites [167]. These resol–VFs composites are usually prepared by compression molding (CM), resin transfer molding (RTM), or hand layup [11,48,76,132,75]; the injection molding and pultrusion process as also typically used by other natural fiber composites have yet to be tried [21,167]. Compared with CM technique, RTM usually renders the composites with better mechanical performances due to less voids in the molded composites which lead to better fiber–matrix interaction [75], although some appropriate organic solvents may be needed [168].

Volatile control is a critical parameter to ensure that the end materials (i.e., VFs-reinforced resol composites) have high density and good properties [26]. The small molecule byproducts of condensation reaction during curing (e.g., water and in some cases accompanied with formaldehyde) can be released and result in a number of voids that may deteriorate the mechanical properties and increase the moisture or water uptake of the composites. By optimized processing (pressure, temperature, and time interval) [57] or formulations, the control of voids can be realized, e.g., vacuum-bag processes with selected vacuum pressure can be utilized to avoid excessive extravasation of resin and void-free quality end materials can be produced [26,169]. Furthermore, higher molding pressure at the gel point (i.e., cure/consolidation) of the resol matrix leads to a reduction of voids in the matrix, and the degree of crosslinking, fiber–matrix interaction, filling, water resistance, and impact strength are therefore improved [26,57].

In addition, there are usually two kinds of mixing method, i.e., direct-mixing and polymerization filling; the latter is an in-situ one, which can render the composites with better mechanical properties [40]. The impregnation method via mats, pulp sheets, or freeze dried micro-/nanofibers can be used to increase the resin content in the composites [42].

3.   Categories and applications

3.1   Cellulosic/lignocellulosic fibers

Cellulosic and lignocellulosic (consisting of 60–80% cellulose, 5–20% lignin and up to 20% moisture) fibers with excellent strength and stiffness have been studied over decades and have demonstrated such advantages in many kinds of composites [44,170,76,106,42,43,171,102]. Lignocellulosic natural fibers, e.g., jute, coir, sisal, hemp, and sugarcane/bagasse, have received significant attention for polymer composites including resols [133,132,22,122]. And with regard to the thermal stability, the cellulose usually shows a significant degradation process (mainly 250–330 °C), lower than that of lignin, but higher than that of hemicellulose [132].

For the macrofibers, we take bagasse as an example. Sugarcane bagasse (main constituents: cellulose, hemicellulose, and lignin), a byproduct of the sugar industry, was usually burned as fuel (incineration for heating or electricity) or fermented to generate second-generation ethanol. It is now been realized that it is an ideal candidate for many value-added applications including reinforcement in composite materials (e.g., plastic, cement), owing to its low density, low cost, and acceptable thermal/mechanical properties [172,48,41]. The original unburned sugarcane bagasse (SBU) with more polar groups on the surface favors intermolecular interactions and consequent adhesion with the resol matrix, compared to that of burned sugarcane bagasse (SBB). The resol–bagasse composites present a higher storage modulus and enhanced impact strength than that of unreinforced thermoset, and thus are showing great potential in many applications such as rigid packaging, nonstructural parts of buildings and automotive vehicles to boost the development of circular bioeconomy based on agro-wastes [41,20,173-177].

And for micro-/nanofibers, the microfibrillated cellulose (MFC) or cellulose nanofiber is set as an example. These alkali-treated MFCs with smaller diameter derived from wood pulps (e.g., kraft) can render the resol resin with 5 times higher toughness (i.e., strain at fracture) than untreated pulp and 2 times higher than untreated MFC for microcomposites and nanocomposites, respectively (both with *ca.* 20 wt.% resol content as binder) [81]. The Young's modulus of the MFC/resol composites is linearly increased (up to 19 GPa) with increasing MFC content, and the coefficient of thermal expansion is decreased [82]. The mechanical properties, especially bending strength (up to 370 MPa), are enhanced stepwise with the higher degree of microfibrillation of pulp fiber. The strength is comparable to that of high-specific-strength magnesium alloy widely used in electronic device casings and an order of magnitude higher than that of ordinary wood fiber-based composites. The microfibrillation can be realized by refining and high pressure homogenization treatments to tune the morphology and exposed surface area. And the enhanced mechanical properties are ascribed to the extremely strong stretched molecule chains of MFC, increased surface area and interfacial bond densities, as well as the continuously interconnected web-like structure in the composites (Fig. 11) [79,83]. The resol composites with bacterial cellulose (BC) could be even stronger than that comprised of MFC (i.e., the bending strength increases to 425 MPa, and particularly, the Young's modulus increases to 28 GPa), owing to the uniform, continuous, and straight nanoscale network of cellulosic elements oriented in-plane during the compression process [84].

3.2 Combined fibers

The combination of two or more natural fibers, or natural and synthetic fibers, in a matrix can produce more effective composites, of which desired properties are attained to overcome some limitations of individual fiber based polymer composites [33,139,178].

These hybrid composites, with combined fibers in a single matrix, can usually achieve better synergistic properties [74,13,133,128].

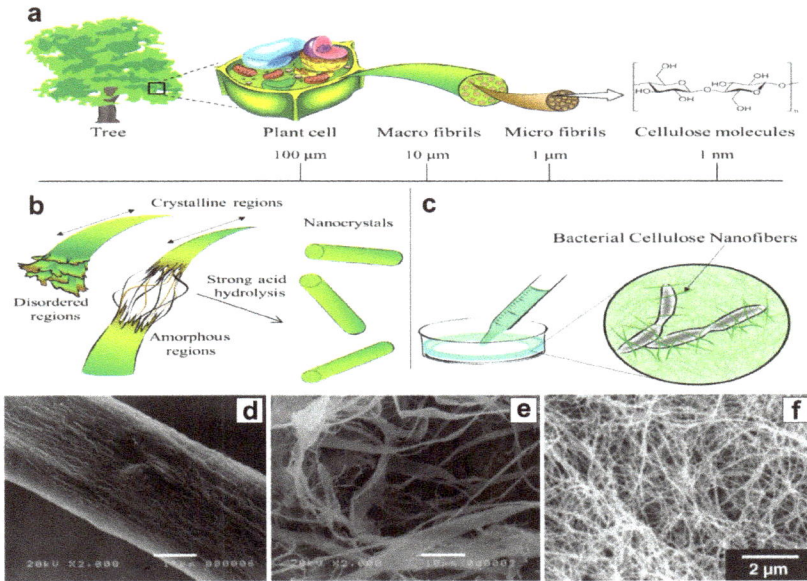

Fig. 11. *(a) Cellulose contained in plants or trees has a hierarchical structure from meter to nanometer scale. (b) Schematic diagram of the preparation of nanocellulose via the reaction between cellulose and strong acid. (c) Bionanocellulose cultured from cellulose-synthesizing bacteria. (d–f) SEM images of kraft pulp single elementary fiber, microfibrillated cellulose (MFC), and bacterial cellulose (BC) pellicle, respectively. Note: d and e scale bar: 10 µm. a–c, Reproduced from MDPI 2020, open access under CC BY 4.0 (Ref. [171]); d and e, Reproduced with permission from Springer-Verlag 2003 (Ref. [83]); f, Reproduced with permission from Springer-Verlag 2004 (Ref. [84]).*

3.2.1 Natrual/natural fibers

There are some typical hybrid composites with resol as matrix based on two kinds of biofibers: sugarcane bagasse lignin/sisal fibers [119], jute/banana fibers [179], jute/cotton hybrid fabrics [178], oil palm empty fruit bunch (OPEFB)/sugarcane bagasse fibers [180], (silane-treated) pineapple leaf fiber (PALF)/kenaf fiber (KF) [33,128,181,161,182,139,163]. The compatibility of the two fibers between themself and the fiber–matrix interface should be taken into consideration [128]. For example, the hybridization of PALF/KF enhanced the mechanical and thermal properties of the composites compared to the single-fiber reinforcement. Because the PALF (with relatively poor interfacial bonding and pull-out mechanism) increases the impact strength while the KF (with good interfacial compatibility) enhances the tensile/flexural strengths and dimensional change restriction (i.e., reduced coefficient of thermal expansion, CTE) [128,139]. For the jute/cotton hybrid fabrics reinforced composites, the jute promotes a higher reinforcing effect and the cotton avoids catastrophic failure. It is found that, for the anisotropy composites, the fabric orientation and roving/fabric characteristics also affect the mechanical properties. Thus it is an effective way to produce composites with higher mechanical properties via hybridizing synergistic fibers and making angle-ply (or cross-ply) laminates of lower anisotropy degree [178].

3.2.2 Natural/synthetic fibers

These typical hybrid composites based on natural/synthetic fibers reinforced resols include jute/rockwool [183], kenaf/Fiberfrax fiber (ceramic fiber) [183], banana fiber/glass fiber [133,152], sisal fiber/glass fiber [101], oil palm fiber/glass fiber [134], coir/glass fiber [104], ridge gourd/glass fiber [109-111], and coconut leaf sheath/jute/glass fabric, *etc.* [184] The synergistic effect will render the hybrid composites with advantages such as better water resistance, thermal and specific mechanical properties, however, some properties such as insulating properties also depend on the circumstances [101,133,115,134,183]. Furthermore, the layering patterns and stacking sequence (e.g., forming interleaving layers, intimate mixture, or periphery/core structure) will affect the mechanical properties and water resistance of the hybrid composites [133,184].

3.3   Foams

Due to the similar density of resol (cured resin, 1.24 g cm^{-3}) to that of VFs fillers (*ca.* 1.3–1.5 g cm^{-3}) [11], the density of resol–VFs composites will not change much with limited incorporation; however, the porous composites or the composite foams would reduce the density quite a lot for ultralight materials [103]. The VFs–resol foam composites are also characterized by excellent flame resistance, low toxic gas emission during combustion, low thermal conductivity, good sound insulation, and high specific strength, which will advance the brittle/fragile phenolic foam as a third-generation polymeric foam and a new kind of environmentally friendly fire-retarding foam composites [103,105,185,37,87,88,162]. For example, a cellulose fiber-reinforced phenolic foam with good interfacial bonding (viscose weight fraction: 2 wt.%) can possess an elevated compressive modulus and strength by 21% and 18%, respectively, compared to the unreinforced foam. Moreover, the incorporation of the cellulose fibers results in a

Materials Research Forum LLC
https://doi.org/10.21741/9781644901854-7

decreased cell size and increased cell density of the foam due to the increased viscosity and bubble nucleation, although the thermal stability decreases slightly. The foam, with a low apparent density of 0.16 g cm^{-3}, is of great interest for insulating and structural applications (Fig. 12) [186].

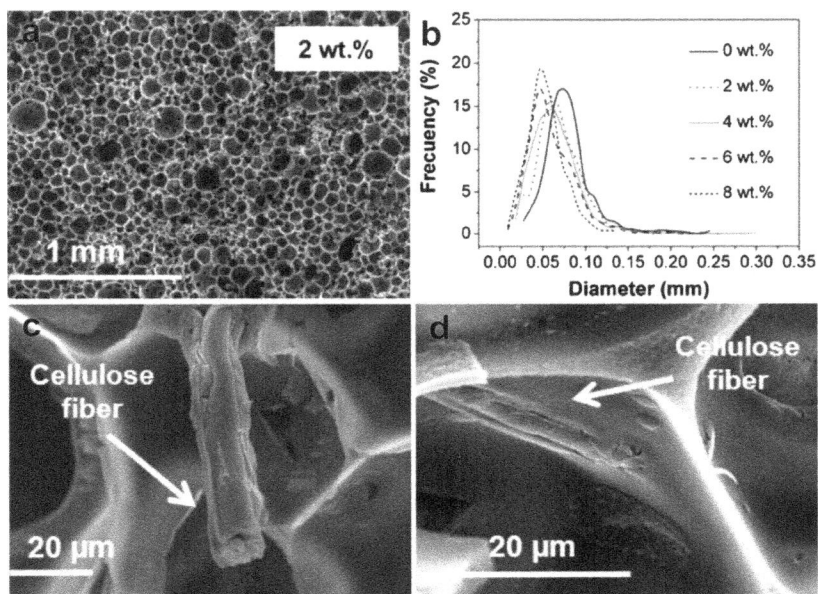

Fig. 12. *(a) SEM image of cellulose fiber-reinforced phenolic foams (CRPFs) with 2 wt.% of cellulose fibers, and (b) cell size distributions of unreinforced phenolic foam and CRPFs with 2, 4, 6 and 8 wt.% of cellulose fibers. SEM images of CRPFs with different fiber weight fractions: (c) 4 wt.% CRPF and (d) 8 wt.% CRPF, showing a good bonding between the cellulose fibers and the phenolic matrix. Reproduced with permission from Elsevier Ltd. 2015 (Ref. [186]).*

3.4   Applications

Vegetable fiber composites have been utilized in various applications including structural and nonstructural materials. Beyond those conventional sectors such as construction, furniture, automobiles, electronics, sporting goods and packaging industries [29,187,16], they are emerging in some exciting high-value-added smart applications.

There are some typical commercial products based on phenol-formaldehyde resin (PF)–vegetable fiber composites, and their trade names include Bakelite (made from PF and

Materials Research Foundations **122** (2022) 154-198   https://doi.org/10.21741/9781644901854-7

wood flour/paper/linen for electrical and mechanical applications), Novotext (randomly oriented cotton fiber-reinforced PF, or cotton textile-phenolic resin once often used for gear wheels of car engines), Tufnol (laminated plastic of high resistance to oils and solvents made from layers of paper or cloth soaked with PF, alternative to unfeasible lubricating oil as lining in loaded bearings), and Ebonol (a paper filled PF for musical instruments) [25,188,189].

Resol-type phenolic resins are also used for making exterior plywood (commonly known as *weather and boil proof*, WBP) because there is no melting point but only a decomposing point of 200 °C and above [25]. For example, lignin-phenolic binder has been synthesized for natural fiber-reinforced composites and their structural material applications thereof [121,125]. Furthermore, by incorporation of some specific components, they may be utilized as functional adhesives, e.g., electromagnetic wave absorbing adhesive made of Fe_3O_4@lignin/phenol-formaldehyde composite (Fig. 13) [190].

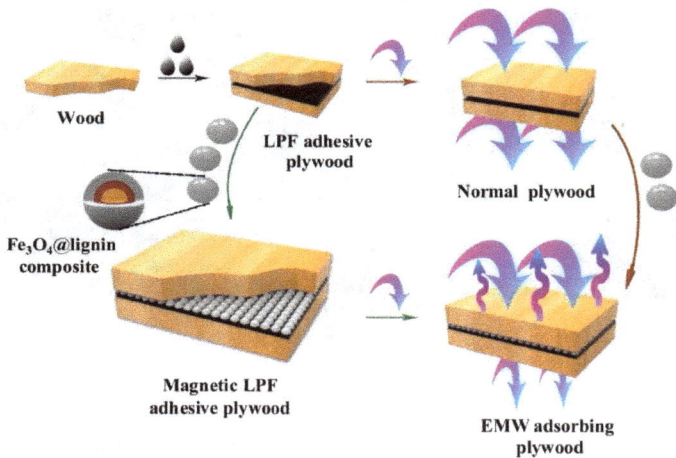

Fig. 13. *Schematic illustration of using lignin as the precursor to synthesize Fe_3O_4@lignin composite for electromagnetic wave (EMW) absorbing lignin-phenol-formaldehyde (LPF) adhesive. Reproduced with permission from Elsevier B.V. 2020 (Ref. [190]).*

Conclusions and perspectives

Phenolic resins, especially resol and its natural fiber composites, are of great advantages when high resistance to fire and elevated temperature is required. By incorporating vegetable fibers (VFs), the brittle nature, elongation, and buckling characteristics of resol resin are considerably improved. From the discussions abovementioned, the following major conclusions can be drawn to advance future investigation and application.

The incorporation of VFs in a resol matrix usually produces stiff structural composite materials that surpass the performances of typical structural materials. Moisture sensitivity of VFs-based polymeric composites is one major factor to hinder the growth prospects of the market, although spiraling demand for lightweight and/or green (eco-friendly) products are among key trends escalating market growth. Different treatment techniques aimed to enhance the mechanical and water-resistant performances of the biofiber-based composites can be realized by improving the compatibility and adhesion between the fibers and the matrix. It is impossible to gather all the advantages through a single method, however, based on the synergistic effect (e.g., VFs selection, surface modification, interfacial improvement, structure and processing optimization), optimized products could be achieved for industrial applications.

Longer VFs are more efficient than short ones due to the enhanced interfacial interactions at the molecular level, however, the increasing dispersion difficulty should also be considered. The content of fibers, i.e., the minimum, critical and maximum, should be determined to evaluate its effect. The void volume fraction usually increases with higher fraction of fiber with random distribution, which leads to reduced density and some typical properties, thus, the molding process should be optimized to diminish void fraction especially at higher fiber percentage. The fiber arrangement (e.g., by forming mats/fabrics, woven, non-woven or hand layup) to enhance the macroscopic entanglement is an alternative approach for higher fiber fraction but with linear rise in some typical mechanical properties. Reduction in diameter of VFs, i.e., by inducing a small amount of nanofiber, can potentially enhance the composite mechanical properties by a lot.

Acknowledgements

This work was supported by National Natural Science Foundation of China (Grant Nos. 51403193, 51403132), Open Project Foundation of the Beijing National Laboratory for Molecular Sciences (2016), Opening Project of Sichuan University of Science and Engineering, Material Corrosion and Protection Key Laboratory of Sichuan Province (2019CL19), Vanadium and Titanium Resource Comprehensive Utilization Key Laboratory of Sichuan Province (2019FTSZ01), Municipal Sci-Tech program of Panzhihua (2019ZD-G-3), and Doctoral innovation fund of Panzhihua University (20190106).

References

[1] Mohanty, A.K., Vivekanandhan, S., Pin, J.-M., Misra, M.: Composites from renewable and sustainable resources: Challenges and innovations. Science **362**(6414), 536-542 (2018). https//doi.org/10.1126/science.aat9072

[2] Dittenber, D.B., GangaRao, H.V.S.: Critical review of recent publications on use of natural composites in infrastructure. Composites Part A: Applied Science and Manufacturing **43**(8), 1419-1429 (2012). https://doi.org/10.1016/j.compositesa.2011.11.019

[3] Milanese, A.C., Cioffi, M.O.H., Voorwald, H.J.C.: Thermal and mechanical behaviour of sisal/phenolic composites. Composites Part B: Engineering **43**(7), 2843-2850 (2012). https://doi.org/10.1016/j.compositesb.2012.04.048

[4] La Mantia, F.P., Morreale, M.: Green composites: A brief review. Composites Part A: Applied Science and Manufacturing **42**(6), 579-588 (2011). https://doi.org/10.1016/j.compositesa.2011.01.017

[5] Koronis, G., Silva, A., Fontul, M.: Green composites: A review of adequate materials for automotive applications. Composites Part B: Engineering **44**(1), 120-127 (2013). https://doi.org/10.1016/j.compositesb.2012.07.004

[6] Zuccarello, B., Marannano, G., Mancino, A.: Optimal manufacturing and mechanical characterization of high performance biocomposites reinforced by sisal fibers. Compos. Struct. **194**, 575-583 (2018). https://doi.org/10.1016/j.compstruct.2018.04.007

[7] Correia, V.C., Santos, S.F., Tonoli, G.H.D., Savastano, H.: 7 - Characterization of vegetable fibers and their application in cementitious composites. In: Harries, K.A., Sharma, B. (eds.) Nonconventional and Vernacular Construction Materials (Second Edition). pp. 141-167. Woodhead Publishing, (2020)

[8] Benaimeche, O., Seghir, N.T., Sadowski, Ł., Mellas, M.: The Utilization of Vegetable Fibers in Cementitious Materials. In: Hashmi, S., Choudhury, I.A. (eds.) Encyclopedia of Renewable and Sustainable Materials. pp. 649-662. Elsevier, Oxford (2020)

[9] Jawaid, M., Sapuan, S.M., Alothman, O.Y.: Green Biocomposites: Manufacturing and Properties, 1st ed. Green Energy and Technology. Springer, Cham, Switzerland (2017)

[10] Sanjay, M.R., Madhu, P., Jawaid, M., Senthamaraikannan, P., Senthil, S., Pradeep, S.: Characterization and properties of natural fiber polymer composites: A comprehensive review. J. Cleaner Prod. **172**, 566-581 (2018). https://doi.org/10.1016/j.jclepro.2017.10.101

[11] Zárate, C.N., Aranguren, M.I., Reboredo, M.M.: Resol–vegetable fibers composites. J. Appl. Polym. Sci. **77**(8), 1832-1840 (2000). https//doi.org/10.1002/1097-4628(20000822)77:8<1832::aid-app21>3.0.co;2-u

[12] Campilho, R.D.S.G.: Introduction to Natural Fiber Composites, in Natural Fiber Composites (1st ed.), Campilho, R.D.S.G. (Ed.), 1st Ed. ed. Natural Fiber Composites (1st ed.). CRC Press, (2015, pp 1-34)

[13] Ashik, K., Sharma, R.: A Review on Mechanical Properties of Natural Fiber Reinforced Hybrid Polymer Composites. Journal of Minerals and Materials Characterization and Engineering **3**, 420-426 (2015). https//doi.org/10.4236/jmmce.2015.35044

[14] Sanjay, M.R., Arpitha, G.R., Naik, L.L., Gopalakrishna, K., Yogesha, B.: Applications of Natural Fibers and Its Composites: An Overview. Natural Resources **7**(3), 108-114 (2016).

[15] Bourmaud, A., Beaugrand, J., Shah, D.U., Placet, V., Baley, C.: Towards the design of high-performance plant fibre composites. Prog. Mater Sci. **97**, 347-408 (2018). https://doi.org/10.1016/j.pmatsci.2018.05.005

[16] Peças, P., Carvalho, H., Salman, H., Leite, M.: Natural Fibre Composites and Their Applications: A Review. Journal of Composites Science **2**(4), 66 (2018).

[17] da Costa Melo, R.Q., Barbosa de Lima, A.G.: Vegetable Fiber-Reinforced Polymer Composites: Fundamentals, Mechanical Properties and Applications. Diffusion Foundations **14**, 1-20 (2018). https//doi.org/10.4028/www.scientific.net/DF.14.1

[18] Kerni, L., Singh, S., Patnaik, A., Kumar, N.: A review on natural fiber reinforced composites. Materials Today: Proceedings **28**, 1616-1621 (2020). https://doi.org/10.1016/j.matpr.2020.04.851

[19] Zimniewska, M., Wladyka-Przybylak, M.: Natural Fibers for Composite Applications. In: Rana, S., Fangueiro, R. (eds.) Fibrous and Textile Materials for Composite Applications. pp. 171-204. Springer Singapore, Singapore (2016)

[20] da Silva, C.G., de Oliveira, F., Frollini, E.: Sugarcane Bagasse Fibers Treated and Untreated: Performance as Reinforcement in Phenolic-Type Matrices Based on Lignosulfonates. Waste and Biomass Valorization **10**(11), 3515-3524 (2019). https//doi.org/10.1007/s12649-018-0365-z

[21] Natural Fiber Composites (NFC) Market Size, Share & Trends Analysis Report By Raw Material, By Matrix, By Technology, By Application, And Segment Forecasts, 2018 - 2024. https://www.grandviewresearch.com/industry-analysis/natural-fiber-composites-market. 12/02/2020

[22] Thakur, V.K., Thakur, M.K.: Processing and characterization of natural cellulose fibers/thermoset polymer composites. Carbohydr. Polym. **109**, 102-117 (2014). https://doi.org/10.1016/j.carbpol.2014.03.039

[23] CompositesWorld. Natural fiber composites: What's holding them back?, https://www.compositesworld.com/articles/natural-fiber-composites-whats-holding-them-back (2019).

[24] M.R, S., Siengchin, S., Parameswaranpillai, J., Jawaid, M., Pruncu, C.I., Khan, A.: A comprehensive review of techniques for natural fibers as reinforcement in composites: Preparation, processing and characterization. Carbohydr. Polym. **207**, 108-121 (2019). https://doi.org/10.1016/j.carbpol.2018.11.083

[25] Wikipedia: Phenol formaldehyde resin. https://en.wikipedia.org/wiki/Phenol_formaldehyde_resin#Biodegradation.

[26. Frollini, E., Silva, C.G., Ramires, E.C.: 2 - Phenolic resins as a matrix material in advanced fiber-reinforced polymer (FRP) composites. In: Bai, J. (ed.) Advanced Fibre-Reinforced Polymer (FRP) Composites for Structural Applications. pp. 7-43. Woodhead Publishing, (2013)

[27] Encyclopædia_Britannica: Phenol formaldehyde. https://www.britannica.com/topic/industrial-polymers-468698/Phenol-formaldehyde.

[28] Haupt, R.A., Sellers, T., Jr.: Characterizations of Phenol-Formaldehyde Resol Resins. Ind. Eng. Chem. Res. **33**(3), 693-697 (1994). https//doi.org/10.1021/ie00027a030

[29] Mohd, A., Naheed, S., Mohammad, J., Mohammad, N., Mohammed, P., Othman, Y.A.: A Review on Phenolic Resin and its Composites. Current Analytical Chemistry **14**(3), 185-197 (2018). http://dx.doi.org/10.2174/1573411013666171003154410

[30] Pilato, L.: Phenolic resins: 100Years and still going strong. React. Funct. Polym. **73**(2), 270-277 (2013). https://doi.org/10.1016/j.reactfunctpolym.2012.07.008

[31] Guedes, J., Florentino, W.M., Mulinari, D.R.: Chapter 4 - Thermoplastics Polymers Reinforced with Natural Fibers. In: Thomas, S., Shanks, R., Chandrasekharakurup, S. (eds.) Design and Applications of Nanostructured Polymer Blends and Nanocomposite Systems. pp. 55-73. William Andrew Publishing, Boston (2016)

[32] Shah, D.U., Schubel, P.J., Clifford, M.J., Licence, P.: Mechanical Property Characterization of Aligned Plant Yarn Reinforced Thermoset Matrix Composites Manufactured via Vacuum Infusion. Polymer-Plastics Technology and Engineering **53**(3), 239-253 (2014). https//doi.org/10.1080/03602559.2013.843710

[33] Asim, M., Jawaid, M., Abdan, K., Ishak, M.R.: Effect of pineapple leaf fibre and kenaf fibre treatment on mechanical performance of phenolic hybrid composites. Fibers and Polymers **18**(5), 940-947 (2017). https//doi.org/10.1007/s12221-017-1236-0

[34] Marliana, M.M., Hassan, A., Yuziah, M.Y.N., Khalil, H.P.S.A., Inuwa, I.M., Syakir, M.I., Haafiz, M.K.M.: Flame retardancy, Thermal and mechanical properties of Kenaf fiber reinforced Unsaturated polyester/Phenolic composite. Fibers and Polymers **17**(6), 902-909 (2016). https//doi.org/10.1007/s12221-016-5888-y

[35] Thakur, V.K., Singha, A.S.: Evaluation of GREWIA OPTIVA Fibers as Reinforcement in Polymer Biocomposites. Polymer-Plastics Technology and Engineering **49**(11), 1101-1107 (2010). https//doi.org/10.1080/03602559.2010.496390

[36] Yu, Y.-L., Huang, X.-A., Yu, W.-J.: High performance of bamboo-based fiber composites from long bamboo fiber bundles and phenolic resins. J. Appl. Polym. Sci. **131**(12), 40371 (2014). https//doi.org/10.1002/app.40371

[37] Tang, Q., Fang, L., Guo, W.: Effects of Bamboo Fiber Length and Loading on Mechanical, Thermal and Pulverization Properties of Phenolic Foam Composites. Journal of Bioresources and Bioproducts **4**(1), 51-59 (2019). https://doi.org/10.21967/jbb.v4i1.184

[38] Das, M., Chakraborty, D.: Effects of alkalization and fiber loading on the mechanical properties and morphology of bamboo fiber composites. II. Resol matrix. J. Appl. Polym. Sci. **112**(1), 447-453 (2009). https//doi.org/10.1002/app.29383

[39] Asim, M., Jawaid, M., Nasir, M., Saba, N.: Effect of Fiber Loadings and Treatment on Dynamic Mechanical, Thermal and Flammability Properties of Pineapple Leaf Fiber and Kenaf Phenolic Composites. Journal of Renewable Materials **6**(4), 383-393 (2018).

[40] Mu, Q., Wei, C., Feng, S.: Studies on mechanical properties of sisal fiber/phenol formaldehyde resin in-situ composites. Polym. Compos. **30**(2), 131-137 (2009). https://doi.org/10.1002/pc.20529

[41] da Silva, C.G., Frollini, E.: Unburned Sugarcane Bagasse: Bio-based Phenolic thermoset Composites as an Alternative for the Management of this Agrowaste. J. Polym. Environ. **28**(12), 3201-3210 (2020). https//doi.org/10.1007/s10924-020-01848-y

[42] Miao, C., Hamad, W.Y.: Cellulose reinforced polymer composites and nanocomposites: a critical review. Cellulose **20**(5), 2221-2262 (2013). https//doi.org/10.1007/s10570-013-0007-3

[43] Joseph, B., K, S.V., Sabu, C., Kalarikkal, N., Thomas, S.: Cellulose nanocomposites: Fabrication and biomedical applications. Journal of Bioresources and Bioproducts **5**(4), 223-237 (2020). https://doi.org/10.1016/j.jobab.2020.10.001

[44] Faruk, O., Sain, M.: Preface. In: Faruk, O., Sain, M. (eds.) Biofiber Reinforcements in Composite Materials. pp. xxvii-xxviii. Woodhead Publishing, (2015)

[45] Ighalo, J.O., Adeyanju, C.A., Ogunniyi, S., Adeniyi, A.G., Abdulkareem, S.A.: An empirical review of the recent advances in treatment of natural fibers for reinforced plastic composites. Compos. Interfaces, 28(9), 925-960 (2021). https//doi.org/10.1080/09276440.2020.1826274

[46] Bledzki, A.K., Reihmane, S., Gassan, J.: Properties and modification methods for vegetable fibers for natural fiber composites. J. Appl. Polym. Sci. **59**(8), 1329-1336 (1996). https//doi.org/10.1002/(sici)1097-4628(19960222)59:8<1329::aid-app17>3.0.co;2-0

[47] Gusse, A.C., Miller, P.D., Volk, T.J.: White-Rot Fungi Demonstrate First Biodegradation of Phenolic Resin. Environ. Sci. Technol. **40**(13), 4196-4199 (2006). https//doi.org/10.1021/es060408h

[48] Paiva, J.M.F., Frollini, E.: Sugarcane bagasse reinforced phenolic and lignophenolic composites. J. Appl. Polym. Sci. **83**(4), 880-888 (2002). https://doi.org/10.1002/app.10085

[49] Sreekala, M.S., Thomas, S., Neelakantan, N.R.: Utilization of short oil palm empty fruit bunch fiber (OPEFB) as a reinforcement in phenol-formaldehyde resins: Studies on mechanical properties. J. Polym. Eng. **16**(4), 265-294 (1997).

[50] Rashid, B., Leman, Z., Jawaid, M., Ghazali, M.J., Ishak, M.R.: The mechanical performance of sugar palm fibres (ijuk) reinforced phenolic composites. International Journal of Precision Engineering and Manufacturing **17**(8), 1001-1008 (2016). https//doi.org/10.1007/s12541-016-0122-9

[51] Rashid, B., Leman, Z., Jawaid, M., Ghazali, M.J., Ishak, M.R., Abdelgnei, M.A.: Dry sliding wear behavior of untreated and treated sugar palm fiber filled phenolic composites using factorial technique. Wear **380-381**, 26-35 (2017). https://doi.org/10.1016/j.wear.2017.03.011

[52] Kashizadeh, R., Esfandeh, M., Rezadoust, A.M., Sahraeian, R.: Physico-mechanical and thermal properties of date palm fiber/phenolic resin composites. Polym. Compos. **40**(9), 3657-3665 (2019). https://doi.org/10.1002/pc.25228

[53] Rashid, B., Leman, Z., Jawaid, M., Ghazali, M.J., Ishak, M.R.: Influence of Treatments on the Mechanical and Thermal Properties of Sugar Palm Fibre Reinforced Phenolic Composites. Bioresources **12**(1), 1447-1462 (2017).

[54] Rashid, B., Leman, Z., Jawaid, M., Ghazali, M.J., Ishak, M.R.: Dynamic Mechanical Analysis of Treated and Untreated Sugar Palm Fibre-based Phenolic Composites. Bioresources **12**(2), 3448-3462 (2017). https//doi.org/10.15376/biores.12.2.3448-3462

[55] Asim, M., Jawaid, M., Khan, A., Asiri, A.M., Malik, M.A.: Effects of Date Palm fibres loading on mechanical, and thermal properties of Date Palm reinforced phenolic composites. Journal of Materials Research and Technology **9**(3), 3614-3621 (2020). https://doi.org/10.1016/j.jmrt.2020.01.099

[56] Kamble, Z., Behera, B.K.: Mechanical properties and water absorption characteristics of composites reinforced with cotton fibres recovered from textile waste. Journal of Engineered Fibers and Fabrics **15**, doi: 10.1177/1558925020901530 (2020). https//doi.org/10.1177/1558925020901530

[57] Megiatto, J.D., Silva, C.G., Ramires, E.C., Frollini, E.: Thermoset matrix reinforced with sisal fibers: Effect of the cure cycle on the properties of the biobased composite. Polym. Test. **28**(8), 793-800 (2009). https://doi.org/10.1016/j.polymertesting.2009.07.001

[58] Barreto, A.C.H., Rosa, D.S., Fechine, P.B.A., Mazzetto, S.E.: Properties of sisal fibers treated by alkali solution and their application into cardanol-based biocomposites. Composites Part A: Applied Science and Manufacturing **42**(5), 492-500 (2011). https://doi.org/10.1016/j.compositesa.2011.01.008

[59] Milanese, A.C., Cioffi, M.O.H., Voorwald, H.J.C.: Flexural behavior of Sisal/Castor oil-Based Polyurethane and Sisal/Phenolic Composites. Materials Research **15**, 191-197 (2012).

[60] Wang, S., Wei, C., Liu, H., Gong, Y., Yang, D., Yang, P., Liu, T.: Studies on Mechanical Properties and Morphology of Sisal Pulp Reinforced Phenolic Composites. Adv. Polym. Tech. **35**(4), 353-360 (2016). https://doi.org/10.1002/adv.21557

[61] Zhong, L., Fu, S., Li, F., Zhan, H.: Chlorine dioxide treatment of sisal fibre: surface lignin and its influences on fibre surface characteristics and interfacial behaviour of sisal fibre/phenolic resin composites. Bioresources **5**(4), 2431-2446 (2010).

[62] Zárate, C.N., Aranguren, M.I., Reboredo, M.M.: Influence of fiber volume fraction and aspect ratio in resol–sisal composites. J. Appl. Polym. Sci. **89**(10), 2714-2722 (2003). https//doi.org/10.1002/app.12404

[63] Razera, I.A.T., Frollini, E.: Composites based on jute fibers and phenolic matrices: Properties of fibers and composites. J. Appl. Polym. Sci. **91**(2), 1077-1085 (2004). https://doi.org/10.1002/app.13224

[64] Yang, Z., Xian, G., Li, H.: Effects of alternating temperatures and humidity on the moisture absorption and mechanical properties of ramie fiber reinforced phenolic plates. Polym. Compos. **36**(9), 1590-1596 (2015). https://doi.org/10.1002/pc.23067

[65] Asim, M., Jawaid, M., Abdan, K., Ishak, M.R.: Dimensional stability of pineapple leaf fibre reinforced phenolic composites. AIP Conference Proceedings **1901**(1), 030016 (2017). https//doi.org/10.1063/1.5010481

[66] De, D., Adhikari, B., De, D.: Grass fiber reinforced phenol formaldehyde resin composite: preparation, characterization and evaluation of properties of composite. Polym. Adv. Technol. **18**(1), 72-81 (2007). https//doi.org/10.1002/pat.854

[67] Singha, A.S., Thakur, V.K.: Synthesis, Characterization and Study of Pine Needles Reinforced Polymer Matrix Based Composites. J. Reinf. Plast. Compos. **29**(5), 700-709 (2010). https//doi.org/10.1177/0731684408100354

[68] Singha, A.S., Thakur, V.K.: Synthesis and Characterization of Pine Needles Reinforced RF Matrix Based Biocomposites. E-Journal of Chemistry **5**, 395827 (2008). https//doi.org/10.1155/2008/395827

[69] Joseph, S., Sreekala, M.S., Oommen, Z., Koshy, P., Thomas, S.: A comparison of the mechanical properties of phenol formaldehyde composites reinforced with banana fibres and glass fibres. Compos. Sci. Technol. **62**(14), 1857-1868 (2002). https://doi.org/10.1016/S0266-3538(02)00098-2

[70] Shah, D.U.: Developing plant fibre composites for structural applications by optimising composite parameters: a critical review. J. Mater. Sci. **48**(18), 6083-6107 (2013). https//doi.org/10.1007/s10853-013-7458-7

[71] Silva, C.G., Benaducci, D., Frollini, E.: Lyocell and cotton fibers as reinforcements for a thermoset polymer. BioResources **7**(1), 78-98 (2012).

[72] Rojo, E., Alonso, M.V., Oliet, M., Del Saz-Orozco, B., Rodriguez, F.: Effect of fiber loading on the properties of treated cellulose fiber-reinforced phenolic composites. Composites Part B: Engineering **68**, 185-192 (2015). https://doi.org/10.1016/j.compositesb.2014.08.047

[73] Maya, M.G., George, S.C., Jose, T., Sreekala, M.S., Thomas, S.: Mechanical Properties of Short Sisal Fibre Reinforced Phenol Formaldehyde Eco-Friendly Composites. Polymers from Renewable Resources **8**(1), 27-42 (2017). https//doi.org/10.1177/204124791700800103

[74] Nunna, S., Chandra, P.R., Shrivastava, S., Jalan, A.: A review on mechanical behavior of natural fiber based hybrid composites. J. Reinf. Plast. Compos. **31**(11), 759-769 (2012). https//doi.org/10.1177/0731684412444325

[75] Cherian, B.M., Leão, A.L., de Morais Chaves, M.R., de Souza, S.F., Sain, M., Narine, S.S.: Environmental ageing studies of chemically modified micro and nanofibril phenol formaldehyde composites. Industrial Crops and Products **49**, 471-483 (2013). https://doi.org/10.1016/j.indcrop.2013.04.033

[76] Thakur, V.K., Singha, A.S., Kaur, I., Nagarajarao, R.P., Yang, L.: Studies on Analysis and Characterization of Phenolic Composites Fabricated from Lignocellulosic Fibres. Polym. Polym. Compos. **19**(6), 505-512 (2011). https//doi.org/10.1177/096739111101900609

[77] Ramires, E.C., Frollini, E.: Tannin–phenolic resins: Synthesis, characterization, and application as matrix in biobased composites reinforced with sisal fibers. Composites Part B: Engineering **43**(7), 2851-2860 (2012). https://doi.org/10.1016/j.compositesb.2012.04.049

[78] Neelamana, I.K., Thomas, S., Parameswaranpillai, J.: Characteristics of banana fibers and banana fiber reinforced phenol formaldehyde composites-macroscale to nanoscale. J. Appl. Polym. Sci. **130**(2), 1239-1246 (2013). https://doi.org/10.1002/app.39220

[79] Nakagaito, A.N., Yano, H.: The effect of morphological changes from pulp fiber towards nano-scale fibrillated cellulose on the mechanical properties of high-strength plant fiber based composites. Appl. Phys. A **78**(4), 547-552 (2004). https//doi.org/10.1007/s00339-003-2453-5

[80] Nishino, T., Takano, K., Nakamae, K.: Elastic modulus of the crystalline regions of cellulose polymorphs. J. Polym. Sci., Part B: Polym. Phys. **33**(11), 1647-1651 (1995). https://doi.org/10.1002/polb.1995.090331110

[81] Nakagaito, A.N., Yano, H.: Toughness enhancement of cellulose nanocomposites by alkali treatment of the reinforcing cellulose nanofibers. Cellulose **15**(2), 323-331 (2008). https//doi.org/10.1007/s10570-007-9168-2

[82] Nakagaito, A.N., Yano, H.: The effect of fiber content on the mechanical and thermal expansion properties of biocomposites based on microfibrillated cellulose. Cellulose **15**(4), 555-559 (2008). https//doi.org/10.1007/s10570-008-9212-x

[83] Nakagaito, A.N., Yano, H.: Novel high-strength biocomposites based on microfibrillated cellulose having nano-order-unit web-like network structure. Appl. Phys. A **80**(1), 155-159 (2005). https//doi.org/10.1007/s00339-003-2225-2

[84] Nakagaito, A.N., Iwamoto, S., Yano, H.: Bacterial cellulose: the ultimate nano-scalar cellulose morphology for the production of high-strength composites. Appl. Phys. A **80**(1), 93-97 (2005). https//doi.org/10.1007/s00339-004-2932-3

[85] Rojo, E., Alonso, M.V., Domínguez, J.C., Saz-Orozco, B.D., Oliet, M., Rodriguez, F.: Alkali treatment of viscose cellulosic fibers from eucalyptus wood: Structural, morphological, and thermal analysis. J. Appl. Polym. Sci. **130**(3), 2198-2204 (2013). https://doi.org/10.1002/app.39399

[86] Thakur, V.K., Singha, A.S., Kaur, I., Nagarajarao, R.P., Liping, Y.: Silane Functionalization of Saccaharum cilliare Fibers: Thermal, Morphological, and Physicochemical Study. Int. J. Polym. Anal. Charact. **15**(7), 397-414 (2010). https//doi.org/10.1080/1023666X.2010.510106

[87] Ma, Y.F., Geng, X., Zhang, X.: Effect of Novel DOPO-g-Coupling Agent Treated Wood Fibers on Properties of Composite Phenolic Foams. Bioresources **13**(3), 6187-6200 (2018).

[88] Ma, Y., Geng, X., Zhang, X., Wang, C., Chu, F.: Synthesis of DOPO-g-GPTS modified wood fiber and its effects on the properties of composite phenolic foams. J. Appl. Polym. Sci. **136**(2), 46917 (2019). https//doi.org/10.1002/app.46917

[89] Trindade, W.G., Hoareau, W., Razera, I.A.T., Ruggiero, R., Frollini, E., Castellan, A.: Phenolic Thermoset Matrix Reinforced with Sugar Cane Bagasse Fibers: Attempt to Develop a New Fiber Surface Chemical Modification Involving Formation of Quinones Followed by Reaction with Furfuryl Alcohol. Macromolecular Materials and Engineering **289**(8), 728-736 (2004). https//doi.org/10.1002/mame.200300320

[90] Thakur, V.K., Singha, A.S., Thakur, M.K.: Graft Copolymerization of Methyl Acrylate onto Cellulosic Biofibers: Synthesis, Characterization and Applications. J. Polym. Environ. **20**(1), 164-174 (2012). https//doi.org/10.1007/s10924-011-0372-7

[91] Chauhan, S.R., Patnaik, A., Kaith, B.S., Satapathy, A., Dwivedy, M.: Analysis of Mechanical Behavior of Phenol Formaldehyde Matrix Composites Using Flax-g-Poly (MMA) as Reinforcing Materials. J. Reinf. Plast. Compos. **28**(16), 1933-1944 (2009). https//doi.org/10.1177/0731684407089131

[92] Kalia, S., Kaith, B.S., Sharma, S., Bhardwaj, B.: Mechanical properties of flax-g-poly(methyl acrylate) reinforced phenolic composites. Fibers and Polymers **9**(4), 416-422 (2008). https//doi.org/10.1007/s12221-008-0067-4

[93] Megiatto Jr., J.D., Oliveira, F.B., Rosa, D.S., Gardrat, C., Castellan, A., Frollini, E.: Renewable Resources as Reinforcement of Polymeric Matrices: Composites Based on Phenolic Thermosets and Chemically Modified Sisal Fibers. Macromolecular Bioscience **7**(9-10), 1121-1131 (2007). https//doi.org/10.1002/mabi.200700083

[94] Kaith, B.S., Kalia, S.: Grafting of Flax Fiber (Linum usitatissimum) with Vinyl Monomers for Enhancement of Properties of Flax-Phenolic Composites. Polym. J. **39**(12), 1319-1327 (2007). https//doi.org/10.1295/polymj.PJ2007073

[95] Megiatto Jr., J.D., Silva, C.G., Rosa, D.S., Frollini, E.: Sisal chemically modified with lignins: Correlation between fibers and phenolic composites properties. Polym. Degrad. Stab. **93**(6), 1109-1121 (2008). https://doi.org/10.1016/j.polymdegradstab.2008.03.011

[96] Tita, S., Medeiros, R., Tarpani, J., Frollini, E., Tita, V.: Chemical modification of sugarcane bagasse and sisal fibers using hydroxymethylated lignin: Influence on impact strength and water absorption of phenolic composites. J. Compos. Mater. **52**(20), 2743-2753 (2018). https//doi.org/10.1177/0021998317753886

[97] Li, C., Wan, J., Pan, Y.-T., Zhao, P.-C., Fan, H., Wang, D.-Y.: Sustainable, Biobased Silicone with Layered Double Hydroxide Hybrid and Their Application in Natural-Fiber Reinforced Phenolic Composites with Enhanced Performance. ACS Sustainable Chem. Eng. **4**(6), 3113-3121 (2016). https//doi.org/10.1021/acssuschemeng.6b00134

[98] Megiatto, J.D., Ramires, E.C., Frollini, E.: Phenolic matrices and sisal fibers modified with hydroxy terminated polybutadiene rubber: Impact strength, water absorption, and morphological aspects of thermosets and composites. Industrial Crops and Products **31**(1), 178-184 (2010). https://doi.org/10.1016/j.indcrop.2009.10.001

[99] Sreekala, M.S., Kumaran, M.G., Thomas, S.: Oil palm fibers: Morphology, chemical composition, surface modification, and mechanical properties. J. Appl. Polym. Sci. **66**(5), 821-835 (1997). https//doi.org/10.1002/(sici)1097-4628(19971031)66:5<821::aid-app2>3.0.co;2-x

[100] Joseph, S., Sreekala, M.S., Thomas, S.: Effect of chemical modifications on the thermal stability and degradation of banana fiber and banana fiber-reinforced phenol formaldehyde composites. J. Appl. Polym. Sci. **110**(4), 2305-2314 (2008). https://doi.org/10.1002/app.27648

[101] Lü, J., Zhong, J.-B., Wei, C.: Studies on the Properties of Sisal Fibre/Phenol Formaldehyde Resin In-situ Composites. Research Journal of Textile and Apparel **10**(3), 51-58 (2006). https//doi.org/10.1108/RJTA-10-03-2006-B007

[101] Pereira, P.H.F., Rosa, M.d.F., Cioffi, M.O.H., Benini, K.C.C.d.C., Milanese, A.C., Voorwald, H.J.C., Mulinari, D.R.: Vegetal fibers in polymeric composites: a review. Polímeros **25**, 9-22 (2015).

[103] Ma, Y., Wang, C., Chu, F.: The structure and properties of eucalyptus fiber/phenolic foam composites under N-β(aminoethyl)-γ-aminopropyl trimethoxy

silane pretreatments. Polish Journal of Chemical Technology **19**(4), 116-121 (2017). https://doi.org/10.1515/pjct-2017-0077

[104] Kumar, N.M., Reddy, G.V., Naidu, S.V., Rani, T.S., Subha, M.C.S.: Mechanical Properties of Coir/Glass Fiber Phenolic Resin Based Composites. J. Reinf. Plast. Compos. **28**(21), 2605-2613 (2009). doi:doi: 10.1177/0731684408093092

[105] Ma, Y.F., Wang, C.P., Chu, F.X.: Effects of Fiber Surface Treatments on the Properties of Wood Fiber-Phenolic Foam Composites. Bioresources **12**(3), 4722-4736 (2017). https//doi.org/10.15376/biores.12.3.4722-4736

[106] Rojo, E., Alonso, M.V., Del Saz-Orozco, B., Oliet, M., Rodriguez, F.: Optimization of the silane treatment of cellulosic fibers from eucalyptus wood using response surface methodology. J. Appl. Polym. Sci. **132**(26), 42157 (2015). https//doi.org/10.1002/app.42157

[107] Surya Rajan, B., Saibalaji, M.A., Rasool Mohideen, S.: Tribological performance evaluation of epoxy modified phenolic FC reinforced with chemically modified Prosopis juliflora bark fiber. Mater. Res. Express **6**(7), 075313 (2019). https//doi.org/10.1088/2053-1591/ab07e6

[108] Rojo, E., Oliet, M., Alonso, M.V., Saz-Orozco, B.D., Rodriguez, F.: Mechanical and interfacial properties of phenolic composites reinforced with treated cellulose fibers. Polymer Engineering & Science **54**(10), 2228-2238 (2014). https://doi.org/10.1002/pen.23772

[109] Varada, A.R., Devi, R.R.: Flexural Properties of Ridge Gourd/ Phenolic Composites and Glass/Ridge Gourd/Phenolic Hybrid Composites. J. Compos. Mater. **42**(6), 593-601 (2008). https//doi.org/10.1177/0021998307086197

[110] Rajulu, A.V., Devi, R.R.: Compressive Properties of Ridge Gourd/Phenolic Composites and Ridge Gourd/Phenolic/Glass Hybrid Composites. J. Reinf. Plast. Compos. **26**(16), 1657-1664 (2007). https//doi.org/10.1177/0731684407081358

[111] Rajulu, A.V., Devi, R.R.: Tensile Properties of Ridge Gourd/Phenolic Composites and Glass/Ridge Gourd/Phenolic Hybrid Composites. J. Reinf. Plast. Compos. **26**(6), 629-638 (2007). https//doi.org/10.1177/0731684407075567

[112] Peng, X., Zhong, L., Ren, J., Sun, R.: Laccase and alkali treatments of cellulose fibre: Surface lignin and its influences on fibre surface properties and interfacial behaviour of sisal fibre/phenolic resin composites. Composites Part A: Applied Science and Manufacturing **41**(12), 1848-1856 (2010). https://doi.org/10.1016/j.compositesa.2010.09.004

[113] Wu, M., Sun, Z., Zhao, X.: Effects of Different Modification Methods on the Properties of Sisal Fibers. Journal of Natural Fibers **17**(7), 1048-1057 (2020). https//doi.org/10.1080/15440478.2018.1554517

[114] Razera, I.A.T., Silva, C.G.d., Almeida, É.V.R.d., Frollini, E.: Treatments of jute fibers aiming at improvement of fiber-phenolic matrix adhesion. Polímeros **24**, 417-421 (2014).

[115] Joseph, S., Thomas, S.: Electrical properties of banana fiber-reinforced phenol formaldehyde composites. J. Appl. Polym. Sci. **109**(1), 256-263 (2008). https://doi.org/10.1002/app.27452

[116] de Oliveira, F., da Silva, C.G., Ramos, L.A., Frollini, E.: Phenolic and lignosulfonate-based matrices reinforced with untreated and lignosulfonate-treated sisal fibers. Industrial Crops and Products **96**, 30-41 (2017). https://doi.org/10.1016/j.indcrop.2016.11.027

[117] da Silva, C.G., Oliveira, F., Ramires, E.C., Castellan, A., Frollini, E.: Composites from a forest biorefinery byproduct and agrofibers: Lignosulfonate-phenolic type matrices reinforced with sisal fibers. Tappi J. **11**(9), 41-49 (2012). https//doi.org/10.32964/tj11.9.41

[118] Tita, S.P.S., Paiva, J.M.F.d., Frollini, E.: Resistência ao Impacto e Outras Propriedades de Compósitos Lignocelulósicos: Matrizes Termofixas Fenólicas Reforçadas com Fibras de Bagaço de Cana-de-açúcar. Polímeros **12**(4), 228-239 (2002). https://doi.org/10.1590/s0104-14282002000400005

[119] Ramires, E.C., Megiatto Jr., J.D., Gardrat, C., Castellan, A., Frollini, E.: Valorization of an industrial organosolv–sugarcane bagasse lignin: Characterization and use as a matrix in biobased composites reinforced with sisal fibers. Biotechnol. Bioeng. **107**(4), 612-621 (2010). https//doi.org/10.1002/bit.22847

[120] Hildebrandt, J., Budzinski, M., Nitzsche, R., Weber, A., Krombholz, A., Thrän, D., Bezama, A.: Assessing the technical and environmental performance of wood-based fiber laminates with lignin based phenolic resin systems. Resources, Conservation and Recycling **141**, 455-464 (2019). https://doi.org/10.1016/j.resconrec.2018.10.029

[121] Mahendran, A.R., Wuzella, G., Aust, N., Müller, U., Kandelbauer, A.: Processing and Characterization of Natural Fibre Reinforced Composites Using Lignin Phenolic Binder. Polym. Polym. Compos. **21**(4), 199-206 (2013). https//doi.org/10.1177/096739111302100401

[122] Oliveira, F.B., Gardrat, C., Enjalbal, C., Frollini, E., Castellan, A.: Phenol–furfural resins to elaborate composites reinforced with sisal fibers—Molecular analysis of resin

and properties of composites. J. Appl. Polym. Sci. **109**(4), 2291-2303 (2008). https://doi.org/10.1002/app.28312

[123] Ramires, E.C., Megiatto, J.D., Gardrat, C., Castellan, A., Frollini, E.: Biobased composites from glyoxal–phenolic resins and sisal fibers. Bioresour. Technol. **101**(6), 1998-2006 (2010). https://doi.org/10.1016/j.biortech.2009.10.005

[124] Barbosa, V., Ramires, E.C., Razera, I.A.T., Frollini, E.: Biobased composites from tannin–phenolic polymers reinforced with coir fibers. Industrial Crops and Products **32**(3), 305-312 (2010). https://doi.org/10.1016/j.indcrop.2010.05.007

[125] Bisanda, E.T.N.: The manufacture of roofing panels from sisal fibre reinforced composites. J. Mater. Process. Technol. **38**(1), 369-379 (1993). https://doi.org/10.1016/0924-0136(93)90209-O

[126] Saha, P., Manna, S., Sen, R., Roy, D., Adhikari, B.: Durability of lignocellulosic fibers treated with vegetable oil–phenolic resin. Carbohydr. Polym. **87**(2), 1628-1636 (2012). https://doi.org/10.1016/j.carbpol.2011.09.070

[127] Li, X., Cai, Z., Winandy, J.E., Basta, A.H.: Selected properties of particleboard panels manufactured from rice straws of different geometries. Bioresour. Technol. **101**(12), 4662-4666 (2010). https://doi.org/10.1016/j.biortech.2010.01.053

[128] Asim, M., Jawaid, M., Paridah, M.T., Saba, N., Nasir, M., Shahroze, R.M.: Dynamic and thermo-mechanical properties of hybridized kenaf/PALF reinforced phenolic composites. Polym. Compos. **40**(10), 3814-3822 (2019). https//doi.org/10.1002/pc.25240

[129] Wikipedia: Young's modulus. https://en.wikipedia.org/wiki/Young's_modulus.

[130] Wikipedia: Ultimate tensile strength. https://en.wikipedia.org/wiki/Ultimate_tensile_strength.

[131] Yeh, M.-K., Tai, N.-H., Lin, Y.-J.: Mechanical properties of phenolic-based nanocomposites reinforced by multi-walled carbon nanotubes and carbon fibers. Composites Part A: Applied Science and Manufacturing **39**(4), 677-684 (2008). https://doi.org/10.1016/j.compositesa.2007.07.010

[132] Indira, K.N., Jyotishkumar, P., Thomas, S.: Thermal stability and degradation of banana fibre/PF composites fabricated by RTM. Fibers and Polymers **13**(10), 1319-1325 (2012). https//doi.org/10.1007/s12221-012-1319-x

[133] Joseph, S., Sreekala, M.S., Koshy, P., Thomas, S.: Mechanical properties and water sorption behavior of phenol–formaldehyde hybrid composites reinforced with banana fiber and glass fiber. J. Appl. Polym. Sci. **109**(3), 1439-1446 (2008). https://doi.org/10.1002/app.27425

[134] Sreekala, M.S., George, J., Kumaran, M.G., Thomas, S.: The mechanical performance of hybrid phenol-formaldehyde-based composites reinforced with glass and oil palm fibres. Compos. Sci. Technol. **62**(3), 339-353 (2002). https://doi.org/10.1016/S0266-3538(01)00219-6

[135] Arpitha, G.R., Yogesha, B.: An Overview on Mechanical Property Evaluation of Natural Fiber Reinforced Polymers. Materials Today: Proceedings **4**(2, Part A), 2755-2760 (2017). https://doi.org/10.1016/j.matpr.2017.02.153

[136] Ticoalu, A., Aravinthan, T., Cardona, F.: A review on the characteristics of gomuti fibre and its composites with thermoset resins. J. Reinf. Plast. Compos. **32**(2), 124-136 (2013). doi:doi: 10.1177/0731684412463109

[137] University_of_Washington: Young's Modulus. https://depts.washington.edu/matseed/mse_resources/Webpage/Biomaterials/young's_ modulus.htm.

[138] Wikipedia: Flexural modulus. https://en.wikipedia.org/wiki/Flexural_modulus.

[139] Asim, M., Jawaid, M., Abdan, K., Ishak, M.R., Alothman, O.Y.: Effect of Hybridization on the Mechanical Properties of Pineapple Leaf Fiber/Kenaf Phenolic Hybrid Composites. Journal of Renewable Materials **6**(1), 38-46 (2018).

[140] Trindade, W.G., Hoareau, W., Megiatto, J.D., Razera, I.A.T., Castellan, A., Frollini, E.: Thermoset Phenolic Matrices Reinforced with Unmodified and Surface-Grafted Furfuryl Alcohol Sugar Cane Bagasse and Curaua Fibers: Properties of Fibers and Composites. Biomacromolecules **6**(5), 2485-2496 (2005). https//doi.org/10.1021/bm058006+

[141] Li, C., Fan, H., Wang, D.-Y., Hu, J., Wan, J., Li, B.: Novel silicon-modified phenolic novolacs and their biofiber-reinforced composites: Preparation, characterization and performance. Compos. Sci. Technol. **87**, 189-195 (2013). https://doi.org/10.1016/j.compscitech.2013.08.016

[142] Chai, L.-L., Chia, C.-H., Zakaria, S., Nabihah, S., Rasid, R.: Morphology and Properties of Polypropylene Blends Containing Phenolic Resin Produced from the Liquefaction of Empty Fruit Bunch Fibres. Polym. Polym. Compos. **19**(8), 669-676 (2011). https//doi.org/10.1177/096739111101900807

[143] Rochardjo, H.S.B., Ridlo, M.: Effects of Fiber Contents on Wear Resistance of Salacca zalacca Frond Fiber Reinforced Phenolic. Mater. Sci. Forum **948**, 181-185 (2019). https//doi.org/10.4028/www.scientific.net/MSF.948.181

[144] Omrani, E., Menezes, P.L., Rohatgi, P.K.: State of the art on tribological behavior of polymer matrix composites reinforced with natural fibers in the green materials

Materials Research Forum LLC
https://doi.org/10.21741/9781644901854-7

world. Engineering Science and Technology, an International Journal **19**(2), 717-736 (2016). https://doi.org/10.1016/j.jestch.2015.10.007

[145] Wei, C., Zeng, M., Xiong, X., Liu, H., Luo, K., Liu, T.: Friction properties of sisal fiber/nano-silica reinforced phenol formaldehyde composites. Polym. Compos. **36**(3), 433-438 (2015). https://doi.org/10.1002/pc.22957

[146] Ma, Y., Liu, Y., Mao, C., Li, J., Yu, J., Tong, J.: Effects of Structured Fibre on Mechanical and Tribological Properties of Phenolic Composites for Application to Friction Brakes. Polym. Polym. Compos. **26**(4), 315-324 (2018). https//doi.org/10.1177/096739111802600406

[147] Xiong, X.-M., Wei, C., Zeng, M.: Study on the Tribological Performance of Sisal Fiber/Polysulfone/Phenolic Composite Friction Material. Advanced Science Letters **4**(3), 1108-1112 (2011). https//doi.org/10.1166/asl.2011.1411

[148] Wang, Z.-Y., Wang, J., Cao, F.-H., Ma, Y.-H., Tej, S., Gusztáv, F.: Influence of banana fiber on physicomechanical and tribological properties of phenolic based friction composites. Mater. Res. Express **6**(7), 075103 (2019). https//doi.org/10.1088/2053-1591/ab160a

[149] Wei, C., Zeng, M., Xiong, X.M., Zhang, F.A.: Thermal and frictional properties of modified sisal fibre/phenolic resin composites. Plastics, Rubber and Composites **39**(2), 61-66 (2010). https//doi.org/10.1179/174328910X12608851832452

[150] Md, J.A., Saibalaji, M.A., B, S.R., Liu, Y.: Characterization of alkaline treated Areva Javanica fiber and its tribological performance in phenolic friction composites. Mater. Res. Express **6**(11), 115307 (2019). https//doi.org/10.1088/2053-1591/ab43ad

[151] Lai, J.C., Ani, F.N., Hassan, A.: Water Absorption of Lignocellulosic Phenolic Composites. Polym. Polym. Compos. **16**(6), 379-387 (2008). https//doi.org/10.1177/096739110801600605

[152] Joseph, S., Oommen, Z., Thomas, S.: Environmental durability of banana-fiber-reinforced phenol formaldehyde composites. J. Appl. Polym. Sci. **100**(3), 2521-2531 (2006). https://doi.org/10.1002/app.23680

[153] Botaro, V.R., Siqueira, G., Megiatto, J.D., Frollini, E.: Sisal fibers treated with NaOH and benzophenonetetracarboxylic dianhydride as reinforcement of phenolic matrix. J. Appl. Polym. Sci. **115**(1), 269-276 (2010). https//doi.org/10.1002/app.31113

[154] Ly, E.B., Lette, M.J., Diallo, A.K., Gassama, A., Takasaki, A., Ndiaye, D.: Effect of Reinforcing Fillers and Fibres Treatment on Morphological and Mechanical Properties of Typha-Phenolic Resin Composites. Fibers and Polymers **20**(5), 1046-1053 (2019). https//doi.org/10.1007/s12221-019-1087-y

[155] Asim, M., Paridah, M.T., Jawaid, M., Nasir, M., Siakeng, R.: Effects of nanoclay on tensile and flexural properties of pineapple leaf fibre reinforced phenolic composite. International Journal of Recent Technology and Engineering **8**(2 Special Issue 4), 473-476 (2019). https//doi.org/10.35940/ijrte.B1092.0782S419

[156] Wang, H., Xian, G., Li, H., Sui, L.: Durability study of a ramie-fiber reinforced phenolic composite subjected to water immersion. Fibers and Polymers **15**(5), 1029-1034 (2014). https//doi.org/10.1007/s12221-014-1029-7

[157] Xian, G., Yin, P., Kafodya, I., Li, H., Wang, W.-l.: Durability study of ramie fiber fabric reinforced phenolic plates under humidity conditions. Science and Engineering of Composite Materials **23**(1), 45-52 (2016). https://doi.org/10.1515/secm-2014-0018

[158] Singh, B., Gupta, M., Verma, A.: The durability of jute fibre-reinforced phenolic composites. Compos. Sci. Technol. **60**(4), 581-589 (2000). https://doi.org/10.1016/S0266-3538(99)00172-4

[159] Azwa, Z.N., Yousif, B.F., Manalo, A.C., Karunasena, W.: A review on the degradability of polymeric composites based on natural fibres. Mater. Des. **47**, 424-442 (2013). https://doi.org/10.1016/j.matdes.2012.11.025

[160] Zárate, C.N., Aranguren, M.I., Reboredo, M.M.: Thermal degradation of a phenolic resin, vegetable fibers, and derived composites. J. Appl. Polym. Sci. **107**(5), 2977-2985 (2008). https//doi.org/10.1002/app.27455

[161] Asim, M., Paridah, M.T., Saba, N., Jawaid, M., Alothman, O.Y., Nasir, M., Almutairi, Z.: Thermal, physical properties and flammability of silane treated kenaf/pineapple leaf fibres phenolic hybrid composites. Compos. Struct. **202**, 1330-1338 (2018). https://doi.org/10.1016/j.compstruct.2018.06.068

[162] Ma, Y., Geng, X., Zhang, X., Wang, C., Chu, F.: A novel DOPO-g-KH550 modification wood fibers and its effects on the properties of composite phenolic foams. Polish Journal of Chemical Technology **20**(2), 47-53 (2018). https://doi.org/10.2478/pjct-2018-0022

[163] Agrebi, F., Hammami, H., Asim, M., Jawaid, M., Kallel, A.: Impact of silane treatment on the dielectric properties of pineapple leaf/kenaf fiber reinforced phenolic composites. J. Compos. Mater. **54**(7), 937-946 (2020). https//doi.org/10.1177/0021998319871351

[164] Agrebi, F., Ghorbel, N., Rashid, B., Kallel, A., Jawaid, M.: Influence of treatments on the dielectric properties of sugar palm fiber reinforced phenolic composites. J. Mol. Liq. **263**, 342-348 (2018). https://doi.org/10.1016/j.molliq.2018.04.130

[165] Huang, Y., Lin, Q., Yang, C., Bian, G., Zhang, Y., Yu, W.: Multi-scale characterization of bamboo bonding interfaces with phenol-formaldehyde resin of different molecular weight to study the bonding mechanism. Journal of The Royal Society Interface **17**(162), 20190755 (2020). doi:https//doi.org/10.1098/rsif.2019.0755

[166] Wang, X., Deng, Y., Li, Y., Kjoller, K., Roy, A., Wang, S.: In situ identification of the molecular-scale interactions of phenol-formaldehyde resin and wood cell walls using infrared nanospectroscopy. RSC Adv. **6**(80), 76318-76324 (2016). https//doi.org/10.1039/C6RA13159J

[167] Lotfi, A., Li, H., Dao, D.V., Prusty, G.: Natural fiber–reinforced composites: A review on material, manufacturing, and machinability. J. Thermoplast. Compos. Mater. **34**(2), 238-284 (2021). https//doi.org/10.1177/0892705719844546

[168] Wang, B., Huang, Y., Liu, L.: Effect of solvents on adsorption of phenolic resin onto γ-aminopropyl-triethoxysilane treated silica fiber during resin transfer molding. J. Mater. Sci. **41**(4), 1243-1246 (2006). https//doi.org/10.1007/s10853-005-4226-3

[169] Hou, T.H., Bai, J.M., Baughman, J.M.: Processing and Properties of a Phenolic Composite System. J. Reinf. Plast. Compos. **25**(5), 495-502 (2006). https//doi.org/10.1177/0731684405058271

[170] Bandyopadhyay-Ghosh, S., Ghosh, S.B., Sain, M.: 19 - The use of biobased nanofibres in composites. In: Faruk, O., Sain, M. (eds.) Biofiber Reinforcements in Composite Materials. pp. 571-647. Woodhead Publishing, (2015)

[171] Miyashiro, D., Hamano, R., Umemura, K.: A Review of Applications Using Mixed Materials of Cellulose, Nanocellulose and Carbon Nanotubes. Nanomaterials **10**(2), 186 (2020).

[172] Hajiha, H., Sain, M.: 17 - The use of sugarcane bagasse fibres as reinforcements in composites. In: Faruk, O., Sain, M. (eds.) Biofiber Reinforcements in Composite Materials. pp. 525-549. Woodhead Publishing, (2015)

[173] Ni, W., Shi, L.: Layer-structured carbonaceous materials for advanced Li-ion and Na-ion batteries: Beyond graphene. J. Vac. Sci. Technol. A **37**(4), 040803 (2019). https//doi.org/10.1116/1.5095413

[174] Xu, Q., Peng, Q., Ni, W., Hou, Z., Li, J., Yu, L.: Study of different effect on foaming process of biodegradable bionolle in supercritical carbon dioxide. J. Appl. Polym. Sci. **100**(4), 2901-2906 (2006). https//doi.org/10.1002/app.23796

[175] Ni, W., Chen, J., Xu, Q.: Synthesis and characterization of hierarchically porous silica with poplar tissue as template with assistance of supercritical CO2. BioResources **3**(2), 461-476 (2008).

[176] Tang, L.-Q., Ni, W., Zhao, H., Xu, Q., Jiao, J.-X.: Preparation of macroporous TiO2 by starch microspheres template with assistance of supercritical CO2. BioResources **4**(1), 38-48 (2009).

[177] Ni, W., Xu, Q., Jiao, J.-X., Liu, X., Ren, C.: Hierarchically porous Fe2O3 and Fe2O3/SiO2 composites prepared by cypress tissue template with assistance of supercritical CO2. BioResources **3**(3), 774-788 (2008).

[178] de Medeiros, E.S., Agnelli, J.A.M., Joseph, K., de Carvalho, L.H., Mattoso, L.H.C.: Mechanical properties of phenolic composites reinforced with jute/cotton hybrid fabrics. Polym. Compos. **26**(1), 1-11 (2005).

[179. Prashanth, B.H.M., Manjunath, T.S., Gouda, P.S.S., Sajjan, S.S., Ramesh, S.: Physico-mechanical response of phenolic resin composites reinforced with jute and banana fibers. AIP Conference Proceedings **2057**(1), 020016 (2019). https//doi.org/10.1063/1.5085587

[180] Ramlee, N.A., Jawaid, M., Zainudin, E.S., Yamani, S.A.K.: Tensile, physical and morphological properties of oil palm empty fruit bunch/sugarcane bagasse fibre reinforced phenolic hybrid composites. Journal of Materials Research and Technology **8**(4), 3466-3474 (2019). https://doi.org/10.1016/j.jmrt.2019.06.016

[181] Asim, M., Paridah, M.T., Jawaid, M., Nasir, M., Saba, N.: Physical and flammability properties of kenaf and pineapple leaf fibre hybrid composites. IOP Conference Series: Materials Science and Engineering **368**, 012018 (2018). https//doi.org/10.1088/1757-899x/368/1/012018

[182] Asim, M., Jawaid, M., Abdan, K., Ishak, M.R.: The Effect of Silane Treated Fibre Loading on Mechanical Properties of Pineapple Leaf/Kenaf Fibre Filler Phenolic Composites. J. Polym. Environ. **26**(4), 1520-1527 (2018). https//doi.org/10.1007/s10924-017-1060-z

[183] Öztürk, B.: Hybrid effect in the mechanical properties of jute/rockwool hybrid fibres reinforced phenol formaldehyde composites. Fibers and Polymers **11**(3), 464-473 (2010). https//doi.org/10.1007/s12221-010-0464-3

[184] Bharath, K., Sanjay, M., Jawaid, M., Harisha, Basavarajappa, S., Siengchin, S.: Effect of stacking sequence on properties of coconut leaf sheath/jute/E-glass reinforced phenol formaldehyde hybrid composites. Journal of Industrial Textiles **49**(1), 3-32 (2019). https//doi.org/10.1177/1528083718769926

[185] Gao, L., Tang, Q., Chen, Y., Wang, Z., Guo, W.: Investigation of novel lightweight phenolic foam-based composites reinforced with flax fiber mats. Polym. Compos. **39**(6), 1809-1817 (2018). https://doi.org/10.1002/pc.24130

[186] Del Saz-Orozco, B., Alonso, M.V., Oliet, M., Domínguez, J.C., Rodriguez, F.: Mechanical, thermal and morphological characterization of cellulose fiber-reinforced phenolic foams. Composites Part B: Engineering **75**, 367-372 (2015). https://doi.org/10.1016/j.compositesb.2015.01.049

[187] Tang, Q., Fang, L., Guo, W.: Investigation into Mechanical, Thermal, Flame-Retardant Properties of Wood Fiber Reinforced Ultra-High-Density Fiberboards. Bioresources **12**(3), 6749-6762 (2017). https//doi.org/10.15376/biores.12.3.6749-6762

[188] Wikipedia: Bakelite. https://en.wikipedia.org/wiki/Bakelite.

[189] Wikipedia: Novotext. https://en.wikipedia.org/wiki/Novotext.

[190] Pei, W., Shang, W., Liang, C., Jiang, X., Huang, C., Yong, Q.: Using lignin as the precursor to synthesize Fe3O4@lignin composite for preparing electromagnetic wave absorbing lignin-phenol-formaldehyde adhesive. Industrial Crops and Products **154**, 112638 (2020). https://doi.org/10.1016/j.indcrop.2020.112638

[191] Ramires, E.C., Megiatto, J.D., Gardrat, C., Castellan, A., Frollini, E.: Biobased composites from glyoxal-phenolic resins and sisal fibers. Bioresour. Technol. **101**(6), 1998-2006 (2010). https//doi.org/10.1016/j.biortech.2009.10.005

Sustainable Natural Fiber Composites   Materials Research Forum LLC
Materials Research Foundations **122** (2022) 199-208   https://doi.org/10.21741/9781644901854-8

Chapter 8

Measurement of Thermal Conductivity of Natural Fiber Composites

K. Ramanaiah[1]*, Srinivas Prasad Sanaka[1], A.V. Ratna Prasad[1], K. Hemachandra Reddy[2]

[1]Department of Mechanical Engineering, V.R.Siddhartha Engineering College, Vijayawada-520007, India

[2]Department of Mechanical Engineering, College of Engineering, JNTUA, Anantapuramu-515002, India

*ramanaiah@vrsiddhartha.ac.in

Abstract

The conventional materials are replaced with natural composites in energy saving applications due to their good insulation properties against heat transfer and the higher specific strength. In this chapter, the extraction of natural fibers, preparation of composites was discussed. The process of measurement of thermal conductivity of composites is elucidated. Based on the test data, natural polymer composites are promising composite materials for energy saving applications.

Keywords

Natural Fibers, Thermal Conductivity, Energy Saving, Composites, Vakka Fiber

Contents

Measurement of Thermal Conductivity of Natural Fiber Composites 199

1.   **Introduction**..**200**

2.   **Preparation of fibers and composites** **200**

3.   **Measurement of thermal conductivity**..**206**

References **208**

Sustainable Natural Fiber Composites   Materials Research Forum LLC
Materials Research Foundations **122** (2022) 199-208   https://doi.org/10.21741/9781644901854-8

1.   Introduction

Green composites have more demand to utilize them in electrical and electronic, automotive, aeronautical, building and construction industries, packing and energy industries. These materials are cost effective, renewable, biodegradable and light weight. The stringent regulations enforced in several countries, the need for green composites attracts the investigators. Thermal conductivity is a significant property in the selection of material for energy saving applications. The thermal conductivity of a material is the rate at which heat transferred by means of conduction per unit area. The influential factors on which the thermo-physical properties of a composite material depends on the resin materials, fiber, fiber content, fiber orientation and the direction of heat flow and operating temperature. Nor Azlina et al. [1,2] studied the potential of oil palm empty fruit bunch (OPEFB) and sugarcane bagasse fibers for thermal insulation application and they concluded that safe, renewable, economic and sustainable thermal insulation materials made from agricultural wastage are the alternatives to wood and reported the thermal conductivities of oil palm empty fruit bunch and sugarcane bagasse fibers. Basim Abu-Jdayil [3] has reported the variation of thermal conductivity from 0.166 to 0.17 W/mK in the temperature range of 0 to 60^0C for Date pit- polyester composite and 0.366 to 0.456 W/mK for Banana-epoxy composite of fiber content from 5 to 20%. H. Takagi et al. [4] adopted the hot wire method to measure thermal conductivity of Bamboo-Poly lactic acid (PLA) composite and concluded that thermal conductivity of bamboo-PLA composite is smaller than glass/carbon fibre reinforced composites. Kim et al. [5] have measured the thermal conductivity of hemp, Kenaf, flax and sisal fibre reinforced polypropylene-composites. The results indicate that the thermal conductivity of composites is in the range of 0.05-0.07 W/mK at 48.5% fiber content. Thermal conductivity, specific heat and thermal diffusivity of borassus fruit, elephant grass and vakka fibers/polyester composites were studied [6-8]. Exhaustive data on mechanical and thermal properties of synthetic fiber based composites is available in literature. However, the data on thermo-physical properties of green materials is scarcely reported in the literature.

2.   Preparation of fibers and composites

Sequence of steps followed in the extraction and preparation of fibers is shown in Fig 1. Water retting method has been applied for the extraction of natural fibers. In this method the microbial compound bonds separated from the stem of plants. Water retting produces homogenous, high-quality fibers.

Sustainable Natural Fiber Composites   Materials Research Forum LLC
Materials Research Foundations **122** (2022) 199-208   https://doi.org/10.21741/9781644901854-8

```
┌─────────────────────┐     ┌─────────────────────┐     ┌─────────────────┐
│  Plant stem cutting │ →   │    Water retting    │ →   │   Rolling of    │
│                     │     │                     │     │     fiber       │
└─────────────────────┘     └─────────────────────┘     └─────────────────┘
                                                                  │
                                                                  ↓
┌─────────────────────┐     ┌─────────────────────┐     ┌─────────────────┐
│   Drying of fiber   │ ←   │      Washing        │ ←   │   Separation    │
└─────────────────────┘     └─────────────────────┘     └─────────────────┘
          │
          ↓
┌─────────────────────┐
│   Fiber Ready for   │
│ fabrication of composite │
└─────────────────────┘
```

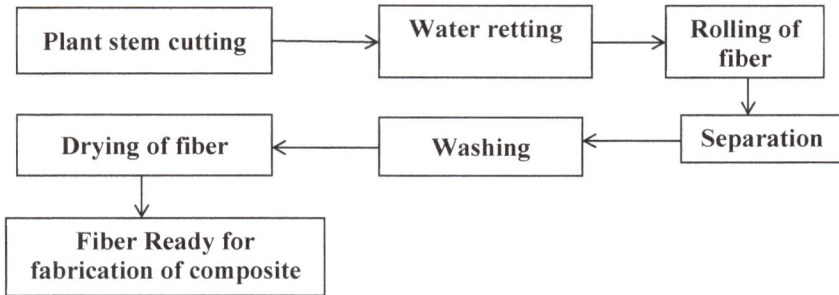

Fig. 1. Sequence of steps in the extraction of fibers

Elephant grass, fishtail palm, typha angastofolia, broom grass, borassus fruit, vakka and sansevieria fibers are extracted from plant/ tree are shown in fig 2-8.

Fig.2. Elephant grass plant and extracted fiber

Sustainable Natural Fiber Composites   Materials Research Forum LLC
Materials Research Foundations **122** (2022) 199-208   https://doi.org/10.21741/9781644901854-8

Fig. 3. Fishtail palm tree and extracted fiber

Fig. 4. Typha Angastofolia plant

Fig. 5. Vakka tree and extracted fiber

Fig. 6. Broom grass and extracted fiber

Fig. 7.. Borassus Tree and extracted fiber

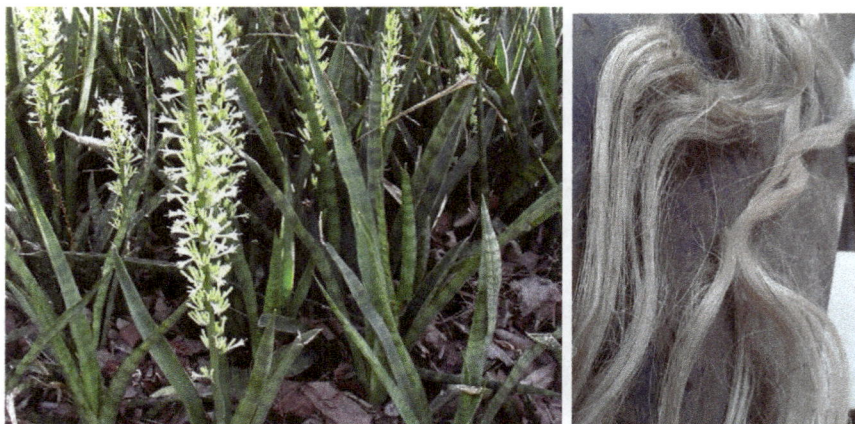

Fig. 8. Sansevieria plant and extracted fiber

The sequence of steps followed in the preparation of samples for testing is shown in Fig. 9.

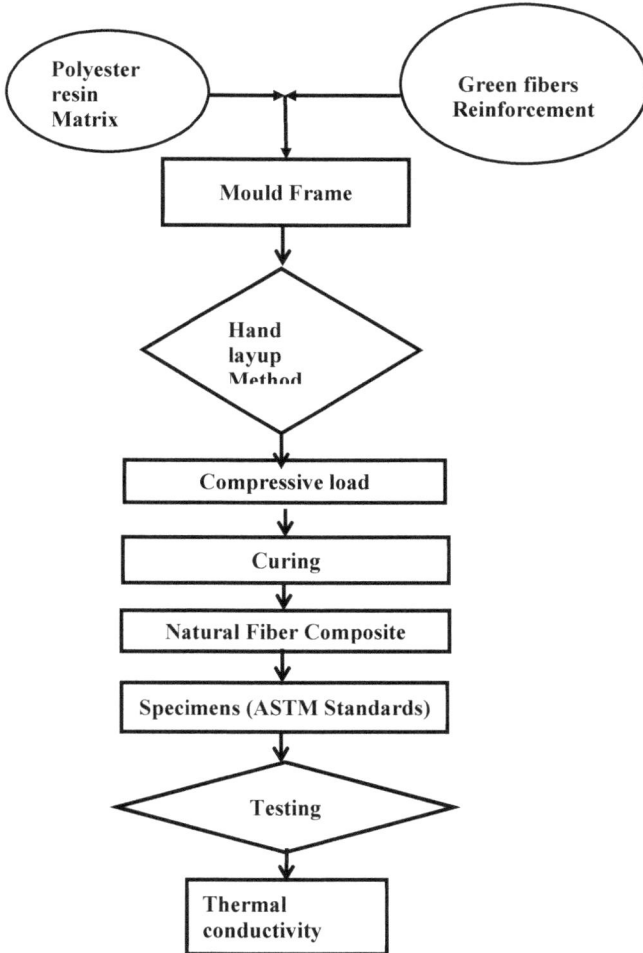

Fig 9. Sequence of steps followed in composite preparation and analysis

Extracted natural fibers are incorporated in unsaturated polyester resin in the preparation of composites using hand lay-up method. Methyl Ethyl Ketone Peroxide (MEKP) catalyst and Cobalt accelerator 1.5% each are mixed to resin for curing at ambient environment.

Sustainable Natural Fiber Composites   Materials Research Forum LLC
Materials Research Foundations **122** (2022) 199-208   https://doi.org/10.21741/9781644901854-8

Layers of fiber and resin were placed until the cavity of mould was completely filled and gently remove any air gap with the by pressing with the roller. Then a thin polyethylene paper of 0.21 mm thick is placed on the rubber mould. A pressure of 0.05 MPa is applied by placing a heavy iron plate and some weights for 24 hours to cure. After that the sample is taken out and filed to get the required sizes. NC thinner is applied on samples and rubbed off to eliminate dust after removing from the mould and is kept in oven for two hours at 800C to remove the moisture content present in it. The prepared samples for thermal conductivity testing as per ASTM E1530 standard are illustrated in fig 10.

Fig 10 Samples used for measuring thermal conductivity

3.   Measurement of thermal conductivity

ASTM E1530-99 standard for the specimen is fulfilled for testing. The diameter of test specimen is 50 mm and 10mm thickness. The DTC 300 model 2022 supplied by TA instruments during testing is shown in Fig 11. Pressure applied on specimen is 0.7 bar. Steady state heat transfer is maintained during testing of the specimen. The tests are carried out by changing the temperature from 30 to 120°C. Once it attains steady state regime, the thermal gradient in terms of temperature between the top and bottom surfaces of sample are measured. Every 30°C rise of temperature is considered as one zone. Total three zones were considered during testing. Therefore, zone one temperature is from 30 to 60 °C and zone2 temperature is from 60-90 °C and zone 3 temperature is from 90 to 120°C. For the calculation of one-dimensional heat transfer through conduction, the following equations are used

$$q = \frac{k(T_1 - T_2)}{L} \tag{1}$$

$$R = \frac{(T_1 - T_2)}{q} \qquad (2)$$

$$k = \frac{L}{R} \qquad (3)$$

Where q is the heat flux (Wm-2), k is the thermal conductivity (W m-1k-1), $T_1 - T_2$ is the difference in temperature (K), L is the thickness of the sample (m), and R is the resistance of sample (m2 K W-1).

Fig. 11. Thermal conductivity test facility

Thermal conductivity test data of natural fiber composites are given in Table 1. Thermal conductivity of the tested green composites is in between the 0.179 to 0.186 W/mK. Based on the quantitative comparison the Vakka is suitable material for energy saving, applications.

Table 1. Thermo-physical properties of all composites

Parameter	Sansevieria sevieria fiber	Broom grass	Vakka fiber	Elephant grass fiber	Borasis fruit fiber	Fishtail palm fiber
Thermal conductivity (W/m.K)	0.183	0.196	0.179	0.183	0.186	0.163

References

[1] Nor Azlina Ramlee, Mohammad Jawaid, Edi Syams Zainudin, Shaikh Abdul Karim Yamani, Modification of oil palm empty fruit bunch and sugarcane bagasse biomass as potential reinforcement for composites panel and thermal insulation materials, J. Bionic Eng. 16 (2019)175-188. https://doi.org/10.1007/s42235-019-0016-5

[2] Nor Azlina Ramlee, Jesuarockiam Naveen, Mohammad Jawaid, Potential of oil palm empty fruit bunch (OPEFB) and sugarcane bagasse fibers for thermal insulation application - A review, Construction and Building Materials 271 (2021) 121519. https://doi.org/10.1016/j.conbuildmat.2020.121519

[3] Basim Abu-Jdayil, Abdel-Hamid Mourad, Waseem Hittini, Muzamil Hassan, Suhaib Hameed, Traditional, state-of-the-art and renewable thermal building insulation materials: An overview, Construction and Building Materials 214 (2019) 709-735. https://doi.org/10.1016/j.conbuildmat.2019.04.102

[4] Hitoshi Takagi, Shuhei Kako, Koji Kusano and Akiharu Ousaka, Thermal conductivity of PLA-bamboo fiber composites, Adv. Composite Mater. 16 (4) (2007) 377-384. https://doi.org/10.1163/156855107782325186

[5] S. W. Kim, S. H. Lee, J. S. Kang and K. H. Kang, Thermal conductivity of thermoplastics reinforced with natural fibers, Int. J. Thermophys. 27 (2006) 1873-1881. https://doi.org/10.1007/s10765-006-0128-0

[6] K.Ramanaiah, A.V.Ratna Prasad, K.Hemachandra Reddy, Experimental study on thermo physical properties of biodegradable borassus fruit fiber-reinforced polyester composites, Materials Today: Proceedings 44 (2021) 1857-1859. https://doi.org/10.1016/j.matpr.2020.12.018

[7] K.Ramanaiah, A.V.Ratna Prasad, K.HemaChandra Reddy, Thermo physical properties of elephant grass fiber reinforced polyester composites, Materials Letters 89 (2012) 156-158. https://doi.org/10.1016/j.matlet.2012.08.070

[8] K.Ramanaiah, A.V.Ratna Prasad, K.HemaChandra Reddy, Thermophysical and fire properties of vakka fiber-reinforced polyester composites, Journal of reinforced plastics and composites, 13 15 (2013)1092-1098. https://doi.org/10.1177/0731684413486366

Sustainable Natural Fiber Composites   Materials Research Forum LLC
Materials Research Foundations **122** (2022) 209-237   https://doi.org/10.21741/9781644901854-9

Chapter 9

Natural Fiber and Biodegradable Plastic Composite

Manoj Mathad[1], Anirudh Kohli[1], Mrutyunjay Adagimath[1], Arun Patil[1]*, Anish Khan[2]

[1]School of Mechanical Engineering, KLE Technological Univeristy, Hubballi, India

[2]Chemistry Department, Faculty of Science, King Abdulaziz University, Jeddah 21589, Saudi Arabia

* patilarun7@gmail.com

Abstract

In today's time, natural fibers are available abundantly and do possess properties such as biodegradability, light weight, good strength etc. due to which more studies are focused in order to develop a composite including natural fibers as their reinforcements. Bio-composites are considered versatile which have gained recognition in industrial material science domain. In this chapter, natural fibers of vegetable and fruit peels are considered as reinforcements and epoxy resin as the matrix to form a novel hybrid composite. The fibers considered in this chapter are onion, potato, carrot, lemon and sweet lime with epoxy resin as their matrix. These natural fibers are reinforced at 10%, 20% and 30% volume fractions. They are subjected to alkali treatment followed by developing the specimen. Then these specimens are subjected to tensile, density, flexural, water absorption rates and also microstructure characterization. Based on the properties obtained for the bio-composite, they are applied for a consumer product device and analysed using ANSYS software. Among all of the developed composites, epoxy lemon yields the optimum results.

Keywords

Bio Composites, Natural Fibers, Biodegradability, Epoxy Resin, Onion Epoxy, Carrot Epoxy, Lemon Epoxy, Potato Epoxy, Sweet Lime Epoxy, Consumer Product

Contents

Natural Fiber and Biodegradable Plastic Composite.....................................**209**

1.   **Introduction**..**210**

2.   **Methodology**...**212**

	2.1	Materials	212
	2.2	Alkali treatment	214
	2.3	Fabrication of composites...	214
	2.4	Density test	214
	2.5	Tensile test..	215
	2.6	Flexural test	215
	2.7	Water absorption test..	215
3.		**Experimental results...**	**216**
	3.1	Density test	216
	3.2	Tensile test..	217
	3.3	Flexural test	220
	3.4	Water absorption test..	223
	3.5	SEM (scanning electron microscope) test	224
4.		**Simulation results ....**	**225**
	4.1	Tensile results	226
	4.2	Flexural results...	228
5.		**Applicative study ....**	**229**
	5.1	Simulation study for consumer product	230
	5.2	Material and mesh criteria	231
	5.3	Boundary and loading condition...	231
	5.4	Structural results	232
Conclusion..			**234**
References			**234**

1. Introduction

With the rise in global pollution and wastage levels, the term biodegradability matters the most. It's desirable to have higher biodegradability so as to decompose the wastage at the highest rate possible. The dumped matter is usually found in terms of organic, medical or commercial waste [1]. The non-biodegradable waste majorly being plastic is dumped in excess every day and to be precise, the amount of biodegradable waste is also dumped in huge quantity. With engineering principles, a small proportion of this biodegradable waste can be used as an alternative for the applications that use non-biodegradable materials. The conversion of organic waste into a suitable form which can be used in applicative field is on the unexplored side [2].

As time passes on, evolution in the domain of engineering innovation enhances and simultaneously, we can see a lot of material resources being utilized for the same purpose. Further, leading to depletion of the naturally available resources [3]. Extinction of natural resources cannot be depicted, but surely there is a decline in availability of useful materials thus there is a need for advanced research & development of materials with an origin dependent on natural matter in the ecosystem [4].

One such case being organic waste which cannot be used directly as obtained for various reasons such as poor shelf life, lower strength and thermal resistance, it's highly effective if used as a composite material. The organic waste includes wastes from kitchen such as vegetables, fruits and wastage found in environment such as dead extract of plants. The end product of the composites involving organic matter as one of its phases is termed as Bio-composites. Composites usually have two main segregations, one being the matrix and the other being reinforcement. There is huge scope developed for bio-compatible materials in sector of product designing & Manufacturing as everyone around is focusing on Biodegradable products. New product development in industrial domain consists of a process called material selection to machine parts, wherein Sustainability, Environmental safety, free from hazardous effect to ecosystem are the major criterions to be fulfilled [5,6].

Natural fiber composites, either biodegradable green composite or hybrid composite material made up of plant fiber reinforcement with epoxy resin, PLA or ABS as polymer matrix, are gaining recognition for various engineering development application because of their low density, bio-degradable design & ecosystem, recyclability, low cost producing & processing and also comparative mechanical properties. Attention on bio-composite is growing in field of aerospace & automotive, structural engineering, packaging industry and other application attributes as mentioned in table 1, with a varied performance efficiency based upon sustainability of loading factor involved [7].

Plant fibers/Green fibers are used as filler or reinforcement in order to increase the biocompatibility of the material, Plant fibers are botanically structural elements of plants and its derivatives like vegetables, fruits, and other by-products. It consists of cellulose, hemicellulose and lignin, which combine together to possess good mechanical strength and stiffness to the tissue of plant [8].

Sustainable Natural Fiber Composites   Materials Research Forum LLC
Materials Research Foundations **122** (2022) 209-237   https://doi.org/10.21741/9781644901854-9

Table 1. Material attributes and applications.

Material attributes	Application attributes
Bio degradable	Single time usable items, safely disposable
Non-Toxic	Human interaction products, toys
Low cost	Consumer goods, coatings
Bio-compatible	Medical applications, polymers
Water absorption	Drying application
Good weight - stiffness	Automotive/aerospace, electronics etc.

2.   Methodology

The Fig. 1 represents the methodology that has been carried out in this chapter.

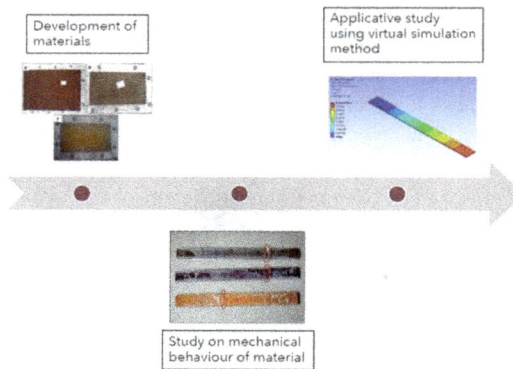

Figure 1. Methodology representation.

2.1   Materials

The composites that have been developed include peels from potato, onion, carrot, lemon and sweet lime as the reinforcements while the matrix is epoxy resin in all the cases (L-12) with hardener (K-6) as shown in fig.2. The natural fibers extracted are in the form of particulate fibers. In order to have the particulate form of fibers, the extracts were initially exposed to sunlight for effective drying in order to remove moisture content and then were grounded to obtain a particle size of 300 microns. Later the composition was tested and the properties as shown in table 2 were obtained.

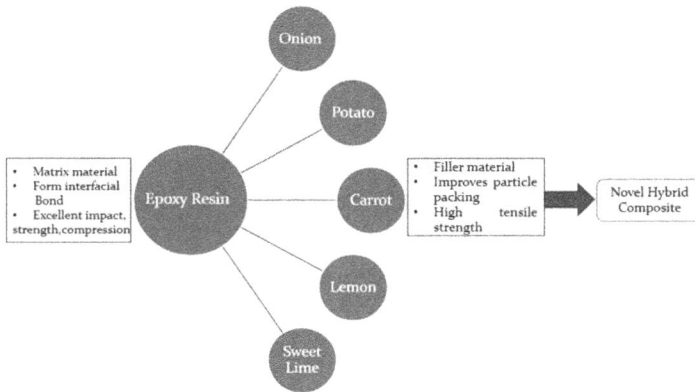

Figure 2. Material description.

Table 2. Natural fiber composition [10].

Sl.no.	Parameters	Unit	Potato fiber	Onion fiber	Carrot fiber	Sweet Lime	Lemon
1	Moisture	%	4.26	4.81	5.14	6.38	7.13
2	Total ash	%	5.34	6.5	7.61	3.70	6.23
3	Crude fiber	%	33.82	32	34.26	32.66	33.15
4	Cellulose	%	42.60	41.54	42.68	41.58	41.80
5	Specific gravity	-	1.075	0.847	0.751	0.725	0.781
6	Lignin	%	44.13	43.11	44.81	38.63	43.51
7	pH	-	6.9	6.6	6.82	7.0	6.8
8	Hemicellulose	%	ND	ND	ND	ND	ND

Note that the following abbreviations are used in the below discussions. Epoxy onion bio composite is abbreviated as EO, epoxy carrot as EC, epoxy potato as EP, epoxy lemon as EL and epoxy sweet lime as ESL bio composites.

2.2   Alkali treatment

The cellulose levels obtained from testing lie in the range of medium-high cellulose levels and alkali treatment is the most suitable for these kinds of fibers. A study revealed that subjecting the natural fibers for alkali treatment enhances its tensile strength, flexural strength and absorption characteristics. After extracting the fibers in particulate form, they were soaked in NaOH solution for 2 hours and then were rinsed with distilled water for 3 consecutive times. In order to avoid clogging of particles when used as reinforcements and remove the moisture content, the fibers were dried in an oven at 50°C for an hour.

2.3   Fabrication of composites

The fabrication of the bio composites was done by hand layup technique. The matrix being epoxy resin of grade L-12 was reinforced with onion, potato, carrot, lemon and sweet lime particles individually. Note that the matrix was mixed with hardener K-6 in the ratio of 10:1. The specimens were fabricated as per ASTM standards with varying volume fractions of the reinforcements for 10%, 20% and 30% [11]. These composites were cured for 24 hours and then carefully detached from the molds. The fabrication of composites is as shown in fig.3.

Figure 3. Composites after hand layup technique.

2.4   Density test

The density of the composites was measured using ASTM D792 standards and also involved Archimedes principle. Initially cantilever setup is positioned on the weighing machine and after attaining the equilibrium, the weight is set to zero [12]. Thereafter, the specimen is placed in air and its weight is recorded. The specimen is then placed in water and weight is recorded. The density was calculated using the Eq. 1.

$$\text{Density } \rho = \frac{W_a}{W_a - W_w} \tag{1}$$

Where ρ, actual density of specimen (g/cc); W_a is weight of specimen in air (g); W_w is weight of specimen in water (g).

2.5 Tensile test

This tensile test is carried out using ASTM D3039 standards. The machine used is universal testing machine (UTM) whose capacity is 10 tons. In this test, the specimen is subjected to a cross head speed of 3mm/min. The standard dimensions of the specimen being 250mm x 25mm x 3mm and the gauge length was 138mm. This test was performed using the Universal Testing Machine.

2.6 Flexural test

This particular test carried out on the specimen is done as per the ASTM D-7264 standards wherein the specimen was subjected to three-point bending test as described in fig. 4. As per the standards, the dimensions of specimen were set to be 154mm x 13mm x 3mm. Note that the cross-head speed was kept same even in case of flexural test i.e., 3mm/min.

Figure 4. Flexural test setup.

2.7 Water absorption test

The test is done as per ASTM D5229 standards. Initially the weight of the bio composite specimens was measured and then submerged in the container of distilled water. After a duration of 24 hours, the specimens are removed from the container and wiped gently with a dry cloth. After the weight is measured, it is then again submerged in water and the weight of the specimen is measured after 24 hours [13]. This is done for a duration of 240 hours. The water absorbed during this duration is equivalent to the increase in the weight of the specimen and is calculated as per the Eq. 2. The fig. 5 shows the experimental tests being conducted and the ouput from each test.

$$\text{Increase in weight \%} = \frac{Final\ weight - Initial\ weight}{Initial\ weight} \times 100 \qquad (2)$$

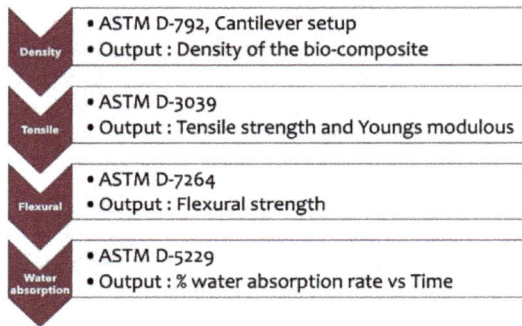

Density	• ASTM D-792, Cantilever setup • Output : Density of the bio-composite
Tensile	• ASTM D-3039 • Output : Tensile strength and Youngs modulous
Flexural	• ASTM D-7264 • Output : Flexural strength
Water absorption	• ASTM D-5229 • Output : % water absorption rate vs Time

Figure 5. Experimental testing standards and output obtained from each test

3.   Experimental results

In this section, the characteristic values obtained from the above-mentioned tests are discussed in detail.

3.1   Density test

The density of the bio composites are obtained in 10%, 20% and 30% of volume fractions of the reinforced biodegradable fibers. The values are listed in table. 3 and the trend observed in the case of density is that with the increase in volume fraction of the reinforcements/fillers, the density decreases. Hence, the highest value of density is observed for epoxy lemon for a volume fraction of 10% and its corresponding value as 1206 Kg/m^3. It is to be noted that the density of epoxy resin is 1300 Kg/m^3. The density of EO is calculated as per Eq. 1. In order to get a visual aspect a graphical reperenstation is shown in fig. 6.

Table 3. Density of bio composites.

Volume fraction (%)	Density (Kg/m³)				
	ESL	EL	EO	EP	EC
10	1192	1206	1138	1153	1121
20	1103	1152	1128	1149	1080
30	1087	1098	1040	1139	1039

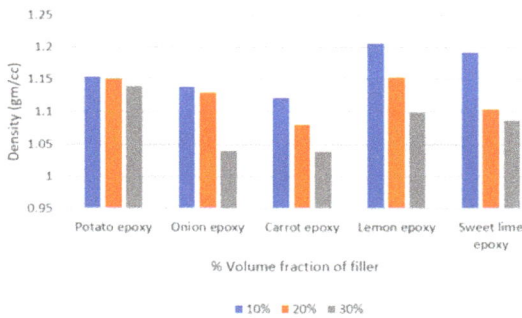

Figure 6. Graphical representation of Density for bio composites

3.2   Tensile test

The results obtained from the tensile test follow a different pattern than that of density test. There are two cases that catch the attention, one being in a way such that tensile strength and young's modulus are in direct correlation with the increase in volume fraction of reinforcements for Epoxy sweet lime and epoxy lemon bio composites. The other being, with increase in volume fractions of the reinforcements the tensile strength and Youngs modulus decreases for Epoxy carrot/potato/onion. The results of tensile test for EO, EP and EC are listed in table 4.

There are certain reasons such as bonding at the interface, particle size and fraction volume percentage of embedded natural fibers which contribute in deciding the property values. From the above table it is quite clear that for volume fractions of 10%, the strength and tensile modulus obtained is highest for Epoxy onion/potato/carrot. The major reason of epoxy/onion showing enhanced results than other two composites is due to the presence of an antibacterial substance called flavonoid which usually blends well with the matrix which

217

in turn make the bonding interface stiff [14]. The fig. 7 shows the failure modes of the specimen under tensile loading and it is quite visible that the failure shape occurs normal to the cross-section which indicates these composites are brittle.

Table 4. Tesnile strength and modulus for EO, EP and EC composites.

Composite	Volume fraction %	Tensile strength (MPa)	Young's modulus (MPa)
Onion/epoxy	10	20.8	659.980
	20	19.13	509.537
	30	16.22	481.510
Potato/epoxy	10	20.75	520
	20	16.79	445.520
	30	9.64	337.105
Carrot/epoxy	10	19.33	515
	20	10.62	366.485
	30	10.09	256.572

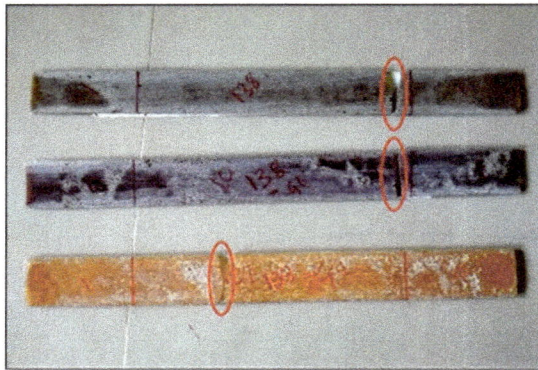

Figure 7. EP, EO and EC specimens failed under tensile load.

Table 5. Tesnile strength and modulus for ESL, EL composites.

Composite	Volume fraction %	Tensile strength (MPa)	Young's modulus (MPa)
Sweet lime/epoxy	10	18.30	743.90
	20	24.04	924.61
	30	35.16	1013.25
Lemon/epoxy	10	28.29	999.64
	20	36.62	1204.61
	30	48.22	1358.30

For the composites, ESL and EL the values of tensile strength and Tensile modulus increase with increase in the volume fraction of the embedded natural fibers. Again, the factors such as strength of natural fibers, interfacial bonding, strength of resin play a vital role for arriving at the specified values. The particle size remains an uncovered area in order to study the variations in tensile strength and modulus with varying particle size. It is to be noted that the tensile strength and modulus is highest in case of lemon/epoxy bio-composite as shown in table. 5. The stress-strain for EO, EC and EP composites are shown in fig. 8. The graphical reprerentation of tensile strength and tensile modulus of all the composites is shown in fig. 9(a) and fig. 9(b). The tensile modulus can be calculated as per the steps and Eq. 3 mentioned below.

Width of the specimen (b)= 25mm

Thickness of the specimen (h)= 3mm

Cross sectional area (A)= $b \times h = 25 \times 3 = 75\text{mm}^2$

Load at failure (F)= 1.56 KN

Maximum allowable stress (σ) = $\dfrac{1.56 \times 1000}{75}$

$\qquad\qquad = 20.8$ MPa

Strain (e)= 0.03151

Tensile modulus (E) = $\dfrac{\sigma}{e}$ $\qquad\qquad\qquad\qquad\qquad\qquad$ (3)

$\qquad\qquad = 659.980$ MPa

Sustainable Natural Fiber Composites   Materials Research Forum LLC
Materials Research Foundations **122** (2022) 209-237   https://doi.org/10.21741/9781644901854-9

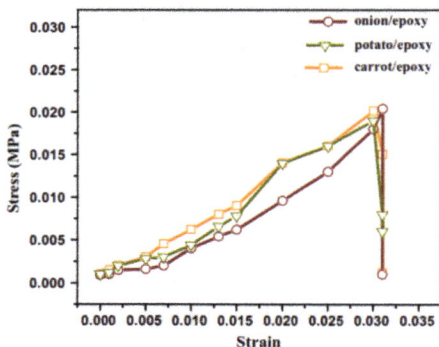

Figure 8. Stress-Strain curve for EO, EP and EC composites.

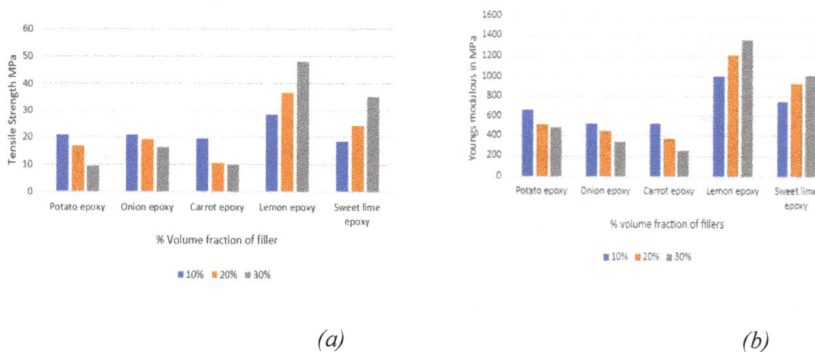

(a)   *(b)*

Figure 9. Graphical representation of (a) Tensile strength and (b) Tensile Modulus.

3.3   Flexural test

The specimens were subjected to three-point bending test and the cross-head speed was 3mm/min which is same as that of tensile test. It is evident from the values obtained for the epoxy/ onion, carrot, potato that with increase in volume fraction of embedded natural fibers there is decrease in the value of flexural strength [15]. While for ESL and EL, with increase in volume fraction of reinforcements the flexural strength also increases.

Sustainable Natural Fiber Composites   Materials Research Forum LLC
Materials Research Foundations **122** (2022) 209-237   https://doi.org/10.21741/9781644901854-9

Table 6. Flexural strength for EO, EP and EC composites

Composite	Volume fraction %	Flexural strength (MPa)
	10	36.06
Onion/epoxy	20	28.85
	30	28.85
	10	35
Potato/epoxy	20	26.25
	30	18.85
	10	34.55
Carrot/epoxy	20	19
	30	17.44

From the table. 6 and table. 7, it is evident that the lemon epoxy composite has the highest flexural strength of 79.32 MPa at 30% volume fraction of the reinforcements owing its credits to high density and good interfacial bonding between the matrix and the filler material. Apart from the citrus fillers embedded bio composites, the highest value is observed for onion epoxy at 10% volume fraction. The fig. 10 represents failure mode of specimens subjected to bending test. While the graphical representation of flexural strength of the composites is shown in fig. 11. The flexural strength is calculated using the Eq. 4 as shown below.

Load at failure (F)= 50N

Gauge length/support span (L)= 100mm

Thickness of specimen (h)= 4mm

Width of the specimen (b) =13mm

$$\text{Flexural strength } (\sigma_f) = \frac{3FL}{2bh^2} \tag{4}$$

$$= \frac{3 \times 50 \times 100}{2 \times 13 \times 4^2}$$

$$= 36.057 \text{ MPa}$$

Materials Research Forum LLC
https://doi.org/10.21741/9781644901854-9

Table 7. Flexural strength for ESL, EL composites

Composite	Volume fraction %	Flexural strength (MPa)
Sweet lime/epoxy	10	50.48
	20	69.40
	30	72.11
Lemon/epoxy	10	57.69
	20	68.50
	30	79.32

Figure 10. EP, EO and EC specimens failed under flexural load

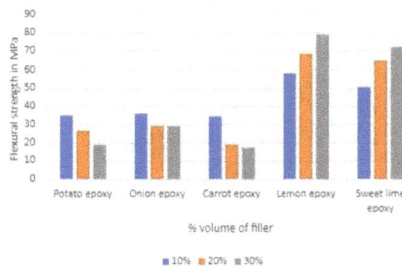

Figure 11. Graphical representation of Flexural strength for bio composites

3.4   Water absorption test

The values obtained are listed in the table. 8. The highest value of water absorption rate is observed for the epoxy sweet lime composite and it is quite generic in all the case that with increase in the volume fraction of the natural fibers, the water absorption rate increases. For ESL and EL, it is seen that the difference is due to the specific gravity i.e., specific gravity for EL is 781 Kg/m³ and that for ESL is 725 Kg/m³ which implies higher the density lesser will be the cavities in the composites and lesser will be the water absorption rate. Hence EL has lesser water absorption rate than ESL.

Table 8. Water absorption rate for bio-composites

Volume fraction (%)	Water absorption after 240 hours (%)				
	ESL	EL	EO	EP	EC
10	4.47	3.68	1.1243	1.2936	1.3301
20	4.82	4.0	1.2681	1.3112	1.3755
30	5.73	4.32	1.2680	1.3277	1.3952

Considering the cases of EO, EP and EC, epoxy carrot shows higher water absorption rate than others because of the hydrophobic nature of the filler material and the matrix. Whenever the filler material absorbs water, it undergoes swelling which leads to crack formation and filling of water takes place in those cracks due to brittle capillary effect. The fig. 12(a) and fig. 12(b) shows graphical analysis of percentage of water absorption rate v/s time for EC and EL.

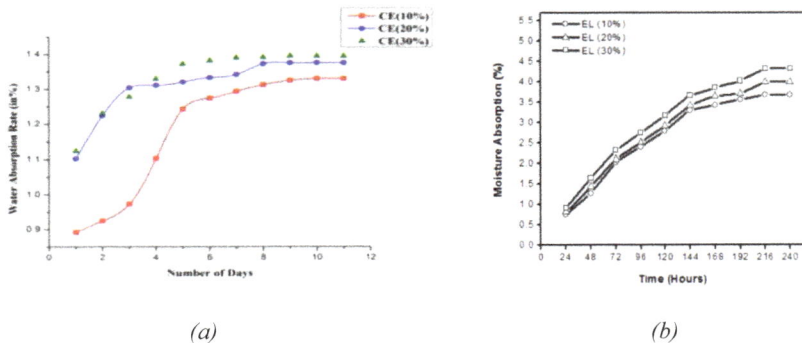

(a)   *(b)*

Figure 12. Water absorption rates of (a) EC composite and (b) EL composite

3.5   SEM (scanning electron microscope) test

In this analysis, the fractured specimens obtained from tensile tests were taken and their cross section was considered for SEM analysis. The micrographs were set at 1000X magnification which shows air voids, matrix/reinforcement bonding etc. The reason for the decline in mechanical properties of EP and EC compared to EO composites can be considered due to large number of air voids. It is also seen that the particle distribution is homogenous in EO, EP and EC composites which indicates good adhesion between filler and the matrix elements. The fig. 13, fig. 14 and fig. 15 show the microstructure of EO, EP and EC composites respectively.

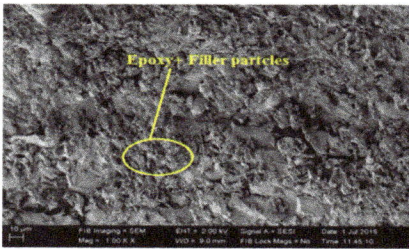

Figure 13. Microstructure characterization   Figure 14. Microstructure characterization
for EO composite   for EP composite

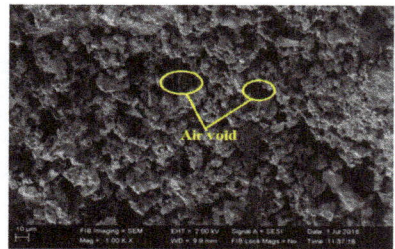

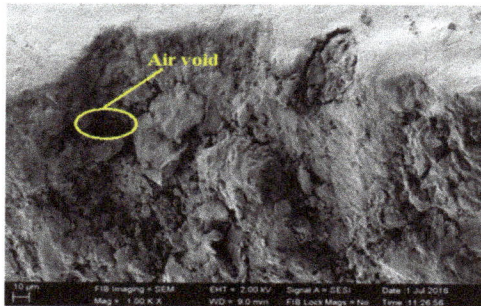

Figure 15. Microstructure characterization for EC composite

For the case of EL and ESL, 30% volume fraction of embedded fibers are studied since the results are superior than for the other volume fractions. Here the micrographs are studied for both 500X and 1000x magnification. The fig. 16(a) and fig. 16(b) shows the microstructure for ESL bio composites and the presence of blow holes may be due to the entrapment of air/gases during fabrication. The fig. 16(c) and fig. 16(d) shows the microstructure of EL composite at 500X and 1000X magnification and it is evident that treatment of particulates form of fibers with NaOH i.e., alkali treatment leads to better matrix-reinforcement adhesion.

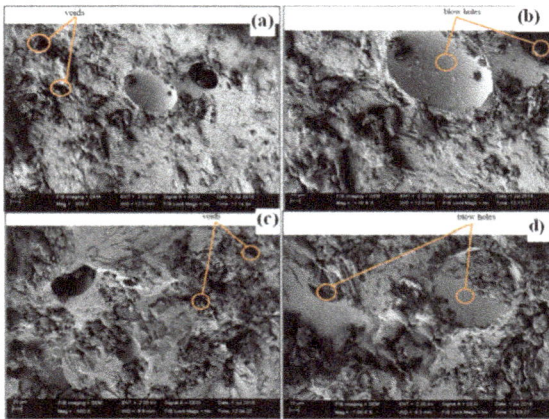

Figure 16. Microstructure characterization (a), (b) for ESL composite and (c), (d) for EL composite

4. Simulation results

After obtaining the tensile, flexural, water absorption results from experimental setups it is now time to demonstrate the same in virtual simulation and compare the results. Since, the optimum values were obtained for 10% volume fraction of embedded natural fibers in epoxy resin, the virtual simulation is done for EP, EO and EC at 10% volume fraction only. Both tensile and flexural results were taken into account for comparison.

The test specimens were prepared in design modeler of ANSYS. For tensile test the specimen was modelled as per ASTM D3039 standards and for flexural test the specimen was modelled as per ASTM 7264 standards. In order to constitute a composite, the modelling was done in such a way that three layers were setup in laminate and not in particulate form of the fillers. The laminates were assigned bonded contact.

Sustainable Natural Fiber Composites   Materials Research Forum LLC
Materials Research Foundations **122** (2022) 209-237   https://doi.org/10.21741/9781644901854-9

4.1 Tensile results

Initially the deformation occurred in the standard specimen of onion epoxy under tensile load is calculated using the Eq. 5 and is similarly done for all the other bio composites. The experimental deformation comes out to be 4.35mm for EO composite. This model is meshed with an element size of 0.5mm and resulting in creation of 939867 nodes and 175000 elements which is evident fig. 18. The model is assigned with onion at 10% volume fraction and epoxy resin properties as shown in fig. 17 and subjected to ultimate tensile load. The tensile test model and its boundary and loading conditions are shown in fig. 19 and fig. 20. The value of deformation is taken out from virtual analysis and is equivalent to 3.69mm.

Deformation $(\Delta) = \frac{FL}{AE}$ (5)

Cross sectional area $(A) = b \times h = 25 \times 3 = 75\text{mm}^2$

Load at failure $(F) = 1.56$ KN

Tensile modulus $(E) = 659.980$ MPa

Effective length $(L) = 138$mm

$(\Delta) = \frac{1560 \times 138}{75 \times 659.98} = 4.35$ mm

Table 9. Material Properties.

Material	Density (g/cm³)	Elastic Modulus (GPa)	Poisson's ratio
Epoxy	1.162	30.79	0.35-0.5
Potato	1.075	0.2	0.26
Onion	0.847	9.71	0.25
Carrot	0.751	1	0.25

The deviation in the results obtained from experimental and virtual simulation is due to the assumption of material whose properties as mentioned in table. 9 are considered to be homogeneous and isotropic in ANSYS software. The values obtained from the software are conservative with the experimental values due to consideration of high factory of safety and model reliability [16]. The analysis results for EC, EP and EO are listed in table. 10.

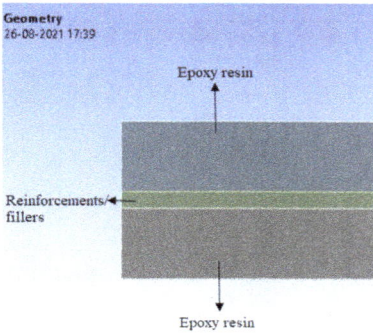

Figure 17. Laminate geometry

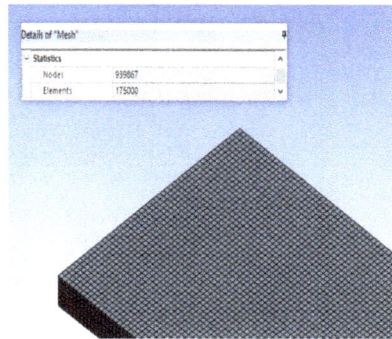

Figure 18. Meshed model

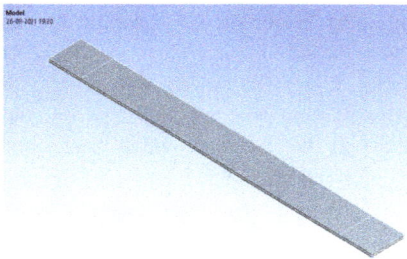

Figure 19. Actual tensile test model

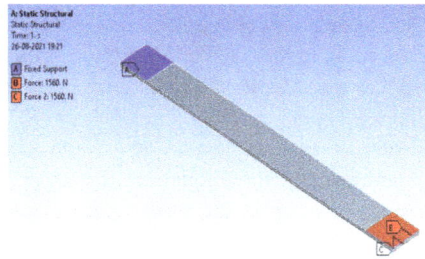

Figure 20. Boundary and loading conditions

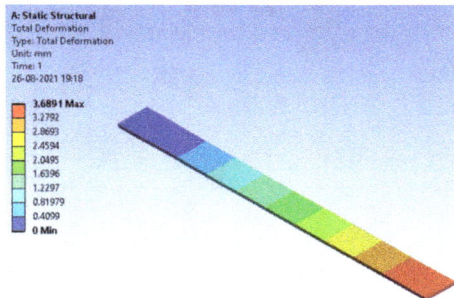

Figure 21. Deformation in EO composite.

Table 10. Experimental and analysis deformation.

Composite	Experimental deformation (mm)	Analysis deformation (mm)	% Error
Onion Epoxy	4.35	3.69	15.17
Carrot epoxy	3.5	3.842	9.77
Potato epoxy	4.2	3.6383	13.37

4.2   Flexural results

In this the specimen is subjected to three-point bending test. The model and the contacts between the laminate layers are shown in fig. 22. A line load of 50N is subjected at the center of the specimen and fixed at two ends which is evident from fig. 23(d) and fig. 23(e). The meshed model is resized and is shown in fig. 23(c). For onion epoxy composite, the flexural strength obtained was 36.06 MPa while with the simulation results, we got 42.82 MPa as listed in table. 11. An error percentage for this composite was equivalent to 15.78 and usually the deviation can clock up to 20% in case of composite material since material reinforcement and behavior of the composite will be different when performed experimentally and through simulation [17]. The deformation obtained in case of flexural test is shown in fig. 23(f).

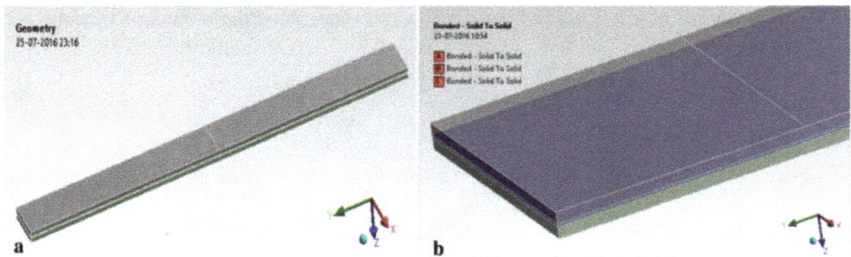

Figure 22. Flexural test geometry and contacts between laminate layers

Table 11. Experimental and analysis flexural strength.

Composite	Experimental flexural strength (MPa)	Virtually obtained flexural strength (MPa)	% Error
EO	36.06	42.82	15.78
EP	35	45.68	23.38
EC	34.55	43.56	20.68

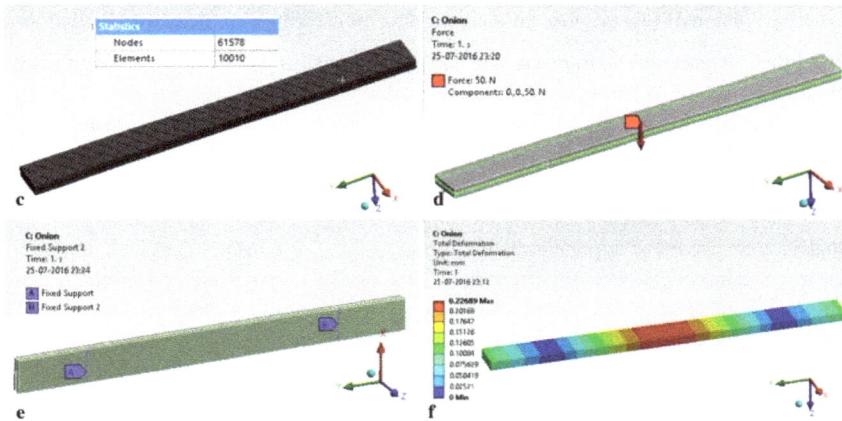

Figure 23. (c) Meshed model, (d), (e) Boundary and loading conditions, (f) Total deformation in EO composite

5. Applicative study

Ecological concerns in innovation development & material science have renewed the interest in bio-composites and green fibers. Green/Bio composite combines resin and plant fibers to form natural composite. Biodegradable polymers found in nature which are part of research study in domain of bio-composite material science are proteins- collagen, albumin, cellulose, polyesters, polymers etc. and matrix or base material being either thermoset or thermoplastic have their specific advantages.

Few of the bio-composites are exceptional with strength weight ratio, considering this mechanical property these materials have great scope for utility in the field of aerospace/automotive like light duty parts, interior structural applications, door frames etc.

Bio-composite material formed with a fusion of plant-based polymers and thermoplastic or thermosetting matrix material make a great commercial product such as floor coating, food industry items, decorative/art products etc. In recent times, 3D printing technology is emerging in manufacturing domain, not just by limiting itself for prototyping but also being utilized to substantial amount in testing various materials [18,19]. Present ongoing study claims that good interfacial bonding material like PLA, Epoxy resin, ABS etc. provides higher feasibility for producing of bio-composites, considering reinforcements as green fibers using the aid of 3D printing technology, specifically FDM (Fused-Deposition modelling) fusion of two or more unlike materials for creating complex structured product.

Hence, there is great scope in producing the bio-composites using additive manufacturing technology, chemical processing and Extrusion technology as we can see tissue engineering integrated with natural composites is being implemented at faster pace in medical field [20]. And utilization of natural fibers in packaging industry has a good value.

5.1   Simulation study for consumer product

Usually, there are lot of applications that products that use non-biodegradable material as their major component. Consumer products such as mouse, joysticks, plugs etc. are most commonly used. The mentioned products use ABS as their outer material which provides higher manufacturing feasibility using additive manufacturing/ 3-d printing [21]. In this simulation an alternative for the use of ABS is analyzed using bio-composites developed in the first half of the discussion. The applicative study involves structural analysis of computer mouse along with the fatigue life estimation for all the composites developed except carrot/epoxy because of high level lignin content. The fig. 24 and fig. 25 shows the model considered for analysis.

Figure 24. Mouse solid model

Figure 25. Hollow model

Sustainable Natural Fiber Composites   Materials Research Forum LLC
Materials Research Foundations **122** (2022) 209-237   https://doi.org/10.21741/9781644901854-9

5.2   Material and mesh criteria

This model is meshed for different element sizes based on the areas of loads application. The fig. 26 represents meshing at the front surface with element size as 0.5mm and the fig. 27 indicates element size of 1.2mm [22]. The rest parts are meshed with medium element size and resulting in the formation of 107277 nodes and 60853 elements. The material as used for the current application is ABS and its properties are mentioned in table. 12.

Table 12. ABS Material properties [23]

Material	ABS
Young's Modulus (MPa)	1470
Density (Kg/m^3)	1010
Poison ratio	0.32

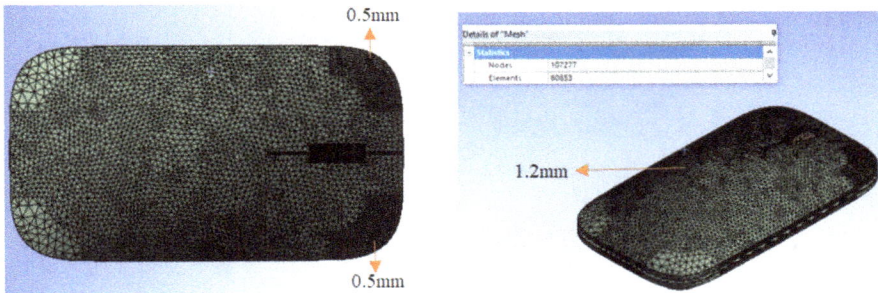

Figure 26. Meshed model with element sizing.   Figure 27. Meshed model with statistics.

5.3   Boundary and loading condition

The locations A and B as shown in fig. 28 are considered to be left and right click button respectively and a pressure equivalent to a load of 10N has been applied individually on both sides [24,25]. The area C is flat and higher contact thus hence a pressure equivalent to 8N has been applied. The lower surface of the mouse is fixed.

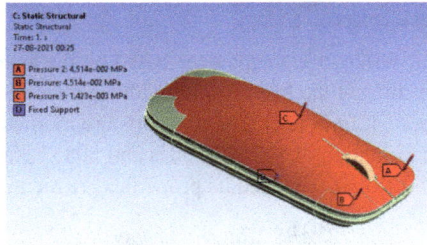

Figure 28. Applied pressure and fixed support

5.4   Structural results

For the specific boundary conditions and the materials selected, this analysis shows equivalent stress the total deformation developed in the consumer product. The fig. 29 and fig. 30 shows that 0.0997 MPa of stress and 0.829 microns of deformation for the ABS material. The deformation is quite negligible in this case. The next case is considered for onion/ epoxy composite and it is evident from the fig. 31 and fig. 32 that the stress is equivalent to 0.1137 MPa and the deformation occurred in the product is 1.82 microns. Considering the same mesh criteria and boundary conditions, this structure is analyzed for EP, ESL and EL.

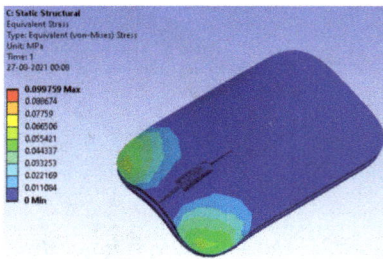

Figure 29. Stress in ABS material   *Figure 30. Deformation in ABS material*

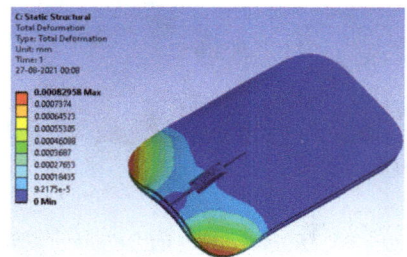

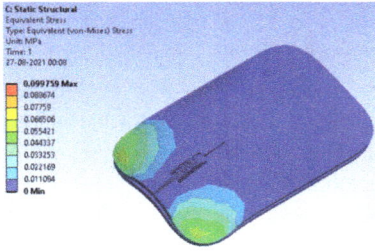

Figure 31. Stress in EO composite

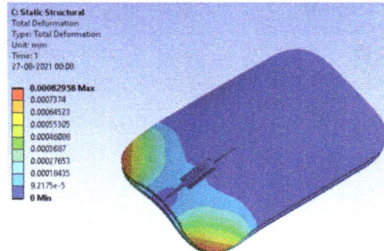

Figure 32. Deformation in EO composite

The structure was also analyzed for fatigue life estimation and it was observed that all the elements had underwent infinite cycles as shown in fig. 33 and hence the developed composites can be used for light weight fatigue applications too [26,27]. The table. 13 shows equivalent stress and deformation values obtained for different natural fiber embedded composites.

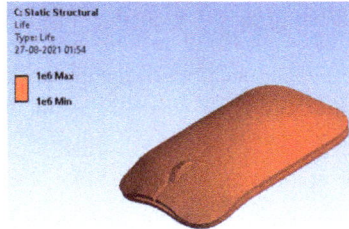

Figure 33. Fatigue life.

Table 13. Structural and fatigue analysis for ABS and developed composites.

Material/Composite	Equivalent stress	Total deformation	Fatigue life (cycles)
ABS	0.0998	0.000829	
Onion/epoxy	0.1137	0.00182	10^6
Potato/epoxy	0.1138	0.00231	
EL	0.1095	0.000891	
ESL	0.1094	0.00119	

Conclusion

The fabricated bio composites were initially tested under experimental setups for density, tensile, flexural and water absorption rates.

i.   The highest density was noted for all the composites at 10% volume fraction among this EL had the highest value of 1206 Kg/m^3 and the least density was observed for EO at 30% volume fraction and was equivalent to 1040 Kg/m^3.

ii.   For the tensile test, the highest value of tensile strength was noted for EL at 30% volume fraction which is 48.22 MPa and tensile or young's modulus was highest for the same which is equivalent to 1358.30 MPa. Note that the values obtained using tensile test were less for EO, EP and EC than the EL and ESL bio composites.

iii.   The same trend as that for tensile test was observed in flexural test too. The maximum flexural was noted for EL at 30% volume fraction which is 79.32 MPa. The water absorption rate increased with increase in reinforcements and was highest for ESL at 30% volume fraction.

Later the same tensile, flexural test were simulated in ANSYS in order to demonstrate the deviation from actual results. The significant factor to be noted here is, the fillers/reinforcements were taken as laminate form and not as particulate form. The deviation recorded for deformation occurring through tensile test at ultimate load was ranging from 0-15% for EO, EP and EC composites, while that in case flexural strength deviation was slightly higher and ranged from 0-20% approximately.

The data obtained for the bio composites was then applied for a consumer product in this case being computer mouse and structural analysis including fatigue life was analyzed. The testing was done for the presently used material i.e., ABS and for all the composites developed except EC due to slightly higher value of lignin content. It was found that the stress developed for ABS was 0.0998 MPa and deformation was 0.829 microns. The results converging to this value were satisfactorily found in EL composite which were 0.1094 MPa of equivalent stress and 1.19 microns of deformation. The deviation for ABS and EL composite in terms of stress developed was 8.86%.

Hence, EL composite showed superior results than the other developed composites and would be an area of future scope for considering additive manufacturing of bio-composites.

References

[1]   Stickler PB (2002) In: Proceedings of the American society for composites 17th technical conference, October 21–24, West Lafayette, IN.

Materials Research Forum LLC
https://doi.org/10.21741/9781644901854-9

[2]   Faruk O, Bledzki AK (2012) Processing of biofiber-reinforced composites. In: Wiley encyclopedia of composites, Wiley Inc., Published Online: 20 Jul 2012. https://doi.org/10.1002/9781118097298.weoc203

[3]   Broutman LJ, Agarwal BD (1974) Polym Eng Sci 14:581–588. https://doi.org/10.1002/pen.760140808

[4]   Bigg DM, Hiscock DF, Preston JR, Bradbury EJ (1988) Polym Compos 9:222–228. https://doi.org/10.1002/pc.750090309

[5]   Bannister MK (2004) Proc Inst Mech Eng Part L J Mater Design Appl 218:87–93.

[6]   M.R. Manshor, H. Anuar, M.N. Nur Aimi, M.I. Ahmad Fitrie, W.B. Wan Nazri, S.M. Sapuan, Y.A. El-Shekeil, M.U. Wahit, Materials and Design 59 (2014) 279-286. https://doi.org/10.1016/j.matdes.2014.02.062

[7]   A. O'Donnell, M.A. Dweib, R.P. Wool. (2004), Bio-composites Science and Technology, 1135-1145. https://doi.org/10.1016/j.compscitech.2003.09.024

[8]   J. F. Silva, J. P. Nunes, A. C. Duro and B. F. Castro. (n.d.), The 19th International Conference on Bio-composite Materials, 8626- 8637.

[9]   Joshi, S.V.; Lawrence, T.D.; Mohanty, A.; Arora, S.: Are natural fiber composites environmentally superior to glass fiber reinforced composites, Compos. Part A Appl. Sci. Manuf. 35, 371–376 (2004). https://doi.org/10.1016/j.compositesa.2003.09.016

[10]   Arun Y.Patil, Umbrajkar Hrishikesh N Basavaraj G D, Krishnaraja G Kodancha, Gireesha R Chalageri 2018. "Influence of Bio-degradable Natural Fiber Embedded in Polymer Matrix", proc.,Materials Today, Elsevier,Vol.5, 7532–7540. https://doi.org/10.1016/j.matpr.2017.11.425

[11]   Arun Y. Patil, N. R. Banapurmath, Jayachandra S.Y., B.B. Kotturshettar, Ashok S Shettar, G. D. Basavaraj, R. Keshavamurthy, T. M. YunusKhan, Shridhar Mathd, 2019. "Experimental and simulation studies on waste vegetable peels as bio-composite fillers for light duty applications", Arabian Journal of Engineering Science, Springer-Nature publications, June, 2019.https://doi.org/10.1007/s13369-019-03951-2. https://doi.org/10.1007/s13369-019-03951-2

[12]   Lau, K.; Ho, M.; Au-Yeung, C.; Cheung, H.: Bio-composites: their multifunctionality. Int. J. Smart Nano Mater. 1(1), 13–27 (2010). https://doi.org/10.1080/19475411003589780

[13]   Foulk, J.D., Akin, D.E., Dodd, R.B.: New low-cost flax fibers for composites. In: SAE Technical Paper Number 2000-01-1133, SAE 2000 World Congress, Detroit, March 6–9, 2000. https://doi.org/10.4271/2000-01-1133

[14]   Pereira, P.H.F.; de Rosa, M.F.; Cioffi, M.O.H.; Benini, K.C.C.; Milanese, A.C.; Voorwald, H.J.C.; Mulinari, D.R.: Vegetal fibers in polymeric composites: a review. Polímeros 25(1), 9–22 (2015). https://doi.org/10.1590/0104-1428.1722

[15]   Pereira, P.H.F.; de Rosa, M.F.; Cioffi, M.O.H.; Benini, K.C.C.; Milanese, A.C.; Voorwald, H.J.C.; Mulinari, D.R.: Vegetal fibers in polymeric composites: a review. Polímeros 25(1), 9–22 (2015). https://doi.org/10.1590/0104-1428.1722

[16]   Arun Y. Patil, N. R. Banapurmath, Shivangi U S, 2020 "Feasibility study of Epoxy coated Poly Lactic Acid as a sustainable replacement for River sand", Journal of Cleaner Production, Elsevier publications, Vol. 267. https://doi.org/10.1016/j.jclepro.2020.121750

[17]   D N Yashasvi , Jatin Badkar, Jyoti Kalburgi, Kartik Koppalkar, 2020 IOP Conf. Ser.: Mater. Sci. Eng. 872 012016. https://doi.org/10.1088/1757-899X/872/1/012016

[18]   Prithviraj Kandekar, Akshay Acharaya, Aakash Chatta, Anup Kamat, 2020 IOP Conf. Ser.: Mater. Sci. Eng. 872 012076. https://doi.org/10.1088/1757-899X/872/1/012076

[19]   Vishalagoud S. Patil, Farheen Banoo, R.V. Kurahatti, Arun Y. Patil, G.U. Raju, Manzoore Elahi M. Soudagar, Ravinder Kumar, C. Ahamed S, A study of sound pressure level (SPL) inside the truck cabin for new acoustic materials: An experimental and FEA approach, Alexandria Engineering Journal, IF: 2.46, Accepted, 2021. (Scopus and Web of Science)

[20]   Arun Y. Patil, Akash Naik, Bhavik Vakani, Rahul Kundu, N. R. Banapurmath, Roseline M, Lekha Krishnapillai, Shridhar N.Mathad, Next Generation material for dental teeth and denture base material: Limpet Teeth (LT) as an alternative reinforcement in Polymethylmethacrylate (PMMA), JOURNAL OF NANO-AND ELECTRONIC PHYSICS, Accepted, Vol. 5 No 4, 04001(7pp) 2021.

[21]   Anirudh Kohli, Ishwar S, Charan M J, C M Adarsha, Arun Y Patil, Basvaraja B Kotturshettar, Design and Simulation study of pineapple leaf reinforced fiber glass as an alternative material for prosthetic limb, IOP Conf. Ser.: Mater. Sci. Eng. Volume 872 012118. https://doi.org/10.1088/1757-899X/872/1/012118

[22]   N. Vijaya Kumar, N. R. Banapurmath, Ashok M. Sajjan, Arun Y. Patil and Sharanabasava V Ganachari, Studies on Hybrid Bio-nanocomposites for Structural applications, Journal of Materials Engineering and Performance, Accepted, 2021.

[23]   Prabhudev S Yavagal, Pavan A Kulkarni, Nikshep M Patil, Nitilaksh S Salimath, Arun Y. Patil, Rajashekhar S Savadi, B B Kotturshettar, Cleaner production of edible

straw as replacement for thermoset plastic, Elsevier, Materials Today Proceedings, March 2020. https://doi.org/10.1016/j.matpr.2020.02.667

[24] Shruti Kiran Totla, Arjun M Pillai, M Chetan, Chetan Warad, Arun Y. Patil, B B Kotturshettar, Analysis of Helmet with Coconut Shell as the Outer Layer, Elsevier, Materials Today Proceedings, March 2020. https://doi.org/10.1016/j.matpr.2020.02.047

[25] Anirudh Kohli, Annika H, Karthik B, Pavan PK, Lohit P A, Prasad B Sarwad, Arun Y Patil and Basvaraja B Kotturshettar, Design and Simulation study of fire-resistant biodegradable shoe, Journal of Physics: Conference Series 1706 (2020) 012185. https://doi.org/10.1088/1742-6596/1706/1/012185

[26] Akshay Kumar, Kiran A R, Mahesh Hombalmath, Manoj Mathad, Siddhi S Rane, Arun Y Patil and B B Kotturshettar, Design and analysis of engine mount for biodegradable and non-biodegradable damping materials, Journal of Physics: Conference Series 1706 (2020) 012182. https://doi.org/10.1088/1742-6596/1706/1/012182

[27] Hombalmath M.M., Patil A.Y., Kohli A., Khan A. (2021) Vegetable Fiber Pre-tensioning Influence on the Composites. In: Jawaid M., Khan A. (eds) Vegetable Fiber Composites and their Technological Applications. Composites Science and Technology. Springer, Singapore. https://doi.org/10.1007/978-981-16-1854-3_6

Sustainable Natural Fiber Composites

Materials Research Foundations **122** (2022) 238-255

Materials Research Forum LLC

https://doi.org/10.21741/9781644901854-10

Chapter 10

Polymer Natural Fibre Composites and its Mechanical Characterization

Manju Sri Anbupalani[1], Chitra Devi Venkatachalam[2]*, Manjula P[1],

[1]Department of Chemical Engineering, Kongu Engineering College, Erode-638060, Tamil Nadu, India

[2]Department of Food Technology, Kongu Engineering College, Erode-638060, India, India

* erchitrasuresh@gmail.com

Abstract

Increasing environmental concern and the dynamic market demand is a driving force for eco-friendly materials, therefore researchers focus on developing biodegradable materials for various applications. Bio- composites are one such eco-friendly material utilizing the advantages of polymer matrix and natural fibre reinforcement. Natural fibre composites are replacing environmentally harmful synthetic materials with better mechanical properties and they have potential applications in different fields like biomedical applications, construction materials, defence, automobiles etc., In addition, they are economically feasible and require low energy for production. Analysing the mechanical characteristics of natural fibre reinforced polymers are crucial to fine tune their utilization and processing techniques. This chapter aims to provide an overview of the mechanical characteristics of NFCs reinforced polymer composites and the details of milestones made with them.

Keyword

Composite Materials, Natural Fibres, Polymers, Mechanical Properties

Contents

Polymer Natural Fibre Composites and its Mechanical Characterization 238

1.   Introduction..239

2.	**Natural fiber**	**240**
2.1	Mechanical behaviour of natural fibers	243
3.	Polymer selection	244
3.1	Interface strength between fiber and polymer	244
3.2	Dispersion of fibre	245
3.3	Fibre orientation	245
4.	**Polymer composites: processing techniques**	**245**
5.	**Mechanical behaviour of natural fiber composites**	**246**
5.1	Tensile properties	246
5.2	Flexural properties	247
5.3	Impact strength	247
5.4	Hardness properties	248
5.5	Dynamic mechanical properties	248
5.6	Thermal properties	248
6.	**Application**	**249**
	Conclusion	**249**
	References	**250**

1.   Introduction

Increasing importance for renewable resource-based materials is due to the environmental awareness over the usage of non-renewable resources like petroleum. Emissions of CO_2 resulting from anthropogenic activities are the constant challenge that the human race is facing. Recently, many industries have adapted environmental friendly production technologies and products. Still, the explorations and adaption of environmental friendly materials are underway due to the growing environmental concerns.

There is broad potential for renewable resources related composite materials and technologies. An extensive research have proved the engineering, environmental and monetary advantages of biodegradable materials. Composites are the combination two or more different constituent phases. The phases are matrix and the reinforcement. matrix material transfers the load to the reinforcement, whereas reinforcement withstand the load and offers strength to the composite[1] Reinforcements and fillers can be in different dimensions like short, long, continuous, discontinuous, or spherical particles. Thus, natural fibers has great scope to produce bio composites. Natural fibers have many advantages and they are discussed in the following sections of this chapter. Material having improved performance, quality, and cost can be attained by the optimised combination of natural fibers, which leads to massive potential for different industries and applications. Sufficient mechanical properties and resilience of natural fiber composites (NFCs) with monetary affordability is an essential factor to commercialize composite materials for large applications. In recent years, comprehensive research are done on NFCs addressing various challenges in the fabrication of composites where the final goal is to achieve NFCs with superior mechanical strength and monetary attributes for certain specific engineering applications.

In common, synthetic polymer matrix materials are chosen to develop natural fiber reinforced composites for several applications such as packing materials, automobile components, electronics and others. The addition of natural fibres to oil-based plastics can offer significant environmental and economic benefits due to depletion of the resources with fluctuating costs. There are plenty of natural fibres and they are relatively low cost. NFC containing polyolefines like polypropylene (PP) and polyethylene are extensively investigated and successfully found in the developing various automobile components [2].

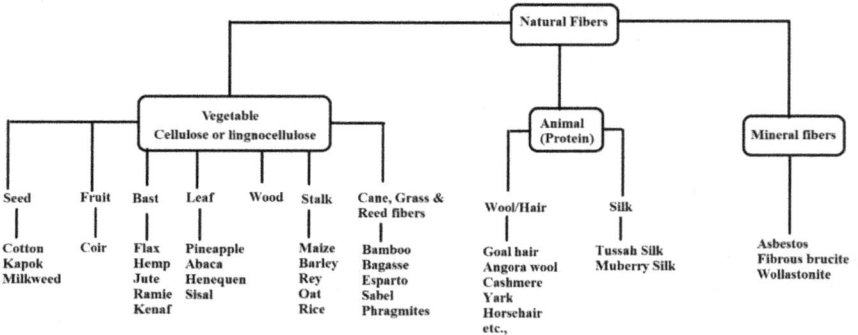

Figure 1. Typical natural fibers [10]

2. Natural fiber

Various types of natural fibers are used for the preparation of composites for industry specific applications [3-6]. These natural fibers are related indirectly or directly in our daily life, which is being used frequently for variety of day-to-day applications. One of the positive assets of such natural fibers is that it can be decomposed by composting or other

Sustainable Natural Fiber Composites   Materials Research Forum LLC
Materials Research Foundations **122** (2022) 238-255   https://doi.org/10.21741/9781644901854-10

methods at the end of their usage[7, 8]. Natural fibers are usually classified into various types (refer with: Fig. 1) based on their origin (e.g. whether they have been obtained from trees, animals or minerals) [9]. Among all these fibers, the plant-based fibers have potential to be utilised for wide ranges of applications, thus having higher commercial importance.

Various natural fibers used for commercial application are shown in figure. (refer with: Fig. 2)[10]. When compared to synthetic fibers, low density is the major advantage of natural fibers [11].

Fiure 2. Typical natural fibers [10]

Many variety of natural fibers like Hibiscus sabdariffa, sisal, jute, henequen, kenaf, Grewiaoptiva, flax, coir, pine needles and many other materials, offer numerous advantages due to their renewable behaviour when compared to synthetic fibers like glass, carbon and aramid fibers,. The distinctive characteristics of natural fibers includes recyclability, low price, acceptable specific strength, biodegradability, low density, ease of separation, good thermal properties, high toughness, nonirritation to the skin, reduced tool wear and improved energy recovery. Thus, Natural fibers are very inexpensive and promising compared to traditional synthetic fibers.

Low density of the natural fibre is due to their reduced weight. Hydrophilic nature of the natural fibre is one of the crucial challenges, which leads to absorption of moisture and it swells. The hydrophilic nature of natural fibers can be reduced through pre-treatment of natural fibre with reagents such as sodium, Sodium chloride and hydroxide. Bleaching and alkali acetylation treatments were also adopted to enhance the adhesion property between fiber and matrix by altering the surface roughness of the fibres [10, 12]. Besides these treatments, environmental friendly treatments were also focused. Sodium bicarbonate is used to remove the hemicellulose substance and surface contaminants of the natural fibre. Table (refer with: table 1) shows chemical composition various natural fibers (plants and animal fibres).

Table 1. Chemical compositions of natural fibers used [10, 13-16]

S. No.	Fiber type	Cellulose wt%	Hemicellulos wt%	Lignin wt%	Pectin wt%	Ash wt%
1.	Hemp	70-74.4	17.9-22.4	3.7-5.7	0.9	8
2.	Jute	45-71.5	12-20.4	12-25	4-10	8
3.	Flax	65-85	16	1-4	5-12	1-2
4.	Kenaf	31-57	21.5	8-19	3-5	-
5.	Ramie	68-76	-	0.6-0.7	1.9	5
6.	Sisal	50-64	6	10-14	10	7
7.	Hardwood	40-50	21-36	20-30	0-1	-
8.	Cotton	85-90	26.9	-	0-1	1
9.	Softwood	40-45	25-30	34-36	1	1
10.	Ricestraw	41-57	33	8-20	8	7
11.	Abaca	56-63	-	12–13	1	1
12.	Bagasse	40-46	24.5-29	12.5-20	-	1.5–2.4
13.	Bamboo	42.3-49.1	24.1-27.7	23.8-26.1	-	1.3-2.0

Sustainable Natural Fiber Composites   Materials Research Forum LLC
Materials Research Foundations **122** (2022) 238-255   https://doi.org/10.21741/9781644901854-10

2.1 Mechanical behaviour of natural fibers

Wide investigation has to be done for the evaluation of the mechanical behaviour and physical characteristics of various natural fibers and existing results from different articles are described in Table (refer with: Table 2).

Table 2. Physiological and mechanical characteristics of prevalent natural fibres [10, 17-21]

S.No.	Type of natural fibers	Density (g/cm³)	Elongation at break (%)	Tensile strength (Mpa)	Young's modulus (Gpa)
1.	Alfa	0.89	5.8	350	22
2.	Abaca	1.5	3–10	400	12
3.	Bamboo	0.6 - 1.1	-	140 - 230	11 - 17
4.	Bagasse	1.25	-	290	17
5.	Coir	1.2	30	175	4 - 6
6.	Banana	1.35	5.9	500	12
7.	Curaua	1.4	3.7 - 4.3	500 - 1150	11.8
8.	Cotton	1.5 - 1.6	7 - 8	287 - 597	5.5 - 12.6
9.	Flax	1.5	2.7–3.2	345 - 1035	27.6
10.	Datepalm	1-1.2	2-4.5	97 - 196	2.5 - 5.4
11.	Henequen	1.2	4.8±1.1	500 ± 70	13.2±3.1
12.	Hemp	1.48	1.6	690	70
13.	Jute	1.3	1.5–1.8	393-773	26.5
14.	Isora	1.2 - 1.3	5–6	500-600	-
15.	Nettle	-	1.7	650	38
16.	Kenaf	-	1.6	930	53
17.	Piassava	1.4	21.9 - 7.8	134–143	1.07 - 4.59
18.	Oilpalm	0.7 - 1.55	25	248	3.2
19.	Ramie	1.5	2.5	560	24.5
20.	Pineapple	0.8 - 1.6	14.5	400 - 627	1.44
21.	Silk	1.3	28 - 30	1300 - 2000	30
22.	Sisal	1.5	2.0 - 2.5	511 - 635	9.4 - 22

3.   Polymer selection

The natural fibre reinforced composite constitute a matrix material, which is an important component. It prevents mechanical abrasion of the fibre surface and passes load into fibres. [22]. The most popular matrices in NFCs are polymeric because of their low weight and can be handled at low temperatures. Along with natural fibre classes of polymers like thermoplastics and thermosets can be used as matrices [23].

The selection of the matrix is restricted by the degradation temperature of natural fibres. Natural fibres are not thermally stable above 200°C but can be processed at higher temperature for limited period of time in certain circumstances [24].   Only thermoplastics like polyethylene (PE), PP, polyolefin, polyvinyl chloride, polystyrene, which can be cured below the same temperature and thermosets are used as matrix because of this restriction [25].

However, it should be noted that PP and PE are frequently used thermoplastic matrices for polymer/natural fibre composites. The most commonly used thermosets are unsaturated polyester (UP), epoxy resin, phenol formaldehyde and VE resins. Thermoplastics have the tendency to be softened repeatedly by applying heat and hardened by cooling and it is possible to be recycled easily, which makes them most favourable in recent commercial application, whereas better comprehension of the natural fibre properties are achieved using thermoset polymers. Replacing petroleum-based polymers are also explored with bioderived matrices. In this regards, PLA is the front-runner from the perspective of mechanical strength and it gives high stiffness in combination with natural fibres [26].

3.1   Interface strength between fiber and polymer

Interfacial adhesion of fibre and matrix has a major influence in ensuring the mechanical behaviour of composites [22]. Since stress is relocated from matrix to fibres through the interface, improved interfacial adhesion is required to attain desired strength. although, it is possible to have a strong adhesion at the interface, enabling crack propagation that can reduce strength and toughness. However, for polymer/natural fibre composites there persist a limitation for the interaction between the hydrophilic plant fibres and hydrophobic polymer matrices, which leads to poor interfacial adhesion. The poor interaction of the plant fibre and matrix material result in the reduced mechanical behaviour as well as high moisture absorption influencing long term properties. For better adhesion, fibre and matrix are brought into contact and the wettability is an essential for good adhesion. The inadequate fibre wetting contributes to interfacial defects that function as stress concentrators [27]. The toughness, flexural and tensile strength of composites have been shown to be influenced by fibre weathering [28]. Physical and chemical processing may increase fibre wettability and thus improve interfacial strength [29, 30].

Sustainable Natural Fiber Composites   Materials Research Forum LLC
Materials Research Foundations **122** (2022) 238-255   https://doi.org/10.21741/9781644901854-10

3.2   Dispersion of fibre

The dispersion of fibres, for short fibre composites, is recognised as a significant factor. The usage of longer fibres will further increase the agglomeration tendency. Good fibre dispersion facilitates good interface adhesion and decreases vacuum by making sure fibres are completely matrix-enclosed [22, 31]. Processing parameters such as temperature and pressure may alter the fibre dispersion. Dispersion and interfacial binding can be changed with additives like stearic acid in PP and PE. MAPP is used as a coupling agent to improve the fibre matrix interaction [32]. Also, fibre modification using grafting process can also be used [32]. Processing techniques has also have a crucial influence on fibre dispersion. Using more intensified processes such as double screw extruder instead of single screw extruder allows increased fibre dispersal, but this is normally leads to fibre damage, which generally depends on the temperature and the screw configuration.

3.3   Fibre orientation

Superior mechanical characteristics are obtained for composites when the fibre is aligned parallel to the direction of the applied load [33, 34]. However, it is more difficult to get alignment with natural fibres than for continuous synthetic fibres. Some alignment is achieved during injection moulding, dependent on matrix viscosity and mould design [35]. Higher levels of fiber alignment are used before impregnation of the matrix with long natural fibre; they are manually mounted in sheets. Modern textile processing of fibers, like spinning, can also be used to manufacture continuous yarn. Wrap spinning is another method to form Aligned fibre yarns. Unidirectional yarn exhibited improved tensile and flexural strength for uniformly aligned fibre yarn than conventional flax epoxy composites [36]. The processing technique employed for synthetic fibre can also be used for natural fibre also that offers better degree of fibre alignment and strength [37].

4.   Polymer composites: processing techniques

Extrusion, injection moulding (IM) and compression moulding are commonly used methods used for NFCs. Resin transfer moulding (RTM) is an another method employed for thermoset matrix materials and for combined flax /PP yarn composites and thermoset matrix composites pultrusion can be successfully employed [22, 38]. Temperature, pressure and processing rate are the crucial factors determining properties of the resulting composites. High temperature may cause fibre degradation that limits the usage of thermoplastic having incompatible. In extrusion process, thermoplastic is used in the pellet form that is softened by applying heat. Then the fibre is mixed with the polymer by means of a single or two rotating screws. Then it is extruded by forcing out of the die at a constant rate. Screw speed determines strength of the material due to air entrapment, excessive melt

Sustainable Natural Fiber Composites   Materials Research Forum LLC
Materials Research Foundations **122** (2022) 238-255   https://doi.org/10.21741/9781644901854-10

temperatures and fibre breakage, whereas low speed leads to poor mixing and insufficient wetting of the fibres. Thus extrusion is used to produce base material for injection moulding. Twin screw extrusion technique gives better dispersion and improved mechanical characteristics comparatively [39].

For thermoplastic matrix materials and thermoset matrix materials Compression moulding (CM) can be employed; the fibre could be chopped fibre or long fibre either with random orientation or aligned orientation. Before pressure and heat is applied, the fibres are stacked alternately with sheets of matrix. The viscosity should to be carefully controlled during pressing and heating, in specific for thick samples. It is necessary to ensure that matrix is impregnated completely in the space between fibres. By controlling the parameters' pressure, temperature, viscosity and holding time good quality composites can be synthesized with considering the type of fibre and matrix [40].

Thermoset in liquid state is inserted into fibre-preform mould in RTM. Here also temperature, injection pressure, resin viscosity are the main variables with this process [40]. Lower processing temperature and reduced thermo-mechanical damages are the benefits of this process [41]. Lower fibre alignment influence the compaction of natural fibres and make them less compatible than glass fibres [41]. This method is suitable for small production units with good component strength.

5.   Mechanical behaviour of natural fiber composites

Tensile test, flexural test, impact test, and hardness tests are used to analyse the mechanical properties of polymer composites reinforced with natural fibre. Those properties of various polymer composites reinforced with natural fibre are described below.

5.1   Tensile properties

Two properties namely tensile strength and Young's modulus are determined through the tensile test [42]. Table (refer with: Table 3) shows the above mentioned two properties of several natural fibre reinforced polymer composites. When the fibre loading is increased tensile strength decreases. Increased fibre content leads to weakening of adhesion between the fibre ad matric, which results in reduced tensile strength. Due to the increased fibre load the interfacial area between the fibre and matrix is becoming weaker, which in turn reduces the tensile strength [43, 44]. Banana fibre-reinforced composite materials had the higher tensile strength whereas composite material reinforced with palm fibres had the lower value. But, the Young's modulus increases with fibre loading. When load is applied in tensile test, partially separated micro spaces are generated that hinders propagation of stress between the fibre and matrix. When fibre loading increases, the degree of obstruction also increases, which can consequently influence the stiffness. The banana fibre and coir fibre reinforced composites had lowest and the highest Young's modulus values, respectively.

Table 3. Variation of mechanical properties of various natural fibre reinforced polymer composites [43, 44]

	Tensile Strength (Mpa)	Young's Modulus (Gpa)	Flexural strength (Mpa)	Flexural Modulus (Mpa)	Impact Strength	Hardness (R1)
Jute-PP	25	1.8	49.5	2.2	40J/m	95
Abaca-PP	24.5	2.1	47	2.1	40J/m	81
Palm-PP	24.5	1.3	50	2	50	34
Bagasse-PP	20	1.3	28	1.1	5.0KJ/mm2	-
Banana-PP	38	0.9	0		11.2	-
Hemp-PP	28	1.5	-	-	-	-
Coir-PP	27.5	2.1	18	2.2	48 J/m	86

5.2 Flexural properties

Flexural property is one of the significant parameter that plays a major role in evaluating the mechanical performance of hybrid green composites in several applications. Flexural properties of various polymer composites reinforced with natural fibre are mentioned in Table (refer with: Table 3). Flexural modulus and Flexural strength increases when the fibre loading is increased. As natural fibres possess high modulus, higher stress is required for the same deformation with high fibre content. Thus, stress transfer could be increased when the fibre matrix adhesion is improved [43-45]. Jute and coir fibre-reinforced composites has the greater flexural strength, whereas bagasse composites has least one.

5.3 Impact strength

The extent of withstanding fracture is known as impact strength, or it is the energy essential to propagate a crack. Table (refer with: Table 3) infers the impact strength of various polymer/natural fibre. Fibre content increases Impact strength of polymer/natural fibre composites. The impact strength value relies on the properties of the natural fibre, matrix material and interfacial adhesion of fibre/matrix. Fibre agglomeration is more for higher fibre content, which can results in stress concentration regions, requiring less energy for crack propagation. Fibre pull-out is a factor of impact failure of composite. While increasing fibre loading and length, increases the force requirement to pull the fibres out. Thus, it consequently improves the impact strength [43-45].

5.4   Hardness properties

Table (refer with: Table 3) shows the hardness property of several natural fibre-reinforced polymer composites. On increasing the fibre loading , hardness of natural fibre reinforced polymer composites increases and this could be justified with increase of stiffness with fibre loading [43-45]. As mentioned in Table (refer with: Table 3), palm reinforcement provides the highest hardness values, wheras the jute fibre reinforcement leads to lowest value.

5.5   Dynamic mechanical properties

The equipment which analyses dynamic mechanical properties of polymer/natural composites is Dynamic mechanical analyzer (DMA). Table (refer with: Table 4) shows the Dynamic mechanical properties of different polymer/natural fibre composites. On the addition of ammonium polyphosphate storage modulus is increased. This can be attributed by improved interfacial adhesion between the matrix and fibre. Alkaline (NaOH) treatment lowers the values of loss factor and the storage modulus is increased. [46]. The same trend is observed in indoum/ PP composites because of the higher time response of the matrix chains. Hence, at high frequencies, the composite behaved like a solid and at low frequencies, the polymer chains have enough time to relax to decreases the loss factor [47]. Kenaf fibre-reinforced polyethylene (PE) composites have high softening temperature, and storage modulus was decreased. In contrast, the same property can be increased with fibre loading because of the composite's stiffening. The loss modulus and loss factor are also enhanced with increase in fibre loading for kenaf fibre reinforced composite.

5.6   Thermal properties

Differential scanning calorimetry (DSC) and Thermogravimetric analysis (TGA) are used to analyse thermal properties of polymer/natural fibre composites. Thermal properties of various polymer/natural fibre composites are briefed in Table (refer with: Table 4).

Sustainable Natural Fiber Composites   Materials Research Forum LLC
Materials Research Foundations **122** (2022) 238-255   https://doi.org/10.21741/9781644901854-10

Table 4. Dynamic mechanical and thermal properties for several polymer/ natural fibre composites

Type of Composite	Storage Modulus (Mpa)	Modulus(Gpa)	Maximum tan delta	Temperature (°C)	Degradation temperature (°C)
Kenaf-PLA	5	1	1.7	67	305
Kenaf-PVC/TPU	-	-	-	-	281
Doum-PP	-	-	0.16	-	-
Kenaf-PE	8	0.7	0.24	-	-
Hemp-PP	-	-	-	-	376

6. Application

In recent couple of decades polymer/natural fibre composites are increasingly used in numbers of car models in door panels, instruments panels, package trays, racks' material, engine covers, boot liners, sun visorsoil/air filters and also used in structural components like seat materials and exterior underfloor panelling. Many internationally the leading automotive industries utilize these materials and the usage is estimated to grow in automotive sector [26]. In developing countries, boards developed from composite can substitute medium density fibreboard in railcars application [17]. NFCs influence the aircraft industry also to make use of it in interior panelling. They are also used in diverse applications such as toys, packaging, funeral articles, railings in marine sector and covers of electronic gadgets like laptops and mobile phones [40, 51]. In sports sector, surfing is a specific application in the context of adopting environmentally friendly materials. Surfboards incorporating NFCs are getting popularised due to its promising mechanical properties and environmental friendly attribute. "Ecoboard" is the earliest one and manufactured by Laminations Ltd, utilizing bio-based resin/hemp fibre.

Conclusion

Mechanical performance of NFCs are being researched widely in recent decades to enhance their utilisation in many engineering sectors. Improvisation occurred because of fine-tuning of fibre selection, methodology of extraction, pre-treatment and addressing interfacial adhesion challenges and processing techniques. This paper has reviewed the various aspects of improving stiffness, strength and impact strength including the effect of weathering and moisture on the mechanical properties. Short and long term performance analysis was also addressed based on the above characteristics. NFCs are compared with GFRPs for its stiffness and cost, tensile sand impact strength, from which it can be concluded that NFCs are approaching GFRFs. The low density values of NFCs gives many advantages for specific properties. Thus, applications of NFCs can also be extended for load bearing, automotive exteriors' underfloor panelling, marine structures and sports apparatus etc., In further, research is essential for addressing moisture absorbing tendency of natural fibres and fire retardance to extend their application scope. Thus, NFCs has a rapid continues uptake and it would approach a positive trend ahead in various engineering/commercial sectors.

References

[1] A.B. Strong, Fundamentals of composites manufacturing: materials, methods and applications, Society of Manufacturing Engineers2008.

[2] M. Zampaloni, F. Pourboghrat, S. Yankovich, B. Rodgers, J. Moore, L. Drzal, A. Mohanty, M. Misra, Kenaf natural fiber reinforced polypropylene composites: A discussion on manufacturing problems and solutions, Composites Part A: Applied Science and Manufacturing 38(6) (2007) 1569-1580. https://doi.org/10.1016/j.compositesa.2007.01.001

[3] V.K. Thakur, M.K. Thakur, R.K. Gupta, Graft copolymers from cellulose: synthesis, characterization and evaluation, Carbohydrate polymers 97(1) (2013) 18-25. https://doi.org/10.1016/j.carbpol.2013.04.069

[4] Y. Xie, C.A. Hill, Z. Xiao, H. Militz, C. Mai, Silane coupling agents used for natural fiber/polymer composites: A review, Composites Part A: Applied Science and Manufacturing 41(7) (2010) 806-819. https://doi.org/10.1016/j.compositesa.2010.03.005

[5] S. Mukhopadhyay, R. Fangueiro, Physical modification of natural fibers and thermoplastic films for composites—a review, Journal of Thermoplastic Composite Materials 22(2) (2009) 135-162. https://doi.org/10.1177/0892705708091860

[6] M.J. John, R.D. Anandjiwala, Recent developments in chemical modification and characterization of natural fiber-reinforced composites, Polymer composites 29(2) (2008) 187-207. https://doi.org/10.1002/pc.20461

[7] V. Thakur, A. Singha, M. Thakur, In-air graft copolymerization of ethyl acrylate onto natural cellulosic polymers, International Journal of Polymer Analysis and Characterization 17(1) (2012) 48-60. https://doi.org/10.1080/1023666X.2012.638470

[8] V. Thakur, A. Singha, M. Thakur, Modification of natural biomass by graft copolymerization, International Journal of Polymer Analysis and Characterization 17(7) (2012) 547-555. https://doi.org/10.1080/1023666X.2012.704561

[9] H. Akil, M. Omar, A. Mazuki, S. Safiee, Z.M. Ishak, A.A. Bakar, Kenaf fiber reinforced composites: A review, Materials & Design 32(8-9) (2011) 4107-4121. https://doi.org/10.1016/j.matdes.2011.04.008

[10] I. Singh, P.K. Rakesh, Processing of Green Composites, Springer2019.

[11] V.K. Thakur, M.K. Thakur, R.K. Gupta, raw natural fiber–based polymer composites, International Journal of Polymer Analysis and Characterization 19(3) (2014) 256-271. https://doi.org/10.1080/1023666X.2014.880016

[12] A. Shalwan, B. Yousif, In state of art: mechanical and tribological behaviour of polymeric composites based on natural fibres, Materials & Design 48 (2013) 14-24. https://doi.org/10.1016/j.matdes.2012.07.014

[13] P.K. Bajpai, I. Singh, J. Madaan, Development and characterization of PLA-based green composites: A review, Journal of Thermoplastic Composite Materials 27(1) (2014) 52-81. https://doi.org/10.1177/0892705712439571

[14] Z. Azwa, B. Yousif, A. Manalo, W. Karunasena, A review on the degradability of polymeric composites based on natural fibres, Materials & Design 47 (2013) 424-442. https://doi.org/10.1016/j.matdes.2012.11.025

[15] F. Yao, Q. Wu, Y. Lei, W. Guo, Y. Xu, Thermal decomposition kinetics of natural fibers: activation energy with dynamic thermogravimetric analysis, Polymer Degradation and Stability 93(1) (2008) 90-98. https://doi.org/10.1016/j.polymdegradstab.2007.10.012

[16] P. Wambua, J. Ivens, I. Verpoest, Natural fibres: can they replace glass in fibre reinforced plastics?, composites science and technology 63(9) (2003) 1259-1264. https://doi.org/10.1016/S0266-3538(03)00096-4

[17] F. Vilaseca, J. Mendez, A. Pelach, M. Llop, N. Canigueral, J. Girones, X. Turon, P. Mutje, Composite materials derived from biodegradable starch polymer and jute strands, Process Biochemistry 42(3) (2007) 329-334. https://doi.org/10.1016/j.procbio.2006.09.004

[18] T. Yu, J. Ren, S. Li, H. Yuan, Y. Li, Effect of fiber surface-treatments on the properties of poly (lactic acid)/ramie composites, Composites Part A: Applied Science and Manufacturing 41(4) (2010) 499-505. https://doi.org/10.1016/j.compositesa.2009.12.006

Sustainable Natural Fiber Composites   Materials Research Forum LLC
Materials Research Foundations **122** (2022) 238-255   https://doi.org/10.21741/9781644901854-10

[19] A.K. Bledzki, A. Jaszkiewicz, D. Scherzer, Mechanical properties of PLA composites with man-made cellulose and abaca fibres, Composites Part A: Applied Science and Manufacturing 40(4) (2009) 404-412. https://doi.org/10.1016/j.compositesa.2009.01.002

[20] A. Bledzki, A. Jaszkiewicz, Mechanical performance of biocomposites based on PLA and PHBV reinforced with natural fibres–A comparative study to PP, Composites science and technology 70(12) (2010) 1687-1696. https://doi.org/10.1016/j.compscitech.2010.06.005

[21] K. Oksman, M. Skrifvars, J.-F. Selin, Natural fibres as reinforcement in polylactic acid (PLA) composites, Composites science and technology 63(9) (2003) 1317-1324. https://doi.org/10.1016/S0266-3538(03)00103-9

[22] K.L. Pickering, M.A. Efendy, T.M. Le, A review of recent developments in natural fibre composites and their mechanical performance, Composites Part A: Applied Science and Manufacturing 83 (2016) 98-112. https://doi.org/10.1016/j.compositesa.2015.08.038

[23] J. Holbery, D. Houston, Natural-fiber-reinforced polymer composites in automotive applications, Jom 58(11) (2006) 80-86. https://doi.org/10.1007/s11837-006-0234-2

[24] J. Summerscales, N.P. Dissanayake, A.S. Virk, W. Hall, A review of bast fibres and their composites. Part 1–Fibres as reinforcements, Composites Part A: Applied Science and Manufacturing 41(10) (2010) 1329-1335. https://doi.org/10.1016/j.compositesa.2010.06.001

[25] P.A. dos Santos, J.C. Giriolli, J. Amarasekera, G. Moraes, Natural fibers plastic composites for automotive applications, 8th Annual automotive composites conference and exhibition (ACCE 2008). Troy, MI: SPE Automotive & Composites Division, 2008, pp. 492-500.

[26] O. Faruk, A.K. Bledzki, H.P. Fink, M. Sain, Progress report on natural fiber reinforced composites, Macromolecular Materials and Engineering 299(1) (2014) 9-26. https://doi.org/10.1002/mame.201300008

[27] P. Chen, C. Lu, Q. Yu, Y. Gao, J. Li, X. Li, Influence of fiber wettability on the interfacial adhesion of continuous fiber-reinforced PPESK composite, Journal of applied polymer science 102(3) (2006) 2544-2551. https://doi.org/10.1002/app.24681

[28] X.-F. Wu, Y.A. Dzenis, Droplet on a fiber: geometrical shape and contact angle, Acta mechanica 185(3) (2006) 215-225. https://doi.org/10.1007/s00707-006-0349-0

Sustainable Natural Fiber Composites   Materials Research Forum LLC
Materials Research Foundations **122** (2022) 238-255   https://doi.org/10.21741/9781644901854-10

[29] Q. Bénard, M. Fois, M. Grisel, Roughness and fibre reinforcement effect onto wettability of composite surfaces, Applied surface science 253(10) (2007) 4753-4758. https://doi.org/10.1016/j.apsusc.2006.10.049

[30] E. Sinha, S. Panigrahi, Effect of plasma treatment on structure, wettability of jute fiber and flexural strength of its composite, Journal of composite materials 43(17) (2009) 1791-1802. https://doi.org/10.1177/0021998309338078

[31] P. Heidi, M. Bo, J. Roberts, N. Kalle, The influence of biocomposite processing and composition on natural fiber length, dispersion and orientation, Journal of Materials Science and Engineering. A 1(2A) (2011) 190. https://doi.org/10.1155/2011/891940

[32] A.R. Sanadi, D.F. Caulfield, R.E. Jacobson, Agro-fiber thermoplastic composites, Paper and composites from agro-based resources (1997) 377-401.

[33] I. Ben Amor, H. Rekik, H. Kaddami, M. Raihane, M. Arous, A. Kallel, Effect of palm tree fiber orientation on electrical properties of palm tree fiber-reinforced polyester composites, Journal of composite materials 44(13) (2010) 1553-1568. https://doi.org/10.1177/0021998309353961

[34] P. Herrera-Franco, A. Valadez-Gonzalez, A study of the mechanical properties of short natural-fiber reinforced composites, Composites Part B: Engineering 36(8) (2005) 597-608. https://doi.org/10.1016/j.compositesb.2005.04.001

[35] P. Joseph, K. Joseph, S. Thomas, Effect of processing variables on the mechanical properties of sisal-fiber-reinforced polypropylene composites, Composites science and Technology 59(11) (1999) 1625-1640. https://doi.org/10.1016/S0266-3538(99)00024-X

[36] J.E. Carpenter, M.H. Miao, P. Brorens, Deformation behaviour of composites reinforced with four different linen flax yarn structures, Advanced Materials Research, Trans Tech Publ, 2007, pp. 263-266. https://doi.org/10.4028/0-87849-466-9.263

[37] M. Khalfallah, B. Abbès, F. Abbès, Y. Guo, V. Marcel, A. Duval, F. Vanfleteren, F. Rousseau, Innovative flax tapes reinforced Acrodur biocomposites: a new alternative for automotive applications, Materials & Design 64 (2014) 116-126. https://doi.org/10.1016/j.matdes.2014.07.029

[38] I. Angelov, S. Wiedmer, M. Evstatiev, K. Friedrich, G. Mennig, Pultrusion of a flax/polypropylene yarn, Composites Part A: Applied Science and Manufacturing 38(5) (2007) 1431-1438. https://doi.org/10.1016/j.compositesa.2006.01.024

[39] R. Malkapuram, V. Kumar, Y.S. Negi, Recent development in natural fiber reinforced polypropylene composites, Journal of reinforced plastics and composites 28(10) (2009) 1169-1189. https://doi.org/10.1177/0731684407087759

[40] M.-p. Ho, H. Wang, J.-H. Lee, C.-k. Ho, K.-t. Lau, J. Leng, D. Hui, Critical factors on manufacturing processes of natural fibre composites, Composites Part B: Engineering 43(8) (2012) 3549-3562. https://doi.org/10.1016/j.compositesb.2011.10.001

[41] G. Francucci, E.S. Rodríguez, A. Vázquez, Experimental study of the compaction response of jute fabrics in liquid composite molding processes, Journal of Composite Materials 46(2) (2012) 155-167. https://doi.org/10.1177/0021998311410484

[42] M.S. Salit, M. Jawaid, N.B. Yusoff, M.E. Hoque, Manufacturing of natural fibre reinforced polymer composites, Springer2015. https://doi.org/10.1007/978-3-319-07944-8

[43] M.R. Rahman, M.M. Huque, M.N. Islam, M. Hasan, Improvement of physico-mechanical properties of jute fiber reinforced polypropylene composites by post-treatment, Composites Part A: Applied Science and Manufacturing 39(11) (2008) 1739-1747. https://doi.org/10.1016/j.compositesa.2008.08.002

[44] R. Karim, M.F. Rahman, M. Hasan, M.S. Islam, A. Hassan, Effect of Fiber Loading and Alkali Treatment on Physical and Mechanical Properties of Bagasse Fiber Reinforced Polypropylene Composites, Journal of Polymer Materials 30(4) (2013).

[45] M. Haque, R. Rahman, N. Islam, M. Huque, M. Hasan, Mechanical properties of polypropylene composites reinforced with chemically treated coir and abaca fiber, Journal of Reinforced Plastics and Composites 29(15) (2010) 2253-2261. https://doi.org/10.1177/0731684409343324

[46] F. Shukor, A. Hassan, M. Hasan, M.S. Islam, M. Mokhtar, PLA/Kenaf/APP biocomposites: effect of alkali treatment and ammonium polyphosphate (APP) on dynamic mechanical and morphological properties, Polymer-Plastics Technology and Engineering 53(8) (2014) 760-766. https://doi.org/10.1080/03602559.2013.869827

[47] H. Essabir, A. Elkhaoulani, K. Benmoussa, R. Bouhfid, F. Arrakhiz, A. Qaiss, Dynamic mechanical thermal behavior analysis of doum fibers reinforced polypropylene composites, Materials & Design 51 (2013) 780-788. https://doi.org/10.1016/j.matdes.2013.04.092

[48] F.M. Salleh, A. Hassan, R. Yahya, A.D. Azzahari, Effects of extrusion temperature on the rheological, dynamic mechanical and tensile properties of kenaf fiber/HDPE

composites, Composites Part B: Engineering 58 (2014) 259-266.
https://doi.org/10.1016/j.compositesb.2013.10.068

[49] F. Shukor, A. Hassan, M.S. Islam, M. Mokhtar, M. Hasan, Effect of ammonium polyphosphate on flame retardancy, thermal stability and mechanical properties of alkali treated kenaf fiber filled PLA biocomposites, Materials & Design (1980-2015) 54 (2014) 425-429. https://doi.org/10.1016/j.matdes.2013.07.095

[50] Y. El-Shekeil, S. Sapuan, M. Jawaid, O. Al-Shuja'a, Influence of fiber content on mechanical, morphological and thermal properties of kenaf fibers reinforced poly (vinyl chloride)/thermoplastic polyurethane poly-blend composites, Materials & Design 58 (2014) 130-135. https://doi.org/10.1016/j.matdes.2014.01.047

[51] K.-Y. Lee, Y. Aitomäki, L.A. Berglund, K. Oksman, A. Bismarck, On the use of nanocellulose as reinforcement in polymer matrix composites, Composites Science and Technology 105 (2014) 15-27. https://doi.org/10.1016/j.compscitech.2014.08.032

Sustainable Natural Fiber Composites

Materials Research Foundations **122** (2022) 256-281

Materials Research Forum LLC

https://doi.org/10.21741/9781644901854-11

Chapter 11

Glass and Natural Fiber Composite: Properties and Applications

Guravtar Singh Mann[1*]. Anish Khan*[2,3], Abdullah M Asiri[2,3]

[1]School of Mechanical Engineering, Lovely Professional University, Phagwara-144411, India

[2]Chemistry Department, Faculty of Science, King Abdulaziz University, Jeddah 21589, Saudi Arabia

[3]Center of Excellence for Advanced Material Research, King Abdulaziz University, Jeddah 21589 Saudi Arabia

* guravtar.14443@lpu.co.in

Abstract

The ecological concerns and issues such as recycling of plastic products and environmental care are increasingly important to handle the present situation of climate change. As a consequence of growing environmental awareness surrounding society a there is a great interest in the research on more eco-friendly materials that are derived from renewable resources. Therefore, natural composites also known as green bio composites have received attention from researchers and industries to develop biodegradable materials by using natural fibres, those possess outstanding degradable and sustainable properties. Because of its superior properties such as high specific strength, low weight, low cost, reasonably good mechanical properties, non-abrasive, eco-friendly, and bio-degradable characteristics, natural fibres these materials are useful to help researchers, scientists and industries to develop different biodegradable and eco-friendly products. This chapter explored the mechanical properties of natural-glass fibre reinforced polymer composites by considering database from google scholar and other relevant sites.

Keywords

Natural Fibers, Bio Composites, Applications

Contents

Glass and Natural Fiber Composite: Properties and Applications 256

1.   **Introduction**..**257**

2.   **Study on natural hybrid composites**...**260**

3.   **Study on hybrid (natural-glass fibre) reinforced polymers** **261**

4.   **Applications**...**262**

        4.1   Biomedical applications 263

        4.2   Packaging..264

        4.3   Automotive applications..266

        4.4   Energy sector 267

        4.5   Architectural applications...268

Conclusion..**269**

Future road map **270**

Acknowledgement **270**

References **270**

1.   Introduction

The pursuit of zero-waste materials is a never-ending quest. The materials having impact with high performance at low costs to meet basic human and environmental concerns have resulted in innovation of "Green or biodegradable composites", a fourth generation engineered composites that can give high-performance in various applications and are environmentally friendly. Therefore, lot of promotion for plant-based 'lignocellulose' products that are sustainable and that can help to reduce carbon foot prints is done these days. Natural reinforced composite (NFRC) materials have risen in popularity as a result of their unique characteristics, such as light weight, high specific properties, biodegradability, and low toxicity [1-4]. Therefore, these materials are used in various applications such as aerospace, automotive and structural applications etc. When strengthened with different fillers/reinforcing materials coming from renewable sources and biodegradable, especially from forest are used for improving their various properties, so that these can be used for various sustainable applications. The bulk of oil-derived packaging is non-recyclable or commercially unviable to recycle, and it does not easily degrade in landfills, resulting in a large amount of no degradable waste. In landfill soils, microorganisms are unable to degrade traditional waste. In terms of other properties, bio composites can play a vital role for the weight reduction of electric vehicles, thereby

Sustainable Natural Fiber Composites   Materials Research Forum LLC
Materials Research Foundations **122** (2022) 256-281   https://doi.org/10.21741/9781644901854-11

compensating for the weight of the electrical batteries [5-8]. These composites also play an important role as biodegradable materials in other fields, such as medical science. The use of suitable technical routes is the most critical part of composite manufacturing [10-17]. The contemporary composites were developed after world war-2 because the world war-2 was mostly fought with fighter planes, which requires material to be lightweight and robust. Therefore, phenolic resin was used for the first time in the fighter planes by the British royal air force in its mosquito bomber aircraft. Further, the use of radar technology resulted in the development of glass fibre reinforced plastics which were used to make the covering of radar equipment. These materials are reinforced by natural resources, known as environmentally friendly composites and bio composites [19-20]. Different types of natural fibres shown in figure 1.

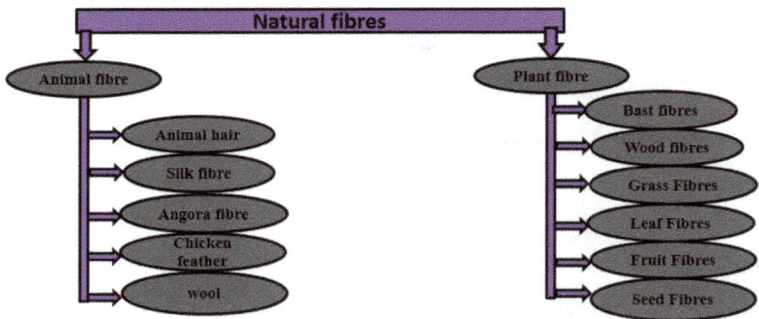

Fig 1. Various types of Natural fibres

Table 1. *Comparison between some mechanical properties of natural and synthetic [21-40]*

Fibre	Density (g/cm³)	Length (mm)	Failure strain (%)	Tensile strength (MPa)	Stiffness/Young's modulus (GPa)	Specific tensile strength (MPa/gCm³)	Specific Young's modulus (GPa/gcm³)
Ramie	1.4	900–1200	2.0–3.8	400–938	44–128	270–620	29–85
Flax	1.4	5–900	1.2-3.2	345–1830	27–80	230–1220	18–53
Hemp	1.4	5–55	1.6	550–1110	58–70	370–740	39–47
Jute	1.4-1.5	1.5–120	1.5-1.8	393–800	10–55	300–610	7.1–39
Harakeke	1.3	4–5	4.2-5.8	440–990	14–33	338–761	11–25
Sisal	1.3-1.5	900	2.0-2.5	507–855	9.4–28	362–610	6.7–20
Alfa	1.4	350	1.5-2.5	188–308	18–25	134–220	13–18

Cotton	1.5-1.6	10–60	3.0-10	287–800	5.5–13	190–530	3.7–8.4
Coir	1.2	20-150	15-30	131–220	4–6	110–180	3.3–5
Silk	1.3	Continuous	15-60	100–1500	5–25	100–1500	4–20
Feather	0.9	10–30	6.9	100–203	3–10	112–226	3.3–11
Wool	1.3	38–152	13.2-35	50–315	2.3–5	38–242	1.8–3.8
E-glass	2.5	Continuous	2.5	2000-3000	70	800–1400	29

One of the most significant developments in the history of materials is the advancement of composite materials and associated design and manufacturing technologies. The focus of research and development has shifted away from monolithic materials and toward natural-glass fibre reinforced polymeric materials [41-42]. Natural fibres as an alternative reinforcement in polymer composites have inspired the interest of many researchers and in recent decades due to their advantages over traditional glass and carbon fibres. In the field of hybrid composites, there is additionally a great deal of work going on to upgrade the mechanical properties of the composites. Therefore composites in which both the grid and the hidden cellulose are obtained from natural assets are recyclable green composites and composites in which one of the parts is separated from common assets are called hybrid composites [43-44]. As a result, plastics [45-46] stay in the atmosphere for a very long time [47]. Until recently, this was not a big cause of concern. Residents living near landfills find them unappealing, and new ones are expensive and difficult to build. Owing to the continued growth of human urban areas and population increases, landfills are being overburdened with waste produced every day [48]. Figure 2 show the various kind of factors influence the natural composites.

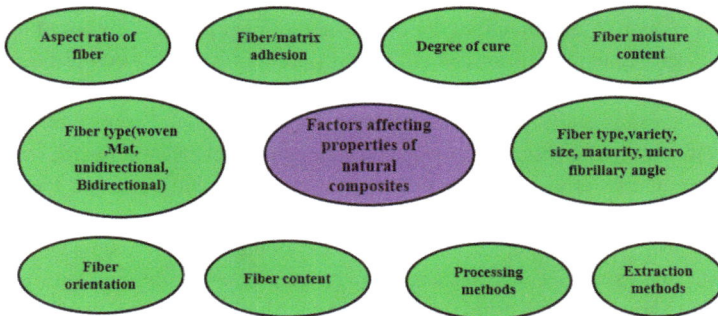

Fig 2. Factors affecting properties of natural composites

Table 2. *Properties comparison of Natural and Glass fibres [49]*

	Natural fibres	Glass fibres
Density	Low	Twice that of natural fibres
Cost	Low	Low, but higher than NF
Renewability	yes	No
Recyclability	Yes	No
Energy consumption	Low	High
CO2 neutral	Yes	No
Abrasion to machines	No	Yes
Health risk when inhaled	No	Yes
Disposal	Biodegradable	Not biodegradable

2. Study on natural hybrid composites

Natural -reinforced composites are becoming more important because of their high degree of recyclability. Moreover these composites have been shown to have great promise to be used in place of glass in some applications. The combination of two or more composites has attracted the attention of many researchers as a way to enhance mechanical properties of composites. The aim of combining two materials in a single composite is to keep the benefits of both while removing some of their drawbacks. For example, replacing carbon fibres in the centre of a laminate with less costly glass fibres will dramatically reduce the cost while maintaining almost all of the flexural properties. When a hybrid composite is loaded in tension in the direction, the more fragile fibres fail first, followed by the more ductile fibres. A number of reviews on natural fibre-reinforced bio composites have been presented. Dixit et al. [50] fabricated coir and sisal reinforced polyester composite (CSRP) hybrid composites using a compression moulding technique. The results revealed that hybridization played an essential role in enhancing the mechanical properties of fabricated composites. Gupta et al. [51] examined the effect of stacking sisal on the mechanical properties of epoxy resin reinforced jute and fabricated sisal /epoxy hybrid composite jute reinforced using hand lay-up to load method. Scanning electron microscopy for surface interfacial studies. In this analysis, a positive consequence of hybridization was observed. The results showed that hybrid composite enhanced mechanical properties (tensile test, flexural test and impact test) of newly fabricated composites.

Sustainable Natural Fiber Composites   Materials Research Forum LLC

Materials Research Foundations **122** (2022) 256-281   https://doi.org/10.21741/9781644901854-11

Table 3. Natural fibers and their chemical composition [52]

Type of	Chemical composition (%)		
	Cellulose	Hemi-cellulose	Lignin
Jute	61-63	13	5-13
Banana	60-65	6—8	5-10
Coir	43	<1	45
Flax	70-72	14	4-5
Mesta	60	15	10
Pineapple leaf	80	–	12
Sisal	60-67	10—15	8-12
Wood	45-50	23	27
Sun hemp	70-80	18-19	4-5
Ramie	80-85	3-4	0.5

Mishra et al. [53] experimentally investigated the mechanical properties of laminated three-ply and five-ply jute/glass hybrids composite and compared to the theoretical values. The hand layup method was used for the fabrication of composites. It was observed that the deflection resistance of the theoretical values of the composite material, particularly for layers 3 (70 %) was much higher than that of layers 5-Taped (2%). The water absorption behaviour was also enhanced with hybridization. In similar study, Berhanu et al. [54] fabricated the jute reinforced polypropylene composites using the compression molding process. The volume fraction of composites was varied as 30, 40 and 50%. Mechanical properties were evaluated using ASTM UTM computerized system. It was observed from scanning electron microscope (SEM), X-ray diffraction (XRD), and thermal analysis (TA) that mechanical properties of composites based on polypropylene were significantly improved by adding the jute reinforcement. M. Sumaila et al. [55] experimentally evaluated the effects of banana length on the substantial and mechanical properties of composite banana and epoxy five distinctive specimens by differentiating the length by weight from 5 mm to 25 mm to 30%. Composites were fabricated using hand layup method. The mechanical properties of fabricated fibers were enhanced significantly by hybridization.

3. Study on hybrid (natural-glass fibre) reinforced polymers

The majority of natural and synthetic glass fibers was combined with natural fibers to reinforce a matrix of hybrid composites. Since glass fibres have a higher mechanical strength, hence, they are preferred. The introduction of glass fibres improved the tensile, flexural, compressive, thermal stability, and water resistance properties of natural fibres. Samal et al. [56] manufactured bamboo and glass fibre hybrid composites using polypropylene as epoxy with an extrusion method followed by injection moulding. Excellent mechanical results for mechanical properties such as tensile, flexural, and impact strength were observed for the composites that were fabricated with 30% loading and 2% MAPP concentration. In another study Singh et al. [57] fabricated partially biodegradable composite using jute as reinforcement and epoxy as matrix material using a compression moulding technique for the understanding effect of curing temperature on various mechanical properties. The temperature range from 80°C to 130°C was considered for investigation. The results revealed that maximum tensile and flexural strength was obtained was 100°C. Abdullah et al. [58] used additives and initiator, for fabricating hybrid resin composites of Reinforced polyester (USP) reinforced by jute fibre (Hessian cloth) and e-glass (mat). The composites were produced using a hand lay-up method. An electronic scanning microscope (SEM) monitors the adhesion of the interface between jute/glass and USP. With the inclusion of different parts of the glass, 25% of the jute showed a more significant improvement in mechanical properties. Moreover, in another study, Ahmed et al. [59] manufactured hand-lay-up isenthalpic polyester composites made of woven jute and glass and studied the impact of hybridization on mechanical characteristics. The composite's performance was evaluated using performance curves. The results indicated that there was a significant improvement in mechanical characteristics such as tensile strength and impact strength with the incorporation of glass fibre in addition water absorption behavior of the composite was enhanced with the addition of glass.

4. Applications

The use of natural as a reinforcement in a polymer matrix drew global attention to the importance of environmental awareness [60]. Application domain of bio composites shown in figure 3.

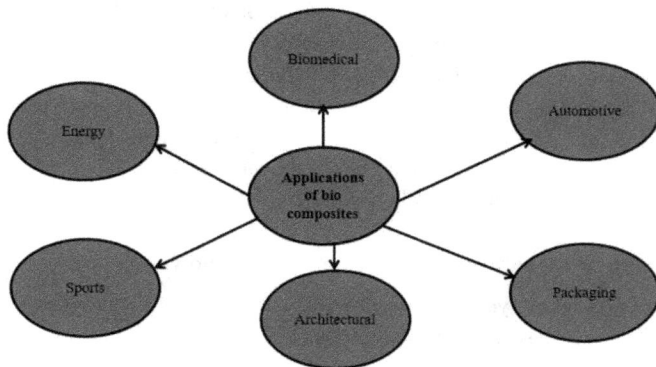

Fig 3. Application domain of Composites

Materials Research Forum LLC
https://doi.org/10.21741/9781644901854-11

In many applications, natural reinforced polymer composites have proved to be a viable alternative to synthetic reinforced polymer composites [61-62]. While many natural composite products are being developed and sold, only a few natural composites have been developed, with the majority of their technologies still in the research and development stages. Cargo racks, door panels, instrument panels, armrests, headrests, and seat shells are all made of natural composites [63]. Decks, docks, window frames, and molded panel parts are only a few of the applications for plastic/wood composites [64]. The kenaf/glass epoxy composite material [65] is used to make the passenger car bumper beam. The creative application of banana in under-floor safety for passenger cars [66] has recently sparked interest in banana reinforced composites. Sisal and Roselle hybrid composites were used to fabricate automobile parts such as rear-view mirrors, visors in two-wheelers, billion seat covers, indicator covers, cover L-side, and name plates [67]. The next section highlights some applications of NFRC.

4.1   Biomedical applications

In the new period, chemistry for modern medicine, in conjunction with the accumulation of natural composites, is expected to be a major theme of chemical activities. As a result, natural polymers and their potential value in biomedicine can be given considerable consideration. As a result, the use of green composites in medical applications has increased dramatically, particularly in the form of implants. For example, in some medical applications, metals such as stainless steel and titanium alloys, polymers (ultrahigh molecular weight polyethylene), and ceramics (hydroxyapatite) were commonly used. However, there has been a paradigm shift in the last two decades from bios stable biomaterials to bio-absorbable or biodegradable (hydrolytically and enzymatically degradable) biomaterials for medical devices that could help the body restore and revive damaged tissues. Santo et al. [68] observed that in the field of tissue engineering (TE), the idea of manufacturing multi-functional scaffolds capable of acting not only as templates for cell transplantation but also of delivering bioactive agents in a controlled manner is an emerging strategy aimed at enhancing tissue regeneration, and developed a complex hybrid release system consisting of a three-dimensional (3D) structure (NPs). Chicken feathers hold the unique advantage of low relative density and good thermal and sound insulating properties. There they can be used in a number of applications, for feather disposal as billions of chickens are culled per year moreover, technologies for the production of chicken feathers into fibrous (feather) and particulate (quill) fractions are developed and patented for biomedical applications. Reddy et al. [69] developed bio-thermoplastics with chicken features for using in tissue engineering with compression moulding process, as the major protein in fibers is biocompatible and have cross-linking properties. The thermoplastic films from features were prepared and investigated the results revealed that feather films were water secure and had good strength. This can be used for fabrication of biomaterials for various biomedical applications. Martelli et al. [70] investigated the influence of polyethylene glycol (PEG) on the films' hygroscopicity and solubility of keratins. The results revealed that these films have lower solubility in water. The coating of silk fibroin with waxes or silicone is done for enhancement material properties of the to be used in biomedical applications. Li et al. [71] studied the behaviour of biomedical textiles such as silk and observed that due to processing, controllable degradability and mechanical properties. this material can be used for extracorporeal implants and soft tissue repair. Chen et al. [72] studied the behaviour of blast by reinforcing them with poly (l-lactic acid) (PLLA) matrix as for fabrication of natural composites for hard tissue repair. Solution blending and freeze drying were carried out for producing PLLA. The results showed that there was a significant improvement in the various properties of the PLLA matrix that showed that these materials have a potential future in promising hard tissue repair applications. Goswami et al. [73] fabricated the foams of the three-components material system consisting of (poly(lactic acid) (PLA), poly(ε&59;-caprolactone) (PCL) and wollastonite (W) for the fabrication of bio composites for biomedical scaffold constructing using compressed CO2. The results showed that newly developed composites showed, show osteoblast cell attachment. Cao et al. [74] used a phase separation technique for studying the behaviour of-of 3D porous poly(lactic-co-glycolic acid) (PLGA) scaffolds. The results revealed that these materials can be used for medical applications. Zhang et al. [75] studied the mechanical properties of composite prepared for bone scaffold prepared using, poly (l-lactic acid) (PLLA), and octadecyl amine functionalized Nano diamond (ND-ODA). The dispersion of nanoparticles increased the hardness and Young's modulus of the composites. It was observed that the properties of newly formed Nano composites close to that of the human cortical bone.

4.2 Packaging

The use of non-biodegradable and non-renewable materials like plastics, glass, and metals in packaging applications has heightened concerns about environmental pollution. As every year, huge amounts of materials are used for packaging applications for the purpose of use and disposal. Burning and landfilling are two old-school methods for dealing with post-consumer plastic waste, all of which pose a health and environmental danger. Traditionally, cellulosic fibres have been used in packaging for a broad range of food types, including frozen or liquid foods, dry food items, and fresh foods [76]. The primary aim of food packaging is to protect and conserve foodstuffs, as well as to maintain their integrity and protection, and to reduce food waste [77]. Cellophane is the most widely used material for food packaging, and it is also known as regenerated cellulose in the film. Cellulose derivatives such as hydroxyethyl cellulose and cellulose acetate were used to make cellophane-based films. Procedures packaging is made up of 100% main, which has strong mechanical properties and elasticity, and is approved for direct contact with package materials (food). It can also be handled with a variety of coating layers to protect foods from moisture, light, bacteria, and other hazards. Traditionally, plant based fibres have been used in packaging for a wide variety of food types, such as frozen food products, liquid foods, new foods and drinks [78]. The major role of food packaging is to safeguard and preserve the food to maintain its quality and protection [79]. The Cellulose-based food packaging, which is most widely used, is cello pane, also known in the film as regenerated cellulose. A number of derivatives of cellulose are such as carboxyl methylcellulose, Methylcellulose, ethyl cellulose, propyl cellulose hydroxide, in the preparation of xyethyl cellulose and cellulose acetate. Cellulose acetate is also widely used as a rigid wrapping film along with cellulose triacetate than other derivatives, since they have low gas and moisture barrier proper- ties [80]. Fibre forms packaging consists of 100% primary fibre that provides high elasticity and strength, and its high purity is certified for direct contact with food, and can be coated with a wide range of coatings to protect food from light, moisture, bacteria and other hazards. Since these energy resources are not recyclable, billions of tons of petroleum products are used to make plastic bags, which can be saved for future uses. As a result, green polymers are playing an increasingly important role in packaging applications in the food industry, as seen in. Srinivasa and Tharanathan addressed the problems surrounding plastics' non-biodegradability [81]. They also classified biopolymers and suggested materials based on chitin/chitosan for use as biopolymers. Youssef et al. [82] examined the strengthening of biopolymer films utilizing nanoparticles and other treatments, concluding that nano composites may minimize flammability while maintaining the transparency of the polymer matrix to some degree. Packaging can also benefit from these composites. Furthermore, the requirement to pack fruits, vegetables, and liquids, among other things, can be met by adding some nanoparticles, as these particles can aid in strong mechanical properties and thermal efficiency. Packaging waste was accounted for, according to Valdes et al. [83] 29.5% of the total municipal solid waste (MSW) in 2009 in the USA and in 2006, 25% of the overall MSW in Europe. Thus, a lot of work still needs to be done to reduce the packaging waste present in MSW considerably. Moreover, the governments around the world have introduced legislation to restrict usage of plastics and to reduce the quantity of waste from packaging [84]. Therefore, focus is shifted towards the plant fibre for the use in packaging applications [85]. The major reason is due to high abundance of these fibre, low weight, high strength, rigidity and biodegradability therefore these fibres can play important role for sustainable packaging [86-87]. The use of nanotechnology in packaging applications is a new field in which packaging materials can be used to improve barrier and mechanical properties, as well as biodegradability. The pure polymer has heat resistance and anti-fire properties [88].

4.3   Automotive applications

Since green composites are lighter in weight than steel and iron parts, they have lower energy consumption as well as vibration dampening benefits in the transportation and automotive industries. In a massive public transportation system composite structural parts are fuel-efficient, environmentally friendly (emits less CO_2), and more versatile than metal, allowing them to be shaped and built into a variety of shapes [89-90]. In recent years, efforts have been made to minimize the use of costly glass, aramid, or carbon s, as well as to lighten the car's body by taking advantage of the lower density and lighter weight of these materials. Natural fibers have been used for several decades, but the invention and subsequent commercialization of synthetics such as glass, aramids, and carbon fiberss in the early and mid-twentieth centuries resulted in a decreased preference for natural fibers in a variety of applications [91]. This was a period of rapid industrial and technological growth, necessitating the use of highly durable materials with consistent properties; a test that the newly manufactured fibers passed with flying colours, while natural fibers failed miserably due to large variations in physical and chemical properties caused by temperature, harvesting methods, transportation, storage, and processing. Since the beginning of time, conventional fibers and, to a lesser degree, glass have been unrivalled as automotive composite reinforcements [92]. In 1953, they began working in the industry. Moulded Glass Company of Ashtabula, Ohio supplied a total of 46 separate glass - reinforced parts for the Chevrolet Corvette [93], where they were manufactured in open moulds by hand rolling polyester resins into glass mats. These were used in the construction of a prototype car, after which a batch of 300 was made. The aim of switching from conventional steel and aluminium parts to glass composites was to reduce weight, improve mechanical properties, and increase processing performance. Since then, glass fibers' positive attributes of great abundance, low cost, good mechanical properties, and consistent performance have earned them a position alongside steel in car body construction, which consumes up to 15% of the world's steel, 25% of the world's glass, and 40% of the world's annual oil production. Despite attempts to minimize the car weight, there has been an overall steady rise over decades, with the weight of an Audi rising by 460 kg between 1972 and 2003 [94]. The use of bio-based materials in vehicle parts was first considered by Ford Motor Company founder Henry Ford in the early 1930s [95]. Inspired by the plight of the farmers, Henry Ford was looking for a way to financially motivate them, and using soil as a realistic source of car parts and fuels seemed like a good idea. Furthermore, in the quest for environmentally friendly light-bodied cars [96], these materials promised to reduce weight and pollution. As a result, the Ford Motor Company embarked on an ambitious

Sustainable Natural Fiber Composites   Materials Research Forum LLC
Materials Research Foundations **122** (2022) 256-281   https://doi.org/10.21741/9781644901854-11

development program into the widespread use of natural fibers in vehicles, culminating in the 1941 unveiling of the concept car [97]. This car had a body made entirely of soy resin-reinforced hemp, sisal, and wheat straw composites, was two-thirds the weight of a standard car, and ran on hemp oil fuel. The abundance of oil at the time, as well as the automotive industry's rapid plunge into WWII, necessitated the development of highly mechanized vehicles, which hampered the progress of this study. The East German Trabant car, which had a monologue construction with the roof, boot lid, bonnet, wings, and doors, was a major breakthrough for bio composites in the automotive industry in 1957. Renewable materials were widely used as reinforcements in composites of interior parts for a variety of passenger and commercial vehicles. Back in 1996, Mercedes-Benz used an epoxy matrix with jute in the door panels of its E-class vehicles [98]. In 2000, when Audi released the A2 midrange car, a new paradigm of green composites application emerged: the door trim panels were made of polyurethane reinforced with a mixed flax/sisal material [99]. Toyota, on the other hand, claims to be the first company to use environmentally friendly materials such as 100% bio plastics. The green composite with natural reinforcement was used in the project. In the spare tire cover, there's a RAUM 2003 model. The component was made of a PLA matrix made from sugar cane and sweet potato, reinforced with kenaf's [100]. Interior components are an example of a later example. Floor mats made from PLA and nylon fibers for Mitsubishi motors, which combine bamboo fibers and a plant-based resin polybutylene succinate (PBS) [101]. Toyota recently created a sugar cane-based eco-plastic that will be used to line the interiors of its vehicles. In reality, it will be used for the first time on the new CT 200's luggage compartment, as announced at the 2011 Automotive World Congress [102]. Furthermore, the availability of green composites is dependent on cost-effective and competitive manufacturing techniques, such as injection-moulded thermoplastics, which are currently used widely by many industries for the fabrication of certain vehicle components. Various material experts predict that an advanced composite body may be 50-67 percent lighter than the current body [103].

4.4 Energy sector

Natural composites, such as Biotex Flax, will play an important role in the energy industry, with rooftop wind turbines being one of the most important applications. The performance characteristics of Biotex flax fibre in technology are used to design and manufacture blades. The main goal of research in this field is to reduce the cost of wind energy (Endowed Chair of Wind Energy) began designing rotor blades for a 1KW wind turbine at Stuttgart University in 2011 [104]. The experiment reviewed a variety of materials for the cost-effective development of turbine blades, but Biotex flax produced the best results for lightweight applications in turbine blade manufacturing. Bamboo, which is one-sixth the weight of steel, is yet another highly productive material of strong compressive strength and elasticity. This substance has the ability to take the place of glass. As compared to glass, the material has strong wind strength and resilience. Bamboo wind blades are 10-12% thinner and more cost-effective than steel blades. Bamboo's small diameter can be used to create artificial panels that can be used in place of wood. Since 2010, China has had about 100 sets of 800KW wind blades in use. The 1.5 MW wind blade passed the static tests of 2MW, 2.5MW, and 3MW. Bamboo can also be considered a potential material in energy sector applications, as demand for wind power blades is projected to increase by 20% in the coming years [105-107].

4.5   Architectural applications

Despite the fact that natural composites have not been regarded as materials for compressive load-bearing applications, the majority of the elements were built for tensile or flexural modes, which is where natural composites are most commonly used. In practice, the components would be subjected to compressive loads to some extent. When the composites are flexed, they undergo a tensile-compressive coupling reaction, and residual compressive forces which exist within the composites due to unbalanced laminate stacking sequences or residual processing stresses. This segment will look at how NFC materials can be used as non-structural and structural components. Natural bast are currently used as reinforcement in a variety of applications. Professor Mizi Fan of Brunel University in the United Kingdom leads NATCOM, a UK government-funded project that has built circular and square NFC rods that can withstand tensile loads. Peng et al. [108] identified a set of mechanical tests that were used to assess and improve NFC reinforced with coarse hemp non twisted yarns. The tensioning of aligned fibers along the rod's axis allows for a higher Vf of reinforcement and improved mechanical properties. This type of component can be used in both indoor and outdoor furniture. In the automotive industry, flat laminated panels reinforced with fabrics or roving's are already in use [109]. Flat laminate panels may be used as wall partitions, roofing components, flooring, or furniture elements in civil engineering. Fabric reinforcements may be used to control properties. The natural appearance and fibrous texture appeal to the eye and match in well with the sustainable design trend. In tensile and flexural modes, NFC has the best properties. Composite properties could be tailored for particular element specifications by controlling the form and arrangement of reinforcement. Rods, tubes, and I-beams made from NFC have been processed to meet precise requirements. Apart from their unique mechanical properties, one of the aesthetic key points in favour of NFC application in civil engineering sustainable design is their "natural" appearance. The use of pigments to reduce UV-induced weathering deterioration could overshadow this effect. Eco-design and energy-efficiency are two main approaches that must be considered in order to achieve the objectives of energy conservation and material reduction. Rosa et al. [110] examined the environmental impacts and thermal conductivity analysis of different building wall materials, as well as performing a life cycle assessment to better understand the effects of environmental changes. Manufacturing was achieved with natural fibers like flax and bio-based epoxy, and the material was tested for thermal conductivity. The findings showed that using eco-sandwiched materials minimized environmental impact, which was long overdue due to the light weight of these materials, which rendered handling and transportation simple. Natural fibers can also be used to create relatively thin building envelopes with high thermal resistance (m^2K/W) and low thermal transmittance, U (W/m2K). Green building concepts are also being discussed these days to reduce infrastructure's environmental effects, so buildings that use the least amount of electricity, water, and materials are being considered. When compared to conventional construction, green buildings are found to have greater benefits in terms of building value (84%), occupant health and well-being (88%), and return on investment (88%). Green composites architecture and constructional applications deliver various advantages, including recycled material, energy efficiency, and durability. Bamboo bio composites have a major influence on interior design, which has its own commercial value. Bio composite is used to make a variety of bamboo items for the exterior and interior that is in high demand on the global market. The majority of users are aware of this. The greatness of this stuff, and to promote efforts to make daily life more sustainable for nature. This is bolstered by its reputation as a revolutionary material, which has garnered praise from a variety of sources, demonstrating that hybrid bamboo can outperform other materials in terms of physical, mechanical, and aesthetic properties [111-112]. Various styles of hybrid bamboo-based items are now available, ranging from ceilings, walls, floors, window frames, doors, and stairs to home decorative accessories. Bamboo joinery can be bent or straightened by heating and clamping, and it can also be used to make special effects. According to previous studies, a standard bamboo stud wall section is formed with distance determined by the thickness of the bamboo boards used in the analysis.

Conclusion

This paper aims to provide a context for understanding the physical and mechanical properties of natural/glass reinforced polymers, while concentrating less on the electrical, thermal, and dynamic properties of hybrid composites consisting of two natural fibers and glass. This study hypothesized that natural composites with glass have stronger mechanical properties than natural composites. In the field of engineering and technology, research on the integration of various natural polymers with synthetic fibers leads to increased applications and the replacement of non-renewable materials. Natural mechanical and

Materials Research Forum LLC
https://doi.org/10.21741/9781644901854-11

physical properties differ from one to the next. Natural reinforced composites are used in a wide range of engineering applications due to their superior properties such as specific strength, low weight, low cost, reasonable mechanical properties, and non-abrasive, environmentally friendly and bio-degradable characteristics. Natural fibers can boost the properties of GFRP and can be used as a replacement material for glass reinforced polymer composites.

Future road map

Green composites should be used more in structural applications in the future. Other applications depend on their continued development and testing. However, a range of issues must be addressed before green composites can be considered completely compatible with synthetic composites. The most significant advancement in recent years has been the establishment of nano composites (i.e., the use of Nano cellulose in the form of crystals or s) are a type of composite material. Natural fibers are used to make this product. Natural fibers make up about 30–40% of the total. Cellulose, with crystalline cellulose accounting for about half of it. This nano cellulose was said to be able to compete with components made of traditional materials. Nanotechnology offers a plethora of possibilities for improving the properties of green composites. Since cellulose nanocrystal and cellulose nanos are stronger than steel and stiffer than aluminum, they are being investigated for a variety of applications. Composites reinforced with cellulose nanocrystals may be available in the near future. Advanced efficiency, longevity, value, service-life, and usefulness while remaining a completely sustainable technology.

Acknowledgement

No conflicts exist for the present work.

References

[1] Bhat, A.H., Dasan, Y.K., Khan, I. and Jawaid, M., 2017. Cellulosic biocomposites: Potential materials for future. In *Green Biocomposites* (pp. 69-100). Springer, Cham. https://doi.org/10.1007/978-3-319-49382-4_4

[2] Khan, A., Rangappa, S.M., Siengchin, S. and Asiri, A.M. eds., 2020. *Bios and Biopolymers for Biocomposites: Synthesis, Characterization and Properties*. Springer Nature. https://doi.org/10.1007/978-3-030-40301-0

[3] Khalil, H.A., Davoudpour, Y., Saurabh, C.K., Hossain, M.S., Adnan, A.S., Dungani, R., Paridah, M.T., Sarker, M.Z.I., Fazita, M.N., Syakir, M.I. and Haafiz, M.K.M.,

Sustainable Natural Fiber Composites   Materials Research Forum LLC
Materials Research Foundations **122** (2022) 256-281   https://doi.org/10.21741/9781644901854-11

2016. A review on nanocellulosic fibres as new material for sustainable packaging: Process and applications. *Renewable and Sustainable Energy Reviews*, *64*, pp.823-836. https://doi.org/10.1016/j.rser.2016.06.072

[4] Jawaid, M.H.P.S. and Khalil, H.A., 2011. Cellulosic/synthetic fibre reinforced polymer hybrid composites: A review. *Carbohydrate polymers*, *86*(1), pp.1-18. https://doi.org/10.1016/j.carbpol.2011.04.043

[5] Vijay, R., Singaravelu, D.L., Vinod, A., Sanjay, M.R., Siengchin, S., Jawaid, M., Khan, A. and Parameswaranpillai, J., 2019. Characterization of raw and alkali treated new natural cellulosic fibers from Tridax procumbens. *International journal of biological macromolecules*, *125*, pp.99-108. https://doi.org/10.1016/j.ijbiomac.2018.12.056

[6] Scott, G. and Wiles, D.M., 2001. Programmed-life plastics from polyolefins: a new look at sustainability. *Biomacromolecules*, *2*(3), pp.615-622. https://doi.org/10.1021/bm010099h

[7] Pandey, J.K. and Singh, R.P., 2005. Green nanocomposites from renewable resources: effect of plasticizer on the structure and material properties of clay-filled starch. *Starch-Stärke*, *57*(1), pp.8-15. https://doi.org/10.1002/star.200400313

[8] Davis, G. and Song, J.H., 2006. Biodegradable packaging based on raw materials from crops and their impact on waste management. *Industrial crops and products*, *23*(2), pp.147-161. https://doi.org/10.1016/j.indcrop.2005.05.004

[9] Rhim, J.W. and Ng, P.K., 2007. Natural biopolymer-based nanocomposite films for packaging applications. *C* Petersen, K., Nielsen, P.V., Bertelsen, G., Lawther, M., Olsen, M.B., Nilsson, N.H. and Mortensen, G., 1999. Potential of biobased materials for food packaging. *Trends in food science & technology*, *10*(2), pp.52-68.*ritical reviews in food science and nutrition*, *47*(4), pp.411-433. https://doi.org/10.1016/S0924-2244(99)00019-9

[10] Khalil, H.A., Davoudpour, Y., Islam, M.N., Mustapha, A., Sudesh, K., Dungani, R. and Jawaid, M., 2014. Production and modification of nanofibrillated cellulose using various mechanical processes: a review. *Carbohydrate polymers*, *99*, pp.649-665. https://doi.org/10.1016/j.carbpol.2013.08.069

[11] Mann, G.S., Singh, L.P., Kumar, P. and Singh, S., 2020. Green composites: A review of processing technologies and recent applications. *Journal of Thermoplastic Composite Materials*, *33*(8), pp.1145-1171. https://doi.org/10.1177/0892705718816354

[12]   Zhao, R., Torley, P. and Halley, P.J., 2008. Emerging biodegradable materials: starch-and protein-based bio-nanocomposites. *Journal of Materials Science*, *43*(9), pp.3058-3071. https://doi.org/10.1007/s10853-007-2434-8

[13]   Lee, S.G. and Xu, X., 2005. Design for the environment: life cycle assessment and sustainable packaging issues. *International Journal of Environmental Technology and Management*, *5*(1), pp.14-41. https://doi.org/10.1504/IJETM.2005.006505

[14]   Kamel, S., 2007. Nanotechnology and its applications in lignocellulosic composites, a mini review. *Express Polymer Letters*, *1*(9), pp.546-575. https://doi.org/10.3144/expresspolymlett.2007.78

[15]   Ramamoorthy, S.K., Skrifvars, M. and Persson, A., 2015. A review of natural fibers used in biocomposites: plant, animal and regenerated cellulose s. *Polymer reviews*, *55*(1), pp.107-162. https://doi.org/10.1080/15583724.2014.971124

[16]   Yan, L., Chouw, N. and Jayaraman, K., 2014. Flax fibre and its composites–A review. *Composites Part B: Engineering*, *56*, pp.296-317. https://doi.org/10.1016/j.compositesb.2013.08.014

[17]   Nirmal, U., Hashim, J. and Ahmad, M.M., 2015. A review on tribological performance of natural fibre polymeric composites. *Tribology International*, *83*, pp.77-104. https://doi.org/10.1016/j.triboint.2014.11.003

[18]   HOBSON, J. and CARUS, M., 2011. Targets for bio-based composites and natural fibres. *JEC composites*, (63), pp.31-32.

[19]   Mohanty, A.K., Misra, M. and Drzal, L.T. eds., 2005. *Natural s, biopolymers, and biocomposites*. CRC press. https://doi.org/10.1201/9780203508206

[20]   Dong C. Review of natural -reinforced hybrid composites. Journal of Reinforced Plastics and Composites. 2018 Mar;37(5):331-48. https://doi.org/10.1177/0731684417745368

[21]   Shah DU, Porter D, Vollrath F. Can silk become an effective reinforcing fibre? A property comparison with flax and glass reinforced composites. Compos Sci Technol 2014;101:173–83. https://doi.org/10.1016/j.compscitech.2014.07.015

[22]   Abdollah MF, Shuhimi FF, Ismail N, Amiruddin H, Umehara N. Selection and verification of kenaf fibres as an alternative friction material using Weighted Decision Matrix method. Materials & Design. 2015 Feb 15;67:577-82. https://doi.org/10.1016/j.matdes.2014.10.091

[23]   De Rosa IM, Kenny JM, Puglia D, Santulli C, Sarasini F. Tensile behavior of New Zealand flax (Phormium tenax) s. Journal of Reinforced Plastics and Composites. 2010 Dec;29(23):3450-4. https://doi.org/10.1177/0731684410372264

[24]   Dittenber DB, GangaRao HVS. Critical review of recent publications on use of natural composites in infrastructure. Composites Part A 2011;43(8):1419–29. https://doi.org/10.1016/j.compositesa.2011.11.019

[25]   Mwaikambo L. Review of the history, properties and application of plant fibres. African Journal of Science and Technology. 2006 Dec;7(2):121.

[26]   Zini E, Scandola M. Green composites: an overview. Polymer composites. 2011 Dec 1;32(12):1905-15. https://doi.org/10.1002/pc.21224

[27]   Brahim SB, Cheikh RB. Influence of fibre orientation and volume fraction on the tensile properties of unidirectional Alfa-polyester composite. Composites Science and Technology. 2007 Jan 1;67(1):140-7. https://doi.org/10.1016/j.compscitech.2005.10.006

[28]   Bos HL, Van Den Oever MJ, Peters OC. Tensile and compressive properties of flax fibres for natural fibre reinforced composites. Journal of Materials Science. 2002 Apr 1;37(8):1683-92.

[29]   Reddy N, Jiang Q, Yang Y. Biocompatible natural silk fibers from Argema mittrei. Journal of Biobased Materials and Bioenergy. 2012 Oct 1;6(5):558-63. https://doi.org/10.1166/jbmb.2012.1255

[30]   Le TM, Pickering KL. The potential of harakeke fibre as reinforcement in polymer matrix composites including modelling of long harakeke fibre composite strength. Composites Part A: Applied Science and Manufacturing. 2015 Sep 1;76:44-53. https://doi.org/10.1016/j.compositesa.2015.05.005

[31]   Carr DJ, Cruthers NM, Laing RM, Niven BE. fibers from three cultivars of New Zealand flax (Phormium tenax). Textile research journal. 2005 Feb;75(2):93-8. https://doi.org/10.1177/004051750507500201

[32]   Pickering KL, Beckermann GW, Alam SN, Foreman NJ. Optimising industrial hemp fibre for composites. Composites Part A: Applied Science and Manufacturing. 2007 Feb 1;38(2):461-8. https://doi.org/10.1016/j.compositesa.2006.02.020

[33]   Pickering K, editor. Properties and performance of natural-fibre composites. Elsevier; 2008 Jun 23. https://doi.org/10.1533/9781845694593

[34] Cheng S, Lau KT, Liu T, Zhao Y, Lam PM, Yin Y. Mechanical and thermal properties of chicken feather /PLA green composites. Composites Part B: Engineering. 2009 Oct 1;40(7):650-4.
https://doi.org/10.1016/j.compositesb.2009.04.011

[35] Huson MG, Bedson JB, Phair NL, Turner PS. Intrinsic strength of wool fibres. Asian Australasian Journal of Animal Sciences. 2000 Jul 1;13:267-.

[36] Gashti MP, Gashti MP. Effect of colloidal dispersion of clay on some properties of wool . Journal of Dispersion Science and Technology. 2013 May 17;34(6):853-8.
https://doi.org/10.1080/01932691.2012.713248

[37] Niu M, Liu X, Dai J, Hou W, Wei L, Xu B. Molecular structure and properties of wool surface-grafted with nano-antibacterial materials. Spectrochimica Acta Part A: Molecular and Biomolecular Spectroscopy. 2012 Feb 1;86:289-93.
https://doi.org/10.1016/j.saa.2011.10.038

[38] Zhan M, Wool RP. Mechanical properties of chicken feather s. Polymer Composites. 2011 Jun 1;32(6):937-44. https://doi.org/10.1002/pc.21112

[39] Efendy MA, Pickering KL. Comparison of harakeke with hemp fibre as a potential reinforcement in composites. Composites Part A: Applied Science and Manufacturing. 2014 Dec 1;67:259-67.
https://doi.org/10.1016/j.compositesa.2014.08.023

[40] Cheung HY, Ho MP, Lau KT, Cardona F, Hui D. Natural fibre-reinforced composites for bioengineering and environmental engineering applications. Composites Part B: Engineering. 2009 Oct 1;40(7):655-63.
https://doi.org/10.1016/j.compositesb.2009.04.014

[41] Xie, Y., Hill, C.A., Xiao, Z., Militz, H. and Mai, C., 2010. Silane coupling agents used for natural /polymer composites: A review. *Composites Part A: Applied Science and Manufacturing*, *41*(7), pp.806-819.
https://doi.org/10.1016/j.compositesa.2010.03.005

[42] Dittenber, D.B. and GangaRao, H.V., 2012. Critical review of recent publications on use of natural composites in infrastructure. *Composites Part A: applied science and manufacturing*, *43*(8), pp.1419-1429.
https://doi.org/10.1016/j.compositesa.2011.11.019

[43] Malkapuram, R., Kumar, V. and Negi, Y.S., 2009. Recent development in natural reinforced polypropylene composites. *Journal of reinforced plastics and composites*, *28*(10), pp.1169-1189. https://doi.org/10.1177/0731684407087759

[44] Shubhra, Q.T., Alam, A.K.M.M. and Quaiyyum, M.A., 2013. Mechanical properties of polypropylene composites: A review. *Journal of thermoplastic composite materials*, *26*(3), pp.362-391. https://doi.org/10.1177/0892705711428659

[45] Meier, M.A., Metzger, J.O. and Schubert, U.S., 2007. Plant oil renewable resources as green alternatives in polymer science. *Chemical Society Reviews*, *36*(11), pp.1788-1802. https://doi.org/10.1039/b703294c

[46] de Espinosa, L.M. and Meier, M.A., 2011. Plant oils: The perfect renewable resource for polymer science?!. *European Polymer Journal*, *47*(5), pp.837-852. https://doi.org/10.1016/j.eurpolymj.2010.11.020

[47] Gassan, J., Chate, A. and Bledzki, A.K., 2001. Calculation of elastic properties of natural s. *Journal of materials science*, *36*(15), pp.3715-3720. https://doi.org/10.1023/A:1017969615925

[48] Charlet, K., Baley, C., Morvan, C., Jernot, J.P., Gomina, M. and Bréard, J., 2007. Characteristics of Hermès flax fibres as a function of their location in the stem and properties of the derived unidirectional composites. *Composites Part A: Applied Science and Manufacturing*, *38*(8), pp.1912-1921. https://doi.org/10.1016/j.compositesa.2007.03.006

[49] Wambua, P., Ivens, J. and Verpoest, I., 2003. Natural fibres: can they replace glass in fibre reinforced plastics?. *Composites science and technology*, *63*(9), pp.1259-1264. https://doi.org/10.1016/S0266-3538(03)00096-4

[50] Dixit, P.S. and Verma, P., 2012. The effect of hybridization on mechanical behaviour of coir/sisal/jute fibres reinforced polyester composite material. *Research Journal of Chemical Sciences ISSN*, *2231*, p.606X.

[51] Gupta, M.K. and Srivastava, R.K., 2015. Effect of sisal fibre loading on mechanical properties of jute fibre reinforced epoxy composite. *Cellulose*, *61*(71), p.65.

[52] Thiruchitrambalam, M., Athijayamani, A., Sathiyamurthy, S. and Thaheer, A.S.A., 2010. A review on the natural -reinforced polymer composites for the development of roselle -reinforced polyester composite. *Journal of Natural s*, *7*(4), pp.307-323. https://doi.org/10.1080/15440478.2010.529299

[53] Mishra, A., 2013. Strength and corrosion testing of jute/glass-epoxy hybrid composite laminates. *Plastic and Polymer Technology*, *2*(2), pp.48-54.

[54] Berhanu, T., Kumar, P. and Singh, I., 2014, December. Mechanical behaviour of jute fibre reinforced polypropylene composites. In *5th International & 25th All India Manufacturing Technology, Design and Research Conference (AIMTDR 2014) December 12th-14th.*

[55] Sumaila, M., Amber, I. and Bawa, M., 2013. Effect of length on the physical and mechanical properties of ramdom oriented, nonwoven short banana (musabalbisiana) /epoxy composite. *Cellulose*, *62*, p.64.

[56] Nayak, S.K., Mohanty, S. and Samal, S.K., 2009. Influence of short bamboo/glass on the thermal, dynamic mechanical and rheological properties of polypropylene hybrid composites. *Materials Science and Engineering: A*, *523*(1-2), pp.32-38. https://doi.org/10.1016/j.msea.2009.06.020

[57] Singh, J.I.P., Singh, S. and Dhawan, V., 2018. Effect of curing temperature on mechanical properties of natural reinforced polymer composites. *Journal of Natural s*, *15*(5), pp.687-696. https://doi.org/10.1080/15440478.2017.1354744

[58] Abdullah-Al-Kafi, Abedin, M.Z., Beg, M.D.H., Pickering, K.L. and Khan, M.A., 2006. Study on the mechanical properties of jute/glass -reinforced unsaturated polyester hybrid composites: Effect of surface modification by ultraviolet radiation. *Journal of Reinforced Plastics and Composites*, *25*(6), pp.575-588. https://doi.org/10.1177/0731684405056437

[59] Ahmed, K.S., Vijayarangan, S. and Rajput, C., 2006. Mechanical behavior of isothalic polyester-based untreated woven jute and glass fabric hybrid composites. *Journal of reinforced plastics and composites*, *25*(15), pp.1549-1569. https://doi.org/10.1177/0731684406066747

[60] Jawaid, M., Khalil, H.A., Hassan, A., Dungani, R. and Hadiyane, A., 2013. Effect of jute fibre loading on tensile and dynamic mechanical properties of oil palm epoxy composites. *Composites Part B: Engineering*, *45*(1), pp.619-624. https://doi.org/10.1016/j.compositesb.2012.04.068

[61] Bharath, K.N. and Basavarajappa, S., 2016. Applications of biocomposite materials based on natural fibers from renewable resources: a review. *Science and Engineering of Composite Materials*, *23*(2), pp.123-133. https://doi.org/10.1515/secm-2014-0088

[62] Ticoalu, A., Aravinthan, T. and Cardona, F., 2010. A review of current development in natural composites for structural and infrastructure applications.

In *Proceedings of the southern region engineering conference (SREC 2010)* (pp. 113-117). Engineers Australia.

[63]   Jarukumjorn, K. and Suppakarn, N., 2009. Effect of glass hybridization on properties of sisal –polypropylene composites. *Composites Part B: Engineering*, *40*(7), pp.623-627. https://doi.org/10.1016/j.compositesb.2009.04.007

[64]   Ganesh, B.N. and Rekha, B., 2015. A comparative study on tensile behavior of plant and animal reinforced composites. *International Journal of Progressive Sciences and Technologies*, *1*(1).

[65]   Nair, L.S. and Laurencin, C.T., 2007. Biodegradable polymers as biomaterials. *Progress in polymer science*, *32*(8-9), pp.762-798. https://doi.org/10.1016/j.progpolymsci.2007.05.017

[66]   Olusegun, D.S., Stephen, A. and Adekanye, T.A., 2012. Assessing mechanical properties of natural fibre reinforced composites for engineering applications. *Journal of Minerals and Materials Characterization and Engineering*, *11*(1), pp.780-784. https://doi.org/10.4236/jmmce.2012.118066

[67]   Harish, S., Michael, D.P., Bensely, A., Lal, D.M. and Rajadurai, A., 2009. Mechanical property evaluation of natural coir composite. *Materials characterization*, *60*(1), pp.44-49. https://doi.org/10.1016/j.matchar.2008.07.001

[68]   de Andrade Silva, F., Toledo Filho, R.D., de Almeida Melo Filho, J. and Fairbairn, E.D.M.R., 2010. Physical and mechanical properties of durable sisal –cement composites. *Construction and building materials*, *24*(5), pp.777-785. https://doi.org/10.1016/j.conbuildmat.2009.10.030

[69]   Santo, V.E., Duarte, A.R.C., Gomes, M.E., Mano, J.F. and Reis, R.L., 2010. Hybrid 3D structure of poly (d, l-lactic acid) loaded with chitosan/chondroitin sulfate nanoparticles to be used as carriers for biomacromolecules in tissue engineering. *The Journal of Supercritical Fluids*, *54*(3), pp.320-327. https://doi.org/10.1016/j.supflu.2010.05.021

[70]   Reddy N, Yang Y. Biocomposites developed using water-plasticized wheat gluten as matrix and jute fibers as reinforcement. Polymer International. 2011 Apr 1;60(4):711-6. https://doi.org/10.1002/pi.3014

[71]   Li, G., Li, Y., Chen, G., He, J., Han, Y., Wang, X. and Kaplan, D.L., 2015. Silk-based biomaterials in biomedical textiles and -based implants. *Advanced healthcare materials*, *4*(8), pp.1134-1151. https://doi.org/10.1002/adhm.201500002

[72] Chen, X., Li, Y. and Gu, N., 2010. A novel basalt -reinforced polylactic acid composite for hard tissue repair. *Biomedical Materials*, *5*(4), p.044104. https://doi.org/10.1088/1748-6041/5/4/044104

[73] Goswami, J., Bhatnagar, N., Mohanty, S. and Ghosh, A.K., 2013. Processing and characterization of poly (lactic acid) based bioactive composites for biomedical scaffold application. *Express Polymer Letters*, *7*(9). https://doi.org/10.3144/expresspolymlett.2013.74

[74] Cao, Y., Groll, T.I., O'Connor, A.J., Stevens, G.W. and Cooper-White, J.J., 2004, December. Systematic selection of solvents for the fabrication of 3D PLGA scaffolds for tissue engineering. In *Transactions-7th World Biomaterials Congress*.

[75] Zhang, Q., Mochalin, V.N., Neitzel, I., Knoke, I.Y., Han, J., Klug, C.A., Zhou, J.G., Lelkes, P.I. and Gogotsi, Y., 2011. Fluorescent PLLA-nanodiamond composites for bone tissue engineering. *Biomaterials*, *32*(1), pp.87-94. https://doi.org/10.1016/j.biomaterials.2010.08.090

[76] Youssef, A.M. and El-Sayed, S.M., 2018. Bionanocomposites materials for food packaging applications: Concepts and future outlook. *Carbohydrate polymers*, *193*, pp.19-27. https://doi.org/10.1016/j.carbpol.2018.03.088

[77] Bradley, E.L., Castle, L. and Chaudhry, Q., 2011. Applications of nanomaterials in food packaging with a consideration of opportunities for developing countries. *Trends in food science & technology*, *22*(11), pp.604-610. https://doi.org/10.1016/j.tifs.2011.01.002

[78] Ludueña, L., Vázquez, A. and Alvarez, V., 2012. Effect of lignocellulosic filler type and content on the behavior of polycaprolactone based eco-composites for packaging applications. *Carbohydrate polymers*, *87*(1), pp.411-421. https://doi.org/10.1016/j.carbpol.2011.07.064

[79] Chauhan, V.S. and Chakrabarti, S.K., 2012. Use of nanotechnology for high performance cellulosic and papermaking products. *Cellulose chemistry and technology*, *46*(5), p.389.

[80] Thielemans, W., Warbey, C.R. and Walsh, D.A., 2009. Permselective nanostructured membranes based on cellulose nanowhiskers. *Green Chemistry*, *11*(4), pp.531-537. https://doi.org/10.1039/b818056c

[81] Srinivasa, P.C. and Tharanathan, R.N., 2007. Chitin/chitosan—Safe, ecofriendly packaging materials with multiple potential uses. *Food reviews international*, *23*(1), pp.53-72. https://doi.org/10.1080/87559120600998163

[82]   Youssef, A.M., 2013. Polymer nanocomposites as a new trend for packaging applications. *Polymer-Plastics Technology and Engineering*, *52*(7), pp.635-660. https://doi.org/10.1080/03602559.2012.762673

[83]   Ashori, A., Babaee, M., Jonoobi, M. and Hamzeh, Y., 2014. Solvent-free acetylation of cellulose nanos for improving compatibility and dispersion. *Carbohydrate polymers*, *102*, pp.369-375. https://doi.org/10.1016/j.carbpol.2013.11.067

[84]   Valdés, A., Mellinas, A.C., Ramos, M., Garrigós, M.C. and Jiménez, A., 2014. Natural additives and agricultural wastes in biopolymer formulations for food packaging. *Frontiers in chemistry*, *2*, p.6. https://doi.org/10.3389/fchem.2014.00006

[85]   Ludueña, L., Vázquez, A. and Alvarez, V., 2012. Effect of lignocellulosic filler type and content on the behavior of polycaprolactone based eco-composites for packaging applications. *Carbohydrate polymers*, *87*(1), pp.411-421. https://doi.org/10.1016/j.carbpol.2011.07.064

[86]   Yam, K.L. and Lee, D.S. eds., 2012. *Emerging food packaging technologies: Principles and practice*. Elsevier. https://doi.org/10.1533/9780857095664

[87]   Zhang, H.C., Kuo, T.C., Lu, H. and Huang, S.H., 1997. Environmentally conscious design and manufacturing: a state-of-the-art survey. *Journal of manufacturing systems*, *16*(5), pp.352-371. https://doi.org/10.1016/S0278-6125(97)88465-8

[88]   Lewis, H., Verghese, K. and Fitzpatrick, L., 2010. Evaluating the sustainability impacts of packaging: the plastic carry bag dilemma. *Packaging Technology and Science: An International Journal*, *23*(3), pp.145-160. https://doi.org/10.1002/pts.886

[89]   Sanjay, M.R. and Yogesha, B., 2017. Studies on natural/glass reinforced polymer hybrid composites: An evolution. *Materials today: proceedings*, *4*(2), pp.2739-2747. https://doi.org/10.1016/j.matpr.2017.02.151

[90]   Sanjay, M.R., Arpitha, G.R. and Yogesha, B., 2015. Study on mechanical properties of natural-glass fibre reinforced polymer hybrid composites: A review. *Materials today: proceedings*, *2*(4-5), pp.2959-2967. https://doi.org/10.1016/j.matpr.2015.07.264

[91]   Holbery, J. and Houston, D., 2006. Natural--reinforced polymer composites in automotive applications. *Jom*, *58*(11), pp.80-86. https://doi.org/10.1007/s11837-006-0234-2

[92]   Koronis, G., Silva, A. and Fontul, M., 2013. Green composites: A review of adequate materials for automotive applications. *Composites Part B: Engineering*, *44*(1), pp.120-127. https://doi.org/10.1016/j.compositesb.2012.07.004

[93]   Mohanty, A.K., Misra, M. and Drzal, L.T. eds., 2005. *Natural s, biopolymers, and biocomposites*. CRC press. https://doi.org/10.1201/9780203508206

[94]   Davies, G., 2012. *Materials for automobile bodies*. Butterworth-Heinemann.

[95]   Drzal, L.T., Misra, M. and Mohanty, A.K. eds., 2005. *Natural s, biopolymers, and biocomposites*. Taylor & Francis.

[96]   Müssig, J., Schmehl, M., Von Buttlar, H.B., Schönfeld, U. and Arndt, K., 2006. Exterior components based on renewable resources produced with SMC technology— Considering a bus component as example. *Industrial Crops and Products*, *24*(2), pp.132-145. https://doi.org/10.1016/j.indcrop.2006.03.006

[97]   Brosius, D., 2006. Natural composites slowly take root. *Composites Technology*, *12*(1), pp.32-37.

[98]   Chen, Y., Sun, L., Chiparus, O., Negulescu, I., Yachmenev, V. and Warnock, M., 2005. Kenaf/ramie composite for automotive headliner. *Journal of Polymers and the Environment*, *13*(2), pp.107-114. https://doi.org/10.1007/s10924-005-2942-z

[99]   Arbelaiz, A., Fernández, B., Cantero, G., Llano-Ponte, R., Valea, A. and Mondragon, I., 2005. Mechanical properties of flax fibre/polypropylene composites. Influence of fibre/matrix modification and glass fibre hybridization. *Composites Part A: applied science and manufacturing*, *36*(12), pp.1637-1644. https://doi.org/10.1016/j.compositesa.2005.03.021

[100] Fatima, S. and Mohanty, A.R., 2011. Acoustical and fire-retardant properties of jute composite materials. *Applied acoustics*, *72*(2-3), pp.108-114. https://doi.org/10.1016/j.apacoust.2010.10.005

[101] Ray, D., 2015. 12 state-of-the-art applications of natural composites in the industry. *Nat Compos*, *5*, p.319. https://doi.org/10.1201/b19062-13

[102] Ashori, A., 2008. Wood–plastic composites as promising green-composites for automotive industries!. *Bioresource technology*, *99*(11), pp.4661-4667. https://doi.org/10.1016/j.biortech.2007.09.043

[103] Akampumuza, O., Wambua, P.M., Ahmed, A., Li, W. and Qin, X.H., 2017. Review of the applications of biocomposites in the automotive industry. *Polymer Composites*, *38*(11), pp.2553-2569. https://doi.org/10.1002/pc.23847

[104] Shahzad, A., 2012. Hemp and its composites–a review. *Journal of Composite Materials*, *46*(8), pp.973-986. https://doi.org/10.1177/0021998311413623

[105] Netravali, A.N. and Chabba, S., 2003. Composites get greener. *Materials today*, *4*(6), pp.22-29. https://doi.org/10.1016/S1369-7021(03)00427-9

[106] Mann, G.S., Singh, L.P., Kumar, P. and Singh, S., 2020. Green composites: A review of processing technologies and recent applications. *Journal of Thermoplastic Composite Materials*, *33*(8), pp.1145-1171. https://doi.org/10.1177/0892705718816354

[107] Van Rijswijk, K., Brouwer, W.D. and Beukers, A., 2001. Application of natural fibre composites in the development of rural societies. *Delft: Delft University of Technology*.

[108] Peng, X., Fan, M., Hartley, J. and Al-Zubaidy, M., 2012. Properties of natural composites made by pultrusion process. *Journal of Composite Materials*, *46*(2), pp.237-246. https://doi.org/10.1177/0021998311410474

[109] Nishino, T., 2004. Natural fibre sources. *Green composites: Polymer composites and the environment*, pp.49-80. https://doi.org/10.1016/B978-1-85573-739-6.50007-5

[110] La Rosa, A.D., Recca, A., Gagliano, A., Summerscales, J., Latteri, A., Cozzo, G. and Cicala, G., 2014. Environmental impacts and thermal insulation performance of innovative composite solutions for building applications. *Construction and Building Materials*, *55*, pp.406-414. https://doi.org/10.1016/j.conbuildmat.2014.01.054

[111] Yudelson, J., 2010. *The green building revolution*. Island Press.

[112] Karus, M., 2004. European hemp industry 2002: cultivation, processing and product lines. *Journal of Industrial Hemp*, *9*(2), pp.93-101. https://doi.org/10.1300/J237v09n02_10

Sustainable Natural Fiber Composites

Materials Research Forum LLC

Materials Research Foundations **122** (2022) 282-302

https://doi.org/10.21741/9781644901854-12

Chapter 12

Pineapple Natural Fibre Composite: Extraction, Mechanical Properties and Application

Anirudh Kohli[1], Manoj Mathad[1], Arun Patil[1*], Anish Khan[2]

[1] School of Mechanical Engineering, [1,2,3] Centre for Material Science, KLE Technological University, Hubballi, India

[2]Chemistry Department, Faculty of Science, King Abdulaziz University, Jeddah 21589, Saudi Arabia

* patilarun7@gmail.com

Abstract

In the present scenario, one out of every five persons on the globe suffers from osteoporosis and other bone-related disorders. This disease has no age restrictions and can strike anybody between the ages of 10 and 100, even newborns. The aforementioned concerns are the consequence of lifestyle disorders as well as damage to the body, which affects the concerned person's body internally, resulting in the aforementioned issues as well as a shortage of Vitamin D. As a result, crutches and other rehabilitation items are in high demand. The above tools are often composed of wood, aluminium, or a steel/aluminum alloy, with Nylon 6/6 cuffs in the case of crutches. Many trees are felt as a result of the creation of such instruments. There is also a significant carbon footprint. As a result, an alternate material for the same purposes is required. Taking cues from the aforementioned issue and waste material in the sphere of residential trash, a PALF composite was identified due to its ease of availability as well as its superior quality and mechanical qualities.

Keywords

Ananas Comosus, PALF-Reinforced Composites, Pineapple Leaf Fibers, Rehabilitation Tool, Biodegradable, Environment-Friendly

Contents

Pineapple Natural Fibre Composite: Extraction, Mechanical Properties and Application **282**

Materials Research Foundations **122** (2022) 282-302   https://doi.org/10.21741/9781644901854-12

1.   **Introductions** **284**

2.   **Material and methodology** **285**

    2.1   Material 285

    2.2   Pineapple leaf fiber (PALF) productions 286

    2.3   De-gumming of PALF 286

    2.3.1 Physical and chemical properties of PALF 287

3.   **Composite preparation** **287**

4.   **Performance of PALF** **288**

    4.1   Tensile strength 288

    4.2   Flexural test 288

    4.3   Impact test 290

    4.4   Fourier transform infrared spectroscopy analysis (FTIR) of
           the composite 290

    4.5   Scanning electron microscopy 290

5.   **Result** **290**

    5.1   Morphological changes in PALF 290

    5.2   Tensile test 290

    5.3   Flexural strength 292

    5.4   Impact strength 293

    5.5   Inter-facial shear strength (IFSS) of the composite 293

    5.6   Fourier transform infrared spectroscopy (FTIR) 293

    5.7   Scanning electron microscopy (SEM) 294

6.   **Application** **294**

    6.1   Virtual simulation of the model 294

    6.1.1 Meshed model of the walking stick 295

    6.2   Simulation results 296

    6.2.1 Total deformation 296

    6.2.2 Equivalent stress 297

Acknowledgment **297**

Reference **297**

1.   Introductions

Since the late twentieth century, there has been a surge in interest in natural fibres as an alternative material in a variety of applications. This approach is not widely investigated for a variety of reasons, including differences in strand physical features and content. Because the same composition and strands with the same physical properties are difficult to obtain, it is not a good candidate for use as an alternative material. However, with the advancement of technology, it has become a viable option because it is replenishable at a faster rate, is biodegradable, is not harmful to the environment, and so on. Because of resource depletion and global warming, companies are shifting toward a more sustainable and environmentally friendly alternative. To achieve the required results, one such endeavour is to employ natural fibres rich in cellulose and lignin.

Their application is easy and quick, allowing us to substitute oil-derived composite materials as well as wood plastic composites[1] and natural fibre reinforced polymers [3]. In this situation, the former choice has been available for a long time, but the latter is gaining popularity in current settings due to interest in natural fibre reinforced polymers from the automotive and aerospace industries. Rice husk, vetiver grass, jute, sisal, pineapple leaf, coir, bamboo, Bagasse, rosella, banana, and other natural fibres are abundant in the agriculture business. Thailand is presently the world's top producer of pineapples[2,7]. The leftover pineapple leaves after harvesting cause a slew of issues for producers. Despite the fact that we have known how to extract fibres from pineapple leaves, this technology is exclusively used in the textile and handcraft paper industries. Despite the fact that several research publications have demonstrated that it is feasible to produce a composite using pineapple leaf and plastics as a base[5-10], it is not frequently used. When compared to glass fibre, it has a substantially greater modulus of stiffness and strength.

After bananas and citrus fruits, pineapple is the third most widely accessible material. It contains 12 distinct species that may be found in tropical places all over the world. Because of its interior chemical makeup of cellulose, lignin, and other components, pineapple is more commonly used in the textile and paper-pulp industries in tropical countries. Despite the fact that it has a diversified composition of numerous components and consequently provides good mechanical and thermal application, the pineapple leaf is always disposed of in this procedure. Pineapple is a tropical fruit that is farmed for its fruit all over the world. The pineapple plant's leaves, which are yet unused, might be easily utilised for commercial purposes [39-42].

Due to a lack of understanding among farmers and the local population about the potential of these leaves, the leaves are burned or left to decay once the fruit is picked [11-15]. In the case of pineapples, the outer skin is dried and used as a scrub or even in nascent form before being used by the textile, paper, and pulp industries to form threads that are then used in these industries to form clothes, papers, and hardboard by mixing it with other materials to achieve the desired result. Let's compare PALF, GF, and polyester in table 1.

Table 1. Properties of materials.

Properties	PALF	GF	Polyester
Density(g/cm^3)	1.527	2.543	1.15
Diameter(μm)	25-65	5-27	-
Tensile Strength (MPa)	415	2550	23
Young's Modulus (GPa)	6.57	72	0.97
Elongation at break(%)	1.62	2.56	1.62

Composites are of two types, one is a filler type as in reinforced types and secondly matrix as in composite. PLA is a very versatile bio-plastic made from agriculture wastes like corn, potato, starch rich vegetables and so on. In spite of its versatile nature still the application of PLA is very much limited as structural reinforcement in the engineering field. PLA is a brittle material with higher strength and a very low elongation of breaking. During the last decade a composite of biodegradable material with bio-fibres is gaining immense popularity compared to synthetic polymer based composites. PALF, a lingo-cellulosic natural fiber which is inexpensive and found in tropical regions provides an opportunity to create an alternative natural fibre based composite for various applications. The primary advantages of using this fiber as reinforcement in polymer composites are listed as follows: (i) nonabrasive nature, (ii) high specific properties (iii) low density, (iv) biodegradability, (v) generation of rural/agricultural-based economy (vi) low energy consumption, and (vii) low cost. These are generally found in matrix material form as it is embedded to form natural fibre composite [25-27,30-32,43].

2.   Material and methodology

2.1   Material

The pineapple leaves utilised in this experiment were cleverly obtained from Indian farmers following their harvest from Assam of Kew and Mauritius Variety. The leaves of the Kew are short and broad, whilst the leaves of the Mauritius variation are long and pointed. The leaves measure 20-60cm long and 2-5cm broad. Sodium hydroxide is an

Sustainable Natural Fiber Composites   Materials Research Forum LLC
Materials Research Foundations **122** (2022) 282-302   https://doi.org/10.21741/9781644901854-12

additional chemical reagent (NaOH). NaOH is a reagent that aids in improving the mechanical characteristics of fibres and removing contaminants from the fibre[4].

2.2   Pineapple leaf fiber (PALF) productions

In this process, a scraping instrument called Ketam is used to separate the fibres while the long bench is laying down. This technique consists of six phases, beginning with washing it under running water after scraping the PALF and drying it in the sun. In this method, we scrap the top layer first, then the bottom layer, and ultimately the fibres to shred the PALF fibre.

The remaining green material accumulated at PALF will be cleaned and eliminated at this stage. Following this stage, the chemicals NaOH are used to clean it, as indicated in Fig 1. The silane treatment comes next. The remaining green material accumulated at PALF will be cleaned and eliminated at this stage. Following this stage, the chemicals NaOH are used to clean it, as indicated in fig 2. The silane treatment comes next [24].

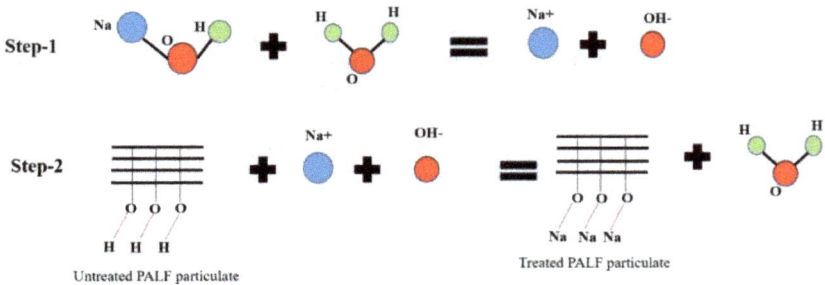

Fig 1 The chemical reaction for above process[15]

2.3   De-gumming of PALF

This is the stage in which the gummy elements from the PALF are eliminated, resulting in the PALF becoming a steep natural fibre. This procedure often removes lignin, pentosan, pectin, and other sticky materials. It is treated with alkaline, silane, or alkaline to complete the process. In this scenario, we treat it with a 3 percent by volume alkaline solution of sodium carbonate (Na_2CO_3). After that, it is cleaned with distilled water and dried in the sun. Following that, it was treated with NaOH solution to improve the mechanical qualities of the natural fibres.

Sustainable Natural Fiber Composites   Materials Research Forum LLC
Materials Research Foundations **122** (2022) 282-302   https://doi.org/10.21741/9781644901854-12

2.3.1 Physical and chemical properties of PALF

According to studies, the length of a PALF bundle is around 1m.

Previously, it was considered that the fibre was round, but this trend altered in 2010[22] when the researchers adopted eclipse as the standard shape and found success since there was a consistent ratio between the major and minor axes. It was discovered that the composites often used finer strands with diameters ranging from 50μm to 100μm. These finer strands are usually found at the bottom layer.

After comparing multiple articles[23,21] on the chemical composition of PALF, it was discovered that its hemi-cellulose composition ranged from 9.45 to 78.80 percent, cellulose composition ranged from 67.12 to 82 percent, hemi-cellulose 80–87.56 percent, pectin 1.2–3 percent, wax along with fat 3.2–4.2 percent, lignin 4.4–15.4 percent, and ash 0.9–2.7 percent.

3.   Composite preparation

To ensure that all of the absorbed moisture was eliminated and that no voids formed, the fibre and PLA resin were dried for 10 hours under vacuum. In this example, PLA pellets were compressed into a 1 mm thick film at 190°C with roughly 18 Bar pressure for 10 minutes, followed by 43 Bar pressure at 190°C for 5 minutes, following which the specimen was removed after cooling to around 100°C. The bed lay ups were completed using the film stacking process in a compression moulder by stacking PLA with PALF fibres alternately. These are done in various weight proportions. To ensure that all of the absorbed moisture was eliminated and that no voids formed, the fibre and PLA resin were dried for 10 hours under vacuum.

In this example, PLA pellets were compressed into a 1 mm thick film at 190°C with roughly 18 Bar pressure for 10 minutes, followed by 43 Bar pressure at 190°C for 5 minutes, following which the specimen was removed after cooling to around 100°C. The bed lay ups were completed using the film stacking process in a compression moulder by stacking PLA with PALF fibres alternately. These are done in various weight proportions [16,19,20,28,29].

4.   Performance of PALF

4.1   Tensile strength

A test sample with dimensions of 150 X 20 X 3 mm was created in accordance with ASTM D3079 requirements, and a test was performed utilising the Enkaky Enterprise, Bangalore-made UTM with a capacity of 10 tonnes. It was run at a cross head speed of 2 mm/min with a humidity of roughly 52% and a temperature of around 263°C. In this situation, we collected three specimens and calculated and published the average value. The standard deviation was also provided. Each specimen was loaded to the point of failure. Tensile test results, tensile strength, percentage of elongation of each specimen, and Young's modulus were also recorded[17,18].

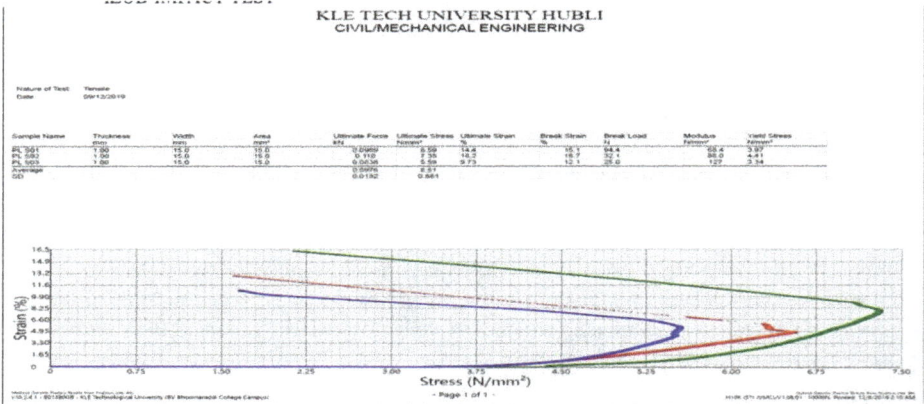

Fig 2 Tensile Test result

4.2   Flexural test

In this, we manufactured specimens with dimensions of 120 X 10 X 4 mm in accordance with ASTM D790 – 03 criteria. We used UTM to perform the three-point bending test. In this scenario, a 1.7 mm/min cross head speed is used to apply a load in the middle of the specimen. The averaged number in this case included the standard deviation and was loaded up to failure, as seen in figure 4.

Sustainable Natural Fiber Composites

Materials Research Forum LLC

Materials Research Foundations **122** (2022) 282-302

https://doi.org/10.21741/9781644901854-12

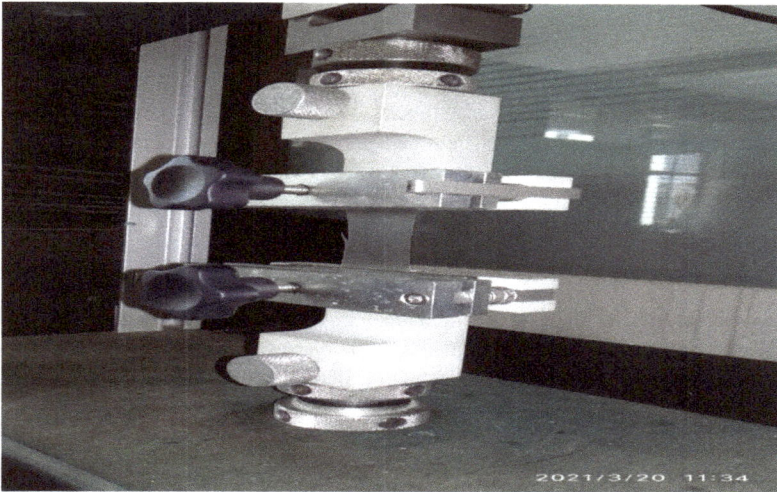

Fig 3 Tensile Test Machine

Fig 4 Flexural testing machine

Materials Research Forum LLC

https://doi.org/10.21741/9781644901854-12

4.3   Impact test

We used ASTM D256 for Izod type testing in this test. In this example, a notch was cut into the sample to generate a stress concentration region, which promotes ductile failure rather than brittle failure[33-35]. It also lowers the energy wasted in the specimen owing to plastic deformation. The following are the parameters for this testing:
Hammer Velocity= 3.45m/s and hammer weight =0.90 5Kilogram
Impact strength was calculated using Equation (1).
Impact Strength= Energy of fracture (Joule)/ Cross-sectional area (m2)   -(1)

4.4   Fourier transform infrared spectroscopy analysis (FTIR) of the composite

To investigate the change in the functional groups of the materials, FTIR analysis is performed on both untreated and treated PALF with NaOH. The samples were investigated in the transmittance mode with a wavelength resolution of 2cm and a range of 600-2200cm.

4.5   Scanning electron microscopy

The fracture morphology of a PLA/PALF composite was investigated using a SEM machine at a 3kV accelerated voltage. The sample was coated with a coating of Platinum before this test to limit the charging impact. In this instance, the sample was seen perpendicular to the fracture surface.

5.   Result

5.1   Morphological changes in PALF

Some morphological changes occur on the surface of PALF after treatment with 6% Na-OH, such as providing a cleaner surface by eliminating cemantic material and other contaminants. In this situation, grooves form as a result of the prolonged soaking period, which leads to fibre surface and polymer manufacturing, which leads to the production of inter-facial formation. As illustrated in Fig 5, the effective surface area of the fibre leads to bond formation with the matrix as a result of matrix construction.

5.2   Tensile test

We can observe that when NaOH treated PALF is compared to untreated PALF, it has a better tensile strength because the treated PALF is free of any fault or discontinuity that leads to failure. We can observe that when the PALF is submerged in NaOH for 5.5-6 hours, its tensile strength is at its peak, but it decreases as the immersion time increases owing to fibre fibrillation. The tensile strength of untreated PALF was 140 MPa, while that of 6 percent NaOH-treated PALF was 165 MPa. There is an approximate 18%

Sustainable Natural Fiber Composites   Materials Research Forum LLC
Materials Research Foundations **122** (2022) 282-302   https://doi.org/10.21741/9781644901854-12

improvement. It demonstrates that by using excessive treatment, we remove far more waxy layers and lignin than was necessary, as well as that by using excessive treatment, it was found damaged and weakened the fibres as shown by fig 6.

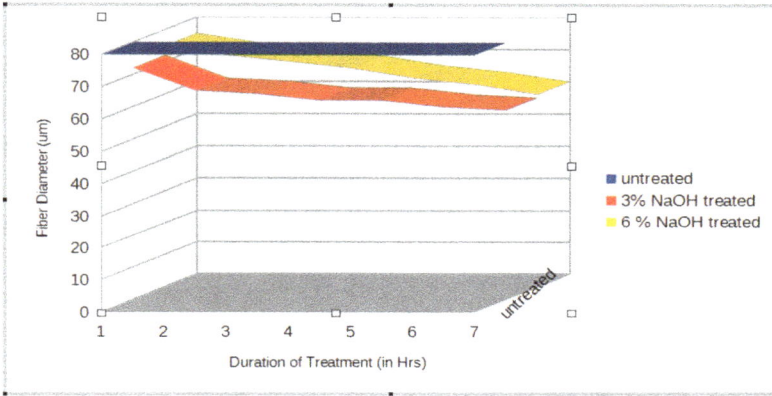

Fig 5 Diameter of untreated and NaOH treated PALF

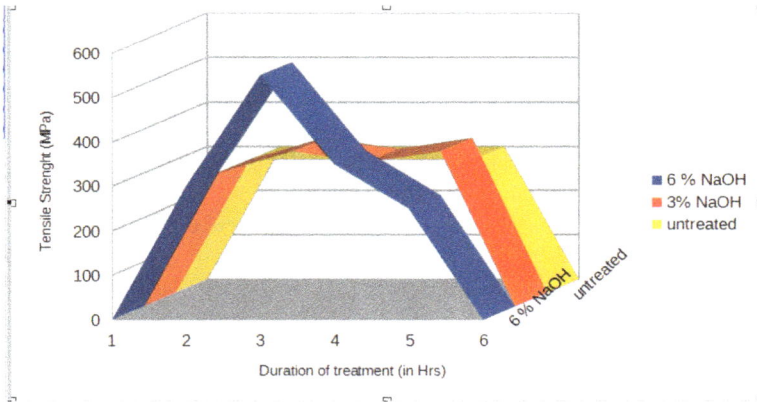

Fig 6. Tensile strength of treated (NaOH) and untreated PALF fibres

Materials Research Forum LLC

https://doi.org/10.21741/9781644901854-12

5.3   Flexural strength

The flexural strength of composites improves as the fibre content increases. The flexural strength of the PALF+PLA composite varied little in this investigation. There was a minor exception between 0 and 20 percent fibre loading. This clearly shows that NaOH-treated bio-composites have increased flexural strength. As demonstrated in Figure 7, the formation of a rougher surface combined with improved adhesion results in greater fibre-matrix interlocking and hence higher mechanical performance of the composite.

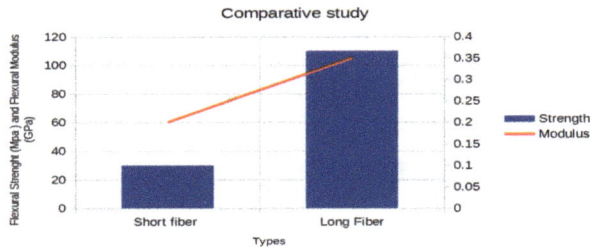

Fig 7. Flexural strength and modulus for short and long fibre composite

In addition, it can be stated that a long fibre composite has better flexural strength and modulus than a short fibre composite in both the treated and untreated cases. Second, the addition of NaOH treatment results in a considerable increase in the flexural characteristics of the long PALF fibre and PLA reinforced composite, with a maximum strength value of 114.03 MPa compared to 33.64 MPa for the short fibre reinforced PALF in PLA matrix composite. Furthermore, the flexural modulus of long fibre composite is 5.70GPa, but short fibre composite has a flexural modulus of just 0.22GPa. These results reveal that long fibre composites have better mechanical properties than short fibre composites in PLA matrix due to a lower degree of fibre alignment in the matrix of the composite. Figure 8 depicts the flexural strength of several fibre orientations. The fibres are I short untreated PALF fibre reinforced bio-composite in PLA matrix (PALFS), (ii) short alkaline treated PALF fibre reinforced bio-composite in PLA matrix (PALFSNA), (iii) long untreated PALF fibre reinforced bio-composite in PLA matrix (PALFLO), and (iv) long alkaline-treated PALF fibre reinforced bio-composite in PLA matrix (PALFLO) (PALFLONA).

Sustainable Natural Fiber Composites   Materials Research Forum LLC
Materials Research Foundations **122** (2022) 282-302   https://doi.org/10.21741/9781644901854-12

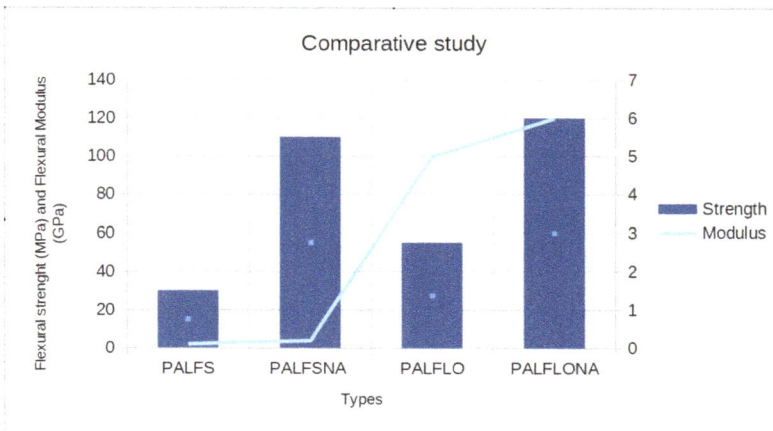

Fig 8. All Flexural properties of bio-composite of PALF reinforced composite in PLA matrix.

5.4   Impact strength

It was observed that with increase in the fibre content of the test sample, it was observed that there was an increase in impact strength. It was observed that for 30% fibre loading, we observe higher impact strength [36-38].

5.5   Inter-facial shear strength (IFSS) of the composite

The inter-facial strength of a composite represents the adhesive bond formation caused by the interlocking of the individual PALF fibre with the PLA matrix. The untreated PALF has an IFSS of 20.6 MPa, but the 6 percent NaOH solution treated PALF has the highest IFSS among the other PALF, with a value of 42.7 MPa. When compared to the nascent condition, this is a 107 percent improvement in the IFSS score.

5.6   Fourier transform infrared spectroscopy (FTIR)

Following the experiment, we can see several changes that have occurred in both untreated and treated bio-composite of PALF reinforced in PLA matrix, such as the vibrational peak at 1174 cm^{-1}, which was caused by stretching of the C-O bond in the lignin component of its acetyl group, as well as a stretched vibration in C=O in the case of ester or carboxylic acid at 1735 cm^{-1}. Both peaks are caused by hemicellulose and cellulose stretching vibrations. We may conclude that the drop in peak is due to the chemical treatment of NaOH, which resulted in the partial elimination of hemicellulose and lignin.

Sustainable Natural Fiber Composites   Materials Research Forum LLC
Materials Research Foundations **122** (2022) 282-302   https://doi.org/10.21741/9781644901854-12

5.7 Scanning electron microscopy (SEM)

In both treated and untreated samples for flexural testing, we detect various micrographs of long and short fibre bio-composite of PALF and PLA. The impact of NaOH on the broken structure of the bio-composite is seen in. The removal of contaminants by the NaOH treatment clearly increases the adhesion between the fibre and the matrix in the case of shot fibres. In the case of Long untreated and treated bio-composite of PALF fibre and PLA matrix, we find voids and breaking. As a result of inadequate wettability or low matrix and fibre adhesion, we witness fibre pull out and fracture in this case. While in case of long fibre treated bio-composite there is better adhesion because of better wettability and increase in surface roughness

6. Application

The rehabilitation sector is one of the fastest expanding industries, and the walking stick is a cornerstone of that industry for both athletes and the elderly. Walking sticks are the most rudimentary yet widely utilised instrument in the rehabilitation business by the general public following any leg-related accident or due to old age. In today's society, sticks are typically composed of wood, PVC pipes, or inexpensive metals. Despite the fact that those are strong and resistant, they still have a number of concerns, such as displacement from their natural course in the case of vertical stick, sliding, and heating issues in the Sahara and Sub-Saharan regions.

6.1 Virtual simulation of the model

We used the mechanical type with coarse meshing to simulate the model. To simulate the model, 800 N and 20 N forces were applied at different specified points to obtain the outcome in terms of the tensile, bending, and torsion forces that the beam will experience. In this case, we viewed the walking stick to be a beam to which force is applied. As seen in Figure 11.

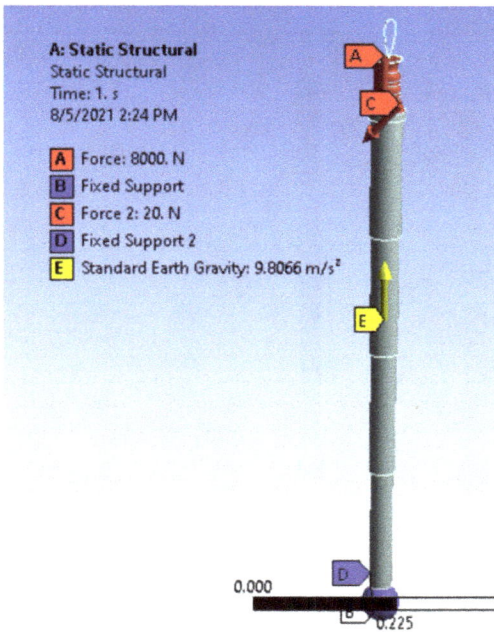

Fig 11 Forces and constraints on virtual model of walking stick.

6.1.1 Meshed model of the walking stick

We have done coarse mechanical meshing in this case to obtain a mesh with 18330 nodes and 9779 elements as shown in Fig 12.

Materials Research Forum LLC

https://doi.org/10.21741/9781644901854-12

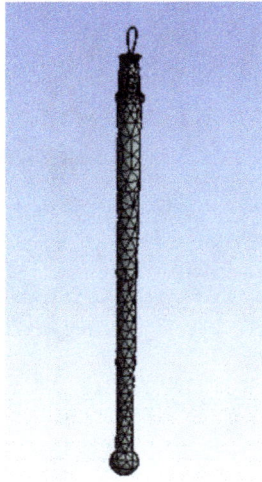

Fig 12. Meshed Model

6.2   Simulation results

6.2.1   Total deformation

In this case, we are attempting to determine the total deformation in relation to the load applied to the body, and after the simulation, we can conclude that the PALF +PLA composite had the highest deformation of 5 mm, followed by wood with a deformation of 2 mm, and steel pipe with a deformation of.1 mm, implying that the steel pipe is the most sturdy and can withstand the greatest amount of load [41-47].

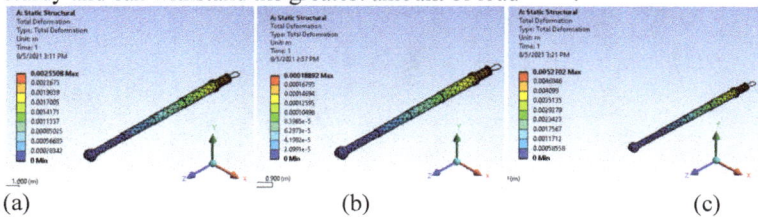

(a)   (b)   (c)

Fig 13 Total deformation in (a) wood (b) Steel (c) PALF+PLA.

6.2.2 Equivalent stress

After performing this test we are trying to compare the stress that the walking stick will experience when it is made up of three different materials which are wood, steel and a composite made up of PALF and PLA. We find that the steel experience the least stress followed by wood and then finally the Composite of PALF +PLA

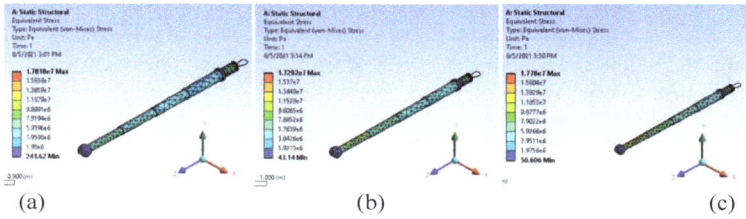

Fig 14 Equivalent stress in (a) wood (b) Steel (c) PALF+PLA.

Acknowledgment

We would like to thank CMS for analyzing mechanical properties of different specimens. We would also like to convey our gratitude to Prof. B.B. Kotturshettar, Dean (Planning and Development) and Head of School of Mechanical Engineering, Prof. P.G. Tewari (Academic dean and Principal) and Prof. B.L Desai (Registrar) KLE Technological University for his logistics support.

Reference

[1] Ashori, A., 2008. Wood–plastic composites as promising green-composites for automotive industries. Bioresour. Technol. 99, 4661–4667. https://doi.org/10.1016/j.biortech.2007.09.043

[2] Ashori, A., Nourbakhsh, A., 2010. Reinforced polypropylene composites: effect of chemical compositions and particle size. Bioresour. Technol. 101, 2515–2519. https://doi.org/10.1016/j.biortech.2009.11.022

[3] Bledzki, A.K., Faruk, O., 2003. Wood fibre reinforced polypropylene composites:effect of fibre geometry and coupling agent on physico-mechanical properties.Appl. Compos. Mater. 10, 365–379.

[4] Bledzki, A.K., Gassan, J., 1999. Composites reinforced with cellulose based fibres.Prog. Polym. Sci. 24, 221–274. https://doi.org/10.1016/S0079-6700(98)00018-5

[5]   Chattopadhyay, S.K., Khandal, R.K., Uppaluri, R., Ghoshal, A.K., 2009. Influence of varying fiber lengths on mechanical, thermal, and morphological properties of MA-g-PP compatibilized and chemically modified short pineapple leaf fiber reinforced polypropylene composite. J. Appl. Polym. Sci. 113, 3750–3756. https://doi.org/10.1002/app.30252

[6]   Chollakup, R., Tantatherdtam, R., Ujjin, S., Sriroth, K., 2011. Pineapple leaf fiber reinforced thermoplastic composites: effect of fiber length and fiber content on their characteristics. J. Appl. Polym. Sci. 119, 1952–1960. https://doi.org/10.1002/app.32910

[7]   Finkenstadt, V.L., Tisserat, B., 2010. Poly(lactic acid) and Osage Orange wood fiber composites for agricultural mulch films. Ind. Crops Prod. 31, 316–320. https://doi.org/10.1016/j.indcrop.2009.11.012

[8]   Hujuri, U., Chattopadhyay, S.K., Uppaluri, R., Ghoshal, A.K., 2007. Effect of maleic anhydride grafted polypropylene on the mechanical and morphological properties of chemically modified short-pineapple-leaf-fiber-reinforced polypropylene composites. J. Appl. Polym. Sci. 107, 1507–1516. https://doi.org/10.1002/app.27156

[9]   Mishra, S., Misra, M., Tripathy, S.S., Nayak, S.K., Mohanty, A.K., 2001. Potentiality of pineapple leaf fiber as reinforcement in PALF–Polyester composite: surface modification and mechanical performance. J. Reinf. Plast. Compos. 20,321–334. https://doi.org/10.1177/073168401772678779

[10]   Mohanty, A.K., Misra, M., Drzal, L.T., Selke, S.E., Harte, B.R., Hinrichsen, G., 2005.

[11]   Natural Fibers, Biopolymers, and Biocomposites. Taylor & Francis, Boca Raton,FL.

[12]   Moran, J.I., Alvarez, V.A., Cyras, V.P., Vazquez, A., 2008. Extraction of cellulose and preparation of nanocellulose from sisal fibers. Cellulose 15, 149–159. https://doi.org/10.1007/s10570-007-9145-9

[13]   Munir, T., Hasan, B., Gerald, T.C., 2009. Preparation and characterization of LDPE and PALF

[14]   PP-wood fiber composites. J. Appl. Polym. Sci. 112, 3095–3102.

[15]   Sam, J.K., Jin, B.M., Gue, H.K., Chang, S.H., 2008. Mechanical properties of polypropylene/naturalfiber composites: comparison of woodfiber and cottonfiber. Polym.Test. 27, 801–806. https://doi.org/10.1016/j.polymertesting.2008.06.002

[16]   Satyanarayana, K.G., Pillai, C.K.S., Pillai, S.G.K., Sukumaran, K., 1982. Structure property studies of fibers from various parts of the coconut tree. J. Mater. Sci. 17,2453–2462. https://doi.org/10.1007/BF00543759

[17]   Susheel, K., Kaith, B.S., Inderjeet, K., 2009. Pretreatments of natural fibers and their application as reinforcing material in polymer composites—a review. Polym. Eng. Sci. 49, 1253–1272. https://doi.org/10.1002/pen.21328

[18]   Stark, N.M., Rowlands, R.E., 2003. Effects of wood fiber characteristics on mechanical properties of wood/polypropylene composites. Wood Fiber Sci. 35, 167–174.

[19]   Threepopnatkul, P., Kaerkitcha, N., Athipongarporn, N., 2009. Effect of surface treatment on performance of pineapple leaf fiber–polycarbonate composites. https://doi.org/10.1016/j.compositesb.2009.04.008

[20]   Composites Part B 40, 628–632.Yao, F., Wu, Q., Lei, Y., Xu, Y., 2008. Rice straw fiber-reinforced high-density polyethylene composite: effect of fiber type and loading. Ind. Crops Prod. 28,63–72. https://doi.org/10.1016/j.indcrop.2008.01.007

[21]   Yusof Y, Yahya SA, Adam A (2015) Novel technology for sustainable pineapple leaf fibers productions. Procedia CIRP 26:756–760. https://doi.org/10.1016/j.procir.2014.07.160

[22]   Mohamed AR, Sapuan SM, Shahjahan M, Khalina A (2010a) Effects of simple abrasive combing and pretreatments on properties of pineapple leaf fibers (PALF) and PALF-vinyl ester composite adhesion. Polym Plast Technol Eng 49:972–978. https://doi.org/10.1080/03602559.2010.482072

[23]   Kalia et al. (eds.), Cellulose Fibers: Bio- and Nano-Polymer Composites, Springer-Verlag Berlin Heidelberg 2011. https://doi.org/10.1007/978-3-642-17370-7

[24]   Ramli, Siti Nur Rabiatutadawiah, Siti Hajar Sheikh Md. Fadzullah, and Zaleha Mustafa. 2017. 'The effect of alkaline treatment and fiber length on pineapple leaf fiber reinforced poly lactic acid biocomposites'. *Jurnal Teknologi* 79(5–2). doi: 10.11113/jt.v79.11293 https://doi.org/10.11113/jt.v79.11293

[25]   Arun Y. Patil, N. R. Banapurmath, Jayachandra S.Y., B.B. Kotturshettar, Ashok S Shettar, G. D. Basavaraj, R. Keshavamurthy, T. M. YunusKhan, Shridhar Mathd, 2019. "Experimental and simulation studies on waste vegetable peels as bio-composite fillers for light duty applications", *Arabian Journal of Engineering Science, Springer-Nature*

publications, June, 2019. https://doi.org/10.1007/s13369-019-03951-2

[26] Arun Y.Patil, Umbrajkar Hrishikesh N Basavaraj G D, Krishnaraja G Kodancha, Gireesha R Chalageri 2018. "Influence of Bio-degradable Natural Fiber Embedded in Polymer Matrix", *proc.,Materials Today*, Elsevier,Vol.**5**, 7532–7540. https://doi.org/10.1016/j.matpr.2017.11.425

[27] Arun Y. Patil, N. R. Banapurmath, Shivangi U S, 2020 "Feasibility study of Epoxy coated Poly Lactic Acid as a sustainable replacement for River sand", *Journal of Cleaner Production, Elsevier publications*, Vol. **267**. https://doi.org/10.1016/j.jclepro.2020.121750

[28] D N Yashasvi , Jatin Badkar, Jyoti Kalburgi, Kartik Koppalkar, 2020 *IOP Conf. Ser.: Mater. Sci. Eng.* 872 012016. https://doi.org/10.1088/1757-899X/872/1/012016

[29] Prithviraj Kandekar, Akshay Acharaya, Aakash Chatta, Anup Kamat, 2020 *IOP Conf. Ser.: Mater. Sci. Eng.* 872 012076. https://doi.org/10.1088/1757-899X/872/1/012076

[30] Vishalagoud S. Patil, Farheen Banoo, R.V. Kurahatti, Arun Y. Patil, G.U. Raju, Manzoore Elahi M. Soudagar, Ravinder Kumar, C. Ahamed S, A study of sound pressure level (SPL) inside the truck cabin for new acoustic materials: An experimental and FEA approach, Alexandria Engineering Journal, 2021. (Scopus and Web of Science)

[31] Arun Y. Patil, Akash Naik, Bhavik Vakani, Rahul Kundu, N. R. Banapurmath, Roseline M, Lekha Krishnapillai, Shridhar N.Mathad, Next Generation material for dental teeth and denture base material: Limpet Teeth (LT) as an alternative reinforcement in Polymethylmethacrylate (PMMA), Journal of nano- and electronic physics, Accepted, Vol. 5 No 4, 04001(7pp) 2021.

[32] Anirudh Kohli, Ishwar S, Charan M J, C M Adarsha, Arun Y Patil, Basvaraja B Kotturshettar, Design and Simulation study of pineapple leaf reinforced fiber glass as an alternative material for prosthetic limb, IOP Conf. Ser.: Mater. Sci. Eng. Volume 872 012118. https://doi.org/10.1088/1757-899X/872/1/012118

[33] Prabhudev S Yavagal, Pavan A Kulkarni, Nikshep M Patil, Nitilaksh S Salimath, Arun Y. Patil, Rajashekhar S Savadi, B B Kotturshettar, Cleaner production of edible straw as replacement for thermoset plastic, Elsevier, Materials Today Proceedings, March 2020. (Scopus and Web of Science). https://doi.org/10.1016/j.matpr.2020.02.667

[34]   Shruti Kiran Totla, Arjun M Pillai, M Chetan, Chetan Warad, Arun Y. Patil, B B Kotturshettar, Analysis of Helmet with Coconut Shell as the Outer Layer, Elsevier, Materials Today Proceedings, March 2020. https://doi.org/10.1016/j.matpr.2020.02.047

[35]   Anirudh Kohli, Annika H, Karthik B, Pavan PK, Lohit P A, Prasad B Sarwad, Arun Y Patil and Basvaraja B Kotturshettar, Design and Simulation study of fire-resistant biodegradable shoe, Journal of Physics: Conference Series 1706 (2020) 012185. https://doi.org/10.1088/1742-6596/1706/1/012185

[36]   Anirudh Kohli, Vrishabh Ghalagi, Manoj Divate, Chetan manakatti, Md. Irshad Karigar, Divakar poojar, Tousif pandugol, Arun Y Patil, Santosh Billur and Basavaraja B Kotturshettar, Journal of Physics: Conference Series 1706 (2020) 012186. https://doi.org/10.1088/1742-6596/1706/1/012186

[37]   Akshay Kumar, Kiran A R, Mahesh Hombalmath, Manoj Mathad, Siddhi S Rane, Arun Y Patil and B B Kotturshettar, Design and analysis of engine mount for biodegradable and non-biodegradable damping materials, Journal of Physics: Conference Series 1706 (2020) 012182. https://doi.org/10.1088/1742-6596/1706/1/012182

[38]   Shivanagouda Mudigoudra, Kiran Ragi, Mahesh Kadennavar, Naveen Danashetty, Prashant Sajjanar, Arun Y Patil, Sridhar M, Suresh H K and Basavaraj B Kotturshettar, Reduction effect of electromagnetic radiation emitted from mobile phones on human head using electromagnetic shielding materials, Journal of Physics: Conference Series 1706 (2020) 012184. https://doi.org/10.1088/1742-6596/1706/1/012184

[39]   Yajnesh M Poojari, Koustubh S Annigeri, Nilesh Bandekar, Kiran U Annigeri, Vinayak badiger, Arun Y Patil, Santosh Billur and Basavaraj B Kotturshettar, An alternative coating material for gas turbine blade for aerospace applications, Journal of Physics: Conference Series 1706 (2020) 012183. https://doi.org/10.1088/1742-6596/1706/1/012183

[40]   Kartik Zalaki, Sharad Patil, Sujay Patil, Aakash Chatta, Shubham Naik, Arun Y Patil and B B Kotturshettar, Alternate material for pressure cooker gasket, Journal of Physics: Conference Series 1706 (2020) 012181. https://doi.org/10.1088/1742-6596/1706/1/012181

[41]   N. Vijaya Kumar, N. R. Banapurmath, Ashok M. Sajjan, Arun Y. Patil and Sharanabasava V Ganachari, Studies on Hybrid Bio-nanocomposites for Structural applications, Journal of Materials Engineering and Performance, Accepted, 2021.

[42]   Sandeep Dhaduti, S. R. Ganachari and Arun Y. Patil, Prediction of injection molding parameters for symmetric spur gear, Journal of Molecular Modeling, Springer Nature, Oct 2020, Accepted, Springer Nature publications. https://doi.org/10.1007/s00894-020-04560-9

[43]   Hombalmath M.M., Patil A.Y., Kohli A., Khan A. (2021) Vegetable Fiber Pre-tensioning Influence on the Composites. In: Jawaid M., Khan A. (eds) Vegetable Fiber Composites and their Technological Applications. Composites Science and Technology. Springer, Singapore. https://doi.org/10.1007/978-981-16-1854-3_6

[44]   Mysore, T.H.M.; **Patil, A.Y.**; Raju, G.U.; Banapurmath, N.R.; Bhovi, P.M.; Afzal, A.; Alamri, S.; Saleel, C.A. Investigation of Mechanical and Physical Properties of Big Sheep Horn As an Alternative Biomaterial for Structural Applications. Materials 2021, 14, x. https://doi.org/10.3390/ma14144039

[45]   N. Vijaya Kumar, N. R. Banapurmath, Ashok M. Sajjan, Arun Y. Patil and Sharanabasava V Ganachari, Studies on Hybrid Bio-nanocomposites for Structural applications, Journal of Materials Engineering and Performance, 2021. https://doi.org/10.1007/s11665-021-05843-9

[46]   Vishalagoud S. Patil, Farheen Banoo, R.V. Kurahatti, Arun Y. Patil, G.U. Raju, Manzoore Elahi M. Soudagar, Ravinder Kumar, C. Ahamed S, A study of sound pressure level (SPL) inside the truck cabin for new acoustic materials: An experimental and FEA approach, Alexandria Engineering Journal, 2021. https://doi.org/10.1016/j.aej.2021.03.074

[47]   Arun Y. Patil, Akash Naik, Bhavik Vakani, Rahul Kundu, N. R. Banapurmath, Roseline M, Lekha Krishnapillai, Shridhar N.Mathad, Next Generation material for dental teeth and denture base material: Limpet Teeth (LT) as an alternative reinforcement in Polymethylmethacrylate (PMMA), JOURNAL OF NANO- AND ELECTRONIC PHYSICS, Sumy State University, Vol. 5 No 4, 04001(7pp) 2021. https://doi.org/10.21272/jnep.13(2).02033

Keyword Index

All-Cellulose Composites and
    Nanocomposites................................... 1
Ananas Comosus 282

Bio Composites...................... 1, 209, 256
Biodegradability 209
Biodegradable.................................... 282

Carrot Epoxy...................................... 209
Cellulose Fibers 96
Cellulose 1
Coconut Fiber 11
Cocos Nucifera 37
Composite Materials.......................... 238
Composites 96, 154, 199
Consumer Product 209

Degradation 128
Dental Material 77
Durability... 128

Eco-Composites................................... 11
Eco-Friendly 96
Energy Saving.................................... 199
Environment-Friendly 282
Epoxy Resin...................................... 209

Fatigue 128
Fiber.. 37
Fibrous 96

Hygroscopicity.................................. 128

Interfacial Interactions........................ 11

Jute.. 37

Lemon Epoxy 209
Lemon Peel 77

Mechanical Properties..................77, 238
Moisture 128
Mortar..96

Natural Fibre Polymer Composites
    (NFCs)...128

Onion Epoxy 209
Onion Peel...77

PALF-Reinforced Composites 282
Phenolic Resins 154
Pineapple Leaf Fibers........................282
PMMA..77
Polymeric Fibers 96
Polymers...238
Potato Epoxy 209
Potato Peel...77
Processing Methods 1

Rehabilitation Tool............................282
Reinforced Composites 37
Reinforcement 154
Resol..154

Surface Treatments..............................11
Sweet Lime Epoxy 209

Thermal Conductivity 199
Thermoplastic Composites...................11
Thermosets 154

Vakka Fiber..199
Vegetable Fibers................................154

About the Editors

Dr. Anish Khan is Currently working as Assistant Professor, at Chemistry Department, Centre of Excellence for Advanced Materials Research (CEAMR), Faculty of Science, King Abdulaziz University, Jeddah, Saudi Arabia. Ph.D. Completed from Aligarh Muslim University, India in 2010. He has 13 years research experience of working in the field of organic-inorganic electrically conducting nano-composites and its applications in making chemical sonsor. He complete Postdoctoral from School of Chemical Sciences, University Sains Malaysia (USM) on electroanalytical chemistry for one year. More than 115 research articles have been published in refereed international journals. More than 10 international conferences/ workshop and 45 books published and 70 Book chapters. Around 20 research project completed. Member of American Nano Society, Field of specialization is polymer nanocomposite/cation-exchanger/chemical sensor/micro biosensor/nanotechnology, application of nanomaterials in electroanalytical chemistry, material chemistry, ion-exchange chromatography and electro-analytical chemistry, dealing with the synthesis, characterization (using different analytical techniques) and derivatization of inorganic ion-exchanger by the incorporation of electrically conducting polymers. Preparation and characterization of hybrid nano composite materials and their applications, Polymeric inorganic cation –exchange materials, Electrically conducting polymeric, materials, Composite material use as Sensors, Green chemistry by remediation of pollution, Heavy metal ion-selective membrane electrode, Biosensor on neurotransmitter.

Dr. A. Manikandan received his early education B.Sc and M.Sc in Chemistry from Bharathidasan University and M. Phill from University of Madras, India. Also, he has received a degree of bachelor of Education (B.Ed.,) from University of Madras and M.Ed., and M.Phil., in Education from Tamil Nadu Teachers Education University, India. He completed his Ph.D. in Chemistry and the thesis entitled "Novel methods of syntheses, structural, morphological and opto-magnetic characterizations of photoelectrochemical nano-catalysts" under the supervision of Dr. S. Arul Antony, Associate Professor, Department of Chemistry, Presidency College, Chennai – 600 005, Tamilnadu, India. He had received 'SRF Fellowship' for the year 2013 from CSIR, New Delhi. His research interest includes the development of magnetically reusable nano-catalyst, modified catalyst for organic pollutant degradation technique (Textile dyes by photo-catalyst) for environmental applications, electrodeposition, catalytic oxidation and reduction process, nano-biomaterials, magnetic nanoparticles, synthesis and characterizations, etc. Currently, he is working as an Associate Professor & Head, Department of Chemistry, Bharath Institute of Higher Education and Research, Chennai -

600073, India. He is active member in Chemical society of India. He has published 185 articles in various reputed international journals with high impact factor. Also, he has been published 7 review articles and 3 book chapters. His Google Scholar Citations 7027, h-index 54 and i-10 index 134. Twenty six of his research students have obtained their M.Phil., degree and six more are pursuing their research for M.Phil., degree. One of his research students completed Ph.D., degree and six more are pursuing their research for Ph.D., degree

Dr. M. Ramesh received M.E degree in Computer Aided Design (CAD) and B.E degree in Mechanical Engineering from University of Madras and Bharathiyar University, Tamil Nadu, India respectively. He is the Madras University Rank holder in M.E. He obtained PhD degree in Mechanical Engineering in the field of eco-friendly composite materials from Jawaharlal Nehru Technological University (JNTU), Andhra Pradesh, India. He has published over 50 research papers in SCI and Scopus indexed Journals, 40 book chapters. He has published two papers in the field of plant fiber based bio-materials in the highest impact factor Journal "Progress in Materials Science" (Impact factor: 31.56) published by Elsevier. His research papers were published by well reputed publishers such as Elsevier, Taylor & Francis, Springer, Wiley, SAGE, Emerald Publishing, etc. He has organized number of Faculty development programs, workshops, short-term courses and has attended over 30 courses conducted by other Institutions and Organizations. He has honored with All India best Research award at Doctorate level and best Publication award by Society for Advancement of Human and Nature (SADHNA), Himachal Pradesh, India. Presently, he is working as Professor & Vice-Principal at KIT-Kalaignarkarunanidhi Institute of Technology, Coimbatore, India. His research interests include bio-materials, polymer nano-composites, hybrid composites and energy efficient materials. More details can be found at (https://scholar.google.co.in/citations?user=YzVpt7AAAAAJ&hl=en).

Dr. Imran Khan is presently working as an Assistant Professor in the Applied Science and Humanities Section, Faculty of Engineering and Technology, University Polytechnic, Aligarh Muslim University, Aligarh-202002, India. Obtained Ph.D. in Applied Physics from Aligarh Muslim University, India in 2015. Selected for the prestigious National Postdoctoral Fellowship (NPDF) from DST-SERB, India in 2016 and completed one year postdoctoral from the Department of Applied Sciences and Humanities, Jamia Milla Islamia, New Delhi-India. More than 24 research papers, 4 book chapters 1 books published and 1 books in progress in referred international Publishers and more than 12 international conferences/ workshop. The current research interests include magnetic nanoparticles, thin films, polymer composites, graphene composites, nanocomposites, optoelectronic materials, hybrid organic-inorganic composites materials, nanoporous nanostructured materials, etc., sol–gel, hydro–/solvo–thermal, microwave methods etc., thin film preparation through PLD,CVD, sputtering, and spin-dip coatings techniques etc.

Prof. Abdullah Mohammed Ahmed Asiri is Professor in Chemistry Department – Faculty of Science -King Abdulaziz University. **Ph.D.** (1995) From University of Walls College of Cardiff, U.K. on Tribochromic compounds and their applications. More than 1000 Research articles and 20 books published. The chairman of the Chemistry Department, King Abdulaziz University currently and also the director of the center of Excellence for Advanced Materials Research. Director of Education Affair Unit– Deanship of Community services. Member of Advisory committee for advancing materials, (National Technology Plan, (King Abdul Aziz City of Science and Technology, Riyadh, Saudi Arabia). Color chemistry. Synthesis of novel photochromic and thermochromic systems,Synthesis of novel colorants and coloration of textiles and plastics, Molecular Modeling, Applications of organic materials into optics such as OEDS, High performance organic Dyes and pigments. New applications of organic photochromic compounds in new novelty. Organic synthesis of heterocyclic compounds as precursor for dyes. Synthesis of polymers functionalized with organic dyes. Preparation of some coating formulations for different applications. Photodynamic thereby using Organic Dyes and Pigments Virtual Labs and Experimental Simulations. He is member of Editorial board of Journal of Saudi Chemical Society, Journal of King Abdul Aziz University,Pigment and Resin Technology Journal, Organic Chemistry Insights, Libertas Academica, Recent Patents on Materials Science, Bentham Science Publishers Ltd. Beside that he has professional membership of International and National Society and Professional bodies.

www.ingramcontent.com/pod-product-compliance
Lightning Source LLC
Chambersburg PA
CBHW071329210326
41597CB00015B/1384